# Multimedia Applications Support for Wireless ATM Networks

ISBN 0-13-021437-X

**Prentice Hall Communications Engineering and Emerging Technologies Series**

*Theodore S. Rappaport, Series Editor*

KIM  *Handbook of CDMA System Design, Engineering, and Optimization*

GARG  *IS-95 CDMA and cdma2000: Cellular/PCS Systems Implementation*

GARG & WILKES  *Principles and Applications of GSM*

HAĆ  *Multimedia Applications Support for Wireless ATM Networks*

LIBERTI & RAPPAPORT  *Smart Antennas for Wireless Communications: IS-95 and Third Generation CDMA Applications*

RAPPAPORT  *Wireless Communications: Principles & Practice*

RAZAVI  *RF Microelectronics*

STARR, CIOFFI & SILVERMAN  *Understanding Digital Subscriber Line Technology*

*FORTHCOMING*

CIMINI & LI  *Orthogonal Frequency Division Multiplexing for Wireless Communication*

MOLISCH  *Wideband Wireless Digital Communication*

POOR & WANG  *Wireless Communication Systems: Advanced Techniques for Signal Reception*

REED & WOERNER  *Software Radio: A Modern Approach to Radio Engineering*

SKLAR  *Digital Communications: Fundamentals and Applications, Second Edition*

TRANTER, KOSBAR, RAPPAPORT & SHANMUGAN  *Simulation of Modern Communications Systems with Wireless Applications*

# Multimedia Applications Support for Wireless ATM Networks

Anna Hać

**Prentice Hall PTR**
Upper Saddle River, NJ 07458
www.phptr.com

**Library of Congress Cataloging-in-Publication Data**

Hać, Anna
   Multimedia applications support for wireless ATM networks / Anna Hac.
     p. cm.
   Includes bibliographical references and index.
   ISBN 0-13-021437-X
     1. Asynchronous transfer mode. 2. Wireless communication systems. 3. Multimedia
systems. I. Title.

TK5105.35. H33 2000
621.382'16--dc21

                                                                                                      00-021806

Editorial/Production Supervision: *Vincent J. Janoski*
Acquisitions Editor: *Bernard Goodwin*
Manufacturing Manager: *Alexis Heydt*
Marketing Manager: *Lisa Konzelmann*
Editorial Assistant: *Diane Spina*
Cover Design Director: *Jerry Volta*
Composition: *Aurelia Scharnhorst*

©2000 by Prentice Hall PTR
Prentice-Hall, Inc.
Upper Saddle River, New Jersey 07458

Prentice Hall books are widely used by corporations and government agencies for training, marketing, and resale.
The publisher offers discounts on this book when ordered in bulk quantities. For more information, contact Corporate Sales Department. Phone: 800-382-3419; Fax: 201-236-7141; E-mail: corpsales@prenhall.com; Or write: Prentice Hall PTR, Corp. Sales Dept., One Lake Street, Upper Saddle River, NJ 07458

Product names mentioned herein are the trademarks of their respective owners.

All rights reserved. No part of this book may be reproduced, in any form or by any means, without permission in writing from the publisher.

Printed in the United States of America

    10  9  8  7  6  5  4  3  2  1

ISBN 0-13-021437-X

Prentice-Hall International (UK) Limited, *London*
Prentice-Hall of Australia Pty. Limited, *Sydney*
Prentice-Hall Canada Inc., *Toronto*
Prentice-Hall Hispanoamericana, S.A., *Mexico*
Prentice-Hall of India Private Limited, *New Delhi*
Prentice-Hall of Japan, Inc., *Tokyo*
Pearson Education Asia Pte. Ltd.
Editora Prentice-Hall do Brasil, Ltda., *Rio de Janeiro*

# Contents

| | |
|---|---|
| **Preface** | **xi** |

| | | |
|---|---|---|
| **1** | **Principles of Telecommunications Network Architecture** | **1** |
| 1.1 | Broadband Networks | 1 |
| | 1.1.1 Basic ATM Protocols | 2 |
| | 1.1.2 Leaky Bucket Model | 8 |
| | 1.1.3 Cooperating Leaky Bucket Model | 10 |
| | 1.1.4 Compression Technology | 11 |
| | 1.1.5 Multimedia Virtual Circuit Model | 14 |
| 1.2 | Rate-Based Congestion Control Schemes for ABR Service in ATM Networks | 17 |
| | 1.2.1 Services in the ATM Network | 18 |
| | 1.2.2 Congestion Control for ABR Service | 18 |
| | 1.2.3 Rate-Based Control Schemes | 20 |
| 1.3 | Multiple Access Protocols for Wireless ATM Networks | 21 |
| | 1.3.1 Wireless Networks | 22 |
| | 1.3.2 Wireless ATM | 24 |
| |     1.3.2.1 ATM Services | 24 |
| |     1.3.2.2 Wireless Integration | 25 |
| | 1.3.3 Multiple-Access Protocol | 26 |
| |     1.3.3.1 Fixed Assignment | 26 |

| | | |
|---|---|---|
| | 1.3.3.2 Random Assignment | 26 |
| | 1.3.3.3 Demand Assignment | 27 |
| | 1.3.4 Demand Assignment Multiple-Access (DAMA) Protocols | 28 |

## 2 Architecture of Telecommunications and Wireless Networks     33

| | | |
|---|---|---|
| 2.1 | Multimedia Network Architecture | 33 |
| | 2.1.1 Design of the Multimedia Conference System | 36 |
| 2.2 | Congestion Control for ABR Traffic in ATM Networks | 41 |
| | 2.2.1 FECN Rate-Control Method | 43 |
| | 2.2.2 BECN Rate-Control Method | 45 |
| | 2.2.3 PRCA | 47 |
| | 2.2.4 Beat Down Problem | 47 |
| | 2.2.5 An Adaptive Congestion Control Scheme | 48 |
| 2.3 | Congestion Control Mechanisms in the Intelligent Network | 52 |
| 2.4 | Guaranteed QoS in Wireless ATM Networks | 59 |
| | 2.4.1 Mobile Quality of Service (M-QoS) | 61 |
| | 2.4.2 Architectures of High-Speed Wireless ATM Networks | 63 |
| |    2.4.2.1 Architecture Based on Distributed Control: AT&T™'s Architecture | 63 |
| |    2.4.2.2 Architecture Based on Hierarchical Organization: NEC™'s Wireless ATM Architecture | 66 |
| |    2.4.2.3 Architecture Based on Intelligent Network: British Telecom™'s Architecture | 66 |
| |    2.4.2.4 Cambridge Wireless ATM LAN Architecture | 69 |
| |    2.4.2.5 LAN Emulation Model for Wireless ATM | 71 |
| |    2.4.2.6 The Mobile Handling Backbone (MHB) | 77 |

## 3 Models of Telecommunications Networks     81

| | | |
|---|---|---|
| 3.1 | Synchronization in Multimedia Systems | 81 |
| | 3.1.1 The Architecture of DMIS | 84 |
| | 3.1.2 Multimedia Synchronization Scheme | 85 |
| |    3.1.2.1 Concept Model Description of Media Synchronization | 86 |
| |    3.1.2.2 Playout Schedule and Synchronization | 88 |
| |    3.1.2.3 Buffer Configuration at the Client Side | 91 |
| |    3.1.2.4 Determination of Control Time in Stored Data Retrieval | 94 |
| |    3.1.2.5 Synchronization in DMIS | 96 |

|   |   |
|---|---|
| 3.1.3 Buffer Control System | 97 |
| 3.2 User Mobility Management in Wireless Networks | 101 |
|    3.2.1 The Architecture of the PCS Network | 103 |
|       3.2.1.1 Access Network | 103 |
|       3.2.1.2 Transport Network | 104 |
|       3.2.1.3 Intelligent Network (IN) | 104 |
|    3.2.2 Location Database Management in the PCS Network | 108 |
|       3.2.2.1 Location Management Strategies Before the PCS Network | 108 |
|       3.2.2.2 Distributed Location Database for the PCS Network | 109 |
|    3.2.3 Placement of a Hierarchical Database | 111 |
| 3.3 Tracking Strategy for Personal Communications Networks | 113 |
|    3.3.1 Structure of PCNs | 113 |
|    3.3.2 Locating Strategies in PCN | 115 |
|    3.3.3 Tracking Strategy | 116 |
| 3.4 Location Management Schemes for Mobile Communications | 125 |
|    3.4.1 The Two-Scope Concept | 128 |
|       3.4.1.1 Global Scope | 128 |
|       3.4.1.2 Local Scope | 130 |
|    3.4.2 Location Management in Local Scope | 130 |
|       3.4.2.1 Method of Static Partitioning and Method of Dynamic Neighbor Zone | 131 |
|       3.4.2.2 The Model | 132 |
|       3.4.2.3 Performance Evaluation of the Dynamic Neighbor Zone Method | 133 |
|       3.4.2.4 Combining the Limited Broadcast and Forwarding Pointer Schemes | 134 |
|       3.4.2.5 Impact of Zone Size on Performance | 136 |
|    3.4.3 Location Management in the Global Scope | 136 |
|       3.4.3.1 The Model | 138 |
|       3.4.3.2 Comparison of Basic Schemes | 139 |
|       3.4.3.3 The Compound Scheme | 140 |
|       3.4.3.4 The Environment of the "Friend Network" | 142 |

# 4 Routing Algorithms 145

| | |
|---|---|
| 4.1 Finding Minimum Cost Multicast Trees with Bounded Path Delay | 145 |
|    4.1.1 DDBMA | 148 |
| 4.2 Multicast and Other Routing Algorithms | 159 |
|    4.2.1 Multicast Routing Algorithms Reducing Congestion | 161 |
|       4.2.1.1 The Algorithm | 162 |
|       4.2.1.2 Cost and Performance Analysis | 169 |
|       4.2.1.3 Applying the Multicast Routing Algorithm to a Multipoint Connection in Broadband Networks | 174 |
| 4.3 Routing in Wireless Personal Networks | 175 |
|    4.3.1 The Architecture and Model of a Wireless Communications Network | 177 |
|    4.3.2 The ESP and DSDSP Routing Algorithms | 182 |
|       4.3.2.1 The ESP Routing Algorithm | 184 |
|       4.3.2.2 The DSDSP Routing Algorithm | 188 |
|       4.3.2.3 A Comparison between the ESP and DSDSP Routing Algorithms | 191 |
| 4.4 MH Protocols for the Internet | 193 |
|    4.4.1 VIP | 193 |
|    4.4.2 LSRIP | 194 |
|    4.4.3 IETF-MIP | 195 |
|    4.4.4 A Comparison between Mobile IPs | 196 |
|    4.4.5 Scalable Mobile Host IP (SMIP) Protocol | 196 |
|       4.4.5.1 HA to FA Inside the Same Area | 198 |
|       4.4.5.2 HA to FA in the Other Area | 199 |
|       4.4.5.3 Moving Inside the Other Area | 202 |
|       4.4.5.4 Moving from the Outside Area to a New Outside Area | 204 |
|       4.4.5.5 SMIP Packet Format | 205 |
|       4.4.5.6 SMIP Notification Message Format | 207 |
|       4.4.5.7 SMIP Authentication | 207 |
|       4.4.5.8 The Popup Method | 208 |
| 4.5 Location Update and Routing Schemes for the Mobile Computing Environment | 209 |
|    4.5.1 Locality Traffic Pattern | 213 |
|    4.5.2 Scheme Description | 215 |
|    4.5.3 Architecture | 217 |

|     |       |                                          |     |
| --- | ----- | ---------------------------------------- | --- |
|     | 4.5.3.1 | Mobile Host (MH)                       | 217 |
|     | 4.5.3.2 | Foreign Agent (FA)                     | 218 |
|     | 4.5.3.3 | Home Agent (HA)                        | 219 |
|     | 4.5.3.4 | Redirection Agent (OSPF Area BR)       | 220 |

    4.5.4 Packet Routing    220
    4.5.5 MH Movements    223

## 5  Resource Allocation in Multimedia Wireless Networks  225

5.1  Dynamic Channel Assignment in Cellular
     Mobile Communications Networks  225
    5.1.1 Cochannel Information Based Dynamic
          Channel Assignment (CDCA) Strategy  232
    5.1.2 Group Dynamic Channel Assignment (GDCA)
          Strategy for Multichannel Traffic  238
5.2  Distributed Dynamic Channel Assignment (DDCA)
     in Cellular Mobile Communication Networks  242
    5.2.1 Centralized DCA Scheme  243
    5.2.2 DDCA  244
    5.2.3 Priority-Based DDCA (PDCA)  246
5.3  The HCA Method for Wireless Communication Networks  251
    5.3.1 Definition of the Problem  252
    5.3.2 Nominal Channel Assignment Strategy and
          Channel Slot Order  255
       5.3.2.1 Frequency Exhaustive and Requirement
              Exhaustive Strategies  255
       5.3.2.2 Importance of Nominal Channel Slot Order  257
    5.3.3 Nominal Channel Allocation in
          the Network Planning Stage  258
    5.3.4 Dynamic Channel Allocation in
          the Network Running Stage  263
5.4  QoS in Wireless Microcellular Networks  265
    5.4.1 Multimedia Traffic in Microcellular Networks  267
    5.4.2 Resource Allocation in Microcellular Networks  268
       5.4.2.1 Admission Threshold Based Scheme  269
       5.4.2.2 Resource Sharing Based Scheme  270
       5.4.2.3 Resource Reservation (RR) Based Scheme  271
    5.4.3 RR with Renegotiation  273

| | |
|---|---|
| **References** | 279 |
| **Glossary of Acronyms** | 293 |
| **Index** | 303 |
| **About the Author** | 311 |

# Preface

Telecommunications networks have been the topic of research for several decades. With the employment of telephone and radio, a new era in communications has begun. Several generations of telecommunications networks have evolved into a worldwide network with countless applications of various media and the interactive users.

This book describes the recent advances in wireless Asynchronous Transfer Mode (ATM) networks. The revolution in networking started with broadband technology, and related applications and services. High-speed networks allow for new applications, which involve new problems and their solutions. Wireless networks are commonly used, and user mobility, network management, and network protocols are important issues. Networks carrying multimedia, voice, data, and video traffic define modern network architecture.

Telecommunications architecture involves broadband and wireless networks, and is suitable for multimedia and mobile applications with bursty traffic. This book reviews network architecture and design principles. Different network architecture and applications are described. Models of telecommunications networks are introduced.

The models of wireless networks include the design of real networks, their architecture, services, and applications. Network management issues, including mobility management, location, and database management are evaluated. Network protocols are discussed. With expanded user services and new applications, network protocols are an important part of network design and performance.

Routing and congestion control is a part of any network architecture. Bursty traffic in high-speed multimedia networks and mobile applications require revision of routing and congestion control techniques and pose new challenges. New solutions to routing and congestion control are introduced. Routing algorithms and flow and congestion control techniques include multicast and other routing algorithms, routing in wireless personal networks, tracking strategy for personal communications networks, location management schemes for mobile communications, and location update and routing schemes for a mobile computing environment.

Multimedia, voice, data and video networks and services combined with wireless and high-speed networks require new solutions to the problems of synchronization in multimedia systems, dynamic channel assignment in cellular mobile communications networks, mobile host protocols, and quality of service (QoS) in wireless microcellular networks.

This book focuses on the newest technology (e.g., high-speed networks, ATM, bursty traffic, multimedia, and wireless), and adapting old and new solutions to the newest technology. The book presents a real applications-oriented approach to solving networking problems. The book introduces and analyzes architecture and design issues in wireless ATM networks.

This book describes the network from the architecture and design perspective. The book includes network design for wireless ATM networks, describes how the network is constructed, and explains the flow of real multimedia traffic in the network. The book combines both mobile and stationary networks with their multimedia applications.

Chapter 1 introduces principles of telecommunications network architecture, and presents examples from existing networks to illustrate these principles. Broadband and wireless ATM networks are described and their protocols are discussed. Congestion control techniques for ATM networks are introduced.

Chapter 2 describes the architecture of telecommunications and wireless networks, including multimedia network architecture, ATM, intelligent network, and various ATM wireless networks.

Network models are studied in Chapter 3. A synchronization model in multimedia systems is introduced. A model of user-mobility management in wireless networks is described. A tracking strategy for personal communications networks is discussed. Location management schemes for mobile communications are introduced.

Routing algorithms are introduced in Chapter 4. The method of finding minimum cost multicast trees with bounded path delay is described. Multicast and other routing algorithms are studied. Routing in wireless personal networks is described. Mobile host protocols for the Internet are discussed. Location update and routing schemes for mobile computing environment are described.

Chapter 5 describes resource allocation in multimedia wireless networks. Dynamic channel assignment, distributed dynamic channel assignment, and hybrid channel allocation method in cellular mobile communications networks are studied. QoS in wireless microcellular networks is discussed.

# CHAPTER 1

# Principles of Telecommunications Network Architecture

A telecommunications network is a collection of nodes and links that communicate by defined sets of formats and protocols. Within the network there are usually three layers: transmission, switching, and service. The transmission layer consists of transmission systems, for example, cables, radio links, and their related technical equipment. The switching layer consists of switching nodes with generic and application software and data. The service layer, distributed among the switching network elements, consists of special hardware, and their application software and data.

## 1.1 Broadband Networks

Broadband integrated services digital network (B-ISDN) based on asynchronous transfer mode (ATM) is the technology of choice for transport of information from multimedia services and applications.

ATM is a cell-based, high-bandwidth, low-delay switching and multiplexing technology that is designed to deliver a variety of high-speed digital communication services. These services include LAN (local area network) interconnection, imaging, and multimedia applications as well as video distribution, video telephony and other video applications.

ATM is asynchronous since the recurrence of cells containing information from an individual customer is not necessarily periodic. ATM handles both connection-oriented and connectionless traffic through the use of adaptation layers and operates at either a constant bit rate (CBR) or variable bit rate (VBR) connection. Each ATM cell sent into the network contains addressing information that establishes a virtual connection from source to destination. All cells are then transferred in sequence over this virtual connection. ATM supports permanent virtual connections (PVC) as well as switched virtual connections (SVC).

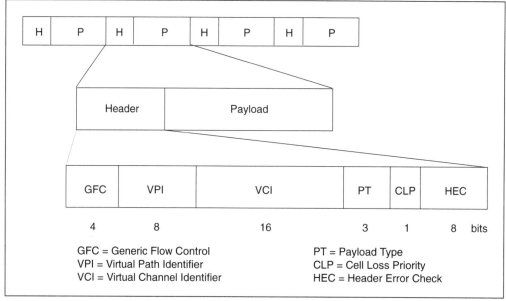

**Figure 1–1**  ATM Cell Format

#### 1.1.1  Basic ATM Protocols

ATM standards define a fixed-size cell with a length of 53 bytes comprised of a 5-byte header and a 48-byte payload as shown in Figure 1–1. Broadband ISDN supports multimedia applications because of its high performance.

Figure 1–2 depicts the ATM Broadband-ISDN protocol reference model as defined in ITU-T Recommendation I.321. The physical layer (PHY) has two sublayers: transmission convergence (TC) and physical medium-dependent (PMD). The PMD sublayer interfaces with the actual physical medium and passes the recovered bit stream to the TC sublayer. The TC sublayer extracts and inserts the ATM cell with the synchronous digital hierarchy (SDH) time division multiplexed (TDM) frame and then passes them to and from the ATM layer, respectively. The ATM layer performs multiplexing, switching, and controlling actions based upon information in the ATM cell header. It passes cells to and accepts cells from the ATM adaptation layer (AAL). The AAL also has two sublayers: segmentation and reassembly (SAR) and convergence sublayer (CS). The AAL passes protocol data units (PDUs) to and accepts them from higher layers. PDUs may differ in variable or fixed length from the ATM cell length. The physical layer (PHY) corresponds to layer 1 of the PHY in the OSI (Open System Interconnection) model while the ATM layer and adaptation layer (AAL) correspond to parts of OSI layer 2 or data-link layer.

The PHY consists of two logical sublayers: the physical medium-dependent (PMD) sublayer and the transmission convergence (TC) sublayer. PMD includes only PMD functions. It provides bit transmission capability, including bit transfer, bit alignment, line coding, and electrical–optical conversion. TC performs functions required to transform a flow of cells into a flow

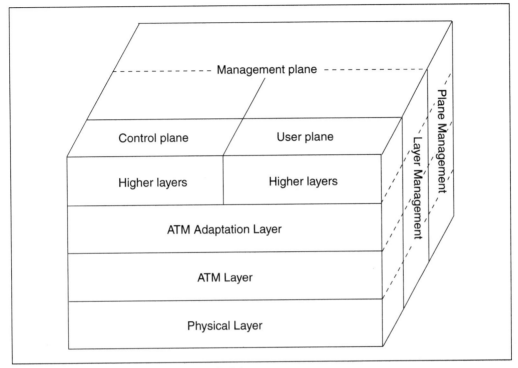

**Figure 1–2** ATM B-ISDN Reference Model

of information that can be transmitted and received over a physical medium. TC functions include (1) transmission frame generation and recovery, (2) transmission frame adaptation, (3) cell delineation, (4) header error control (HEC) sequence generation and cell-header verification, and (5) cell-rate decoupling.

The transmission frame adaptation function performs the actions that are necessary to structure the cell flow according to the payload structure of the transmission frame (transmit direction) and to extract this cell flow out of the transmission frame (receive direction). In the United States, the transmission frame requires Synchronous Optical Network (SONET) envelopes for transmission rates higher than 45 Mbps.

SONET, a synchronous transmission structure, is often used for framing and synchronization at the PHY. The basic time unit of a SONET frame is 125 microseconds. The SONET frame structure is depicted in Figure 1–3. The basic building block of SONET is the synchronous transport signal level 1 (STS-1) with a bit rate of 51.84 Mbps. The STS-1 frame structure can be drawn as 90 columns and 9 rows of 8-bit bytes. The first 3 columns of STS-1 contain section and line overhead bytes used for error monitoring, system maintenance functions, synchronization, and identification of payload type. The remaining 87 columns and 9 rows are used to carry the STS-1 synchronous payload envelope. Higher-rate SONET signals are obtained by byte-interleaving $n$ frame-aligned STS-1's to form an STS-$n$ (e.g., STS-3 has a bit rate of 155.52 Mbps).

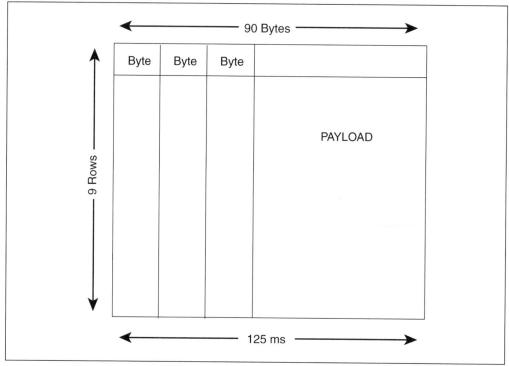

**Figure 1-3** SONET STS-1 Payload Envelope

Cell delineation prepares the cell flow to enable the receiver side to recover cell boundaries. In the transmit direction, cell boundaries are identified and confirmed, and the cell flow is descrambled. The HEC mechanism covers the entire cell header, which is available to this layer by the time the cell is passed down to it. The code used for this function is capable of either single-bit correction or multiple-bit error detection. The transmitting side computes the HEC field value.

Due to the PHY framing overhead, the transfer capacity at the user-network interface (UNI) is 155.52 Mbps with a cell-fill capacity of 149.76 Mbps. Since the ATM cell has 5 bytes of overhead, the 48 bytes information field allows for a maximum of 135.631 Mbps of actual user information. A second UNI interface is defined at 622.08 Mbps with the service bit rate of approximately 600 Mbps. Access at these rates requires a fiber-based loop.

For information transport, an ATM uses virtual connections that are divided into two levels: the virtual-path level and the virtual-channel level. ATM layer functions include 1) generic flow control, 2) cell header generation and extraction, 3) cell virtual-path identifier/virtual-channel identifier (VPI/VCI) translation, and 4) cell multiplexing and demultiplexing.

The generic flow control (GFC) function is used only at the UNI. It may assist the customer network in controlling the cell flow towards the network, but it does not perform flow

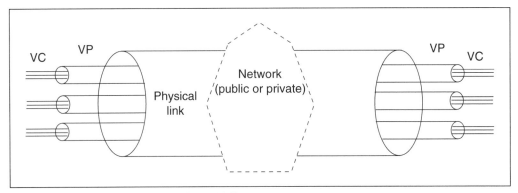

**Figure 1–4** Relation Between VCs and VPs

control of traffic from the network. The GFC can also be used within a user's premises to share ATM capacity among the workstations.

Cell-header generation and extraction functions apply at points where the ATM layer is terminated. In the transmit direction, the cell-header generation function receives a cell-information field from a higher layer and generates an appropriate ATM cell header, except for the HEC sequence which is calculated and inserted by the PHY. In the receive direction, the cell-header extraction function removes the ATM cell header and passes the cell information field to the ATM adaptation layer.

The virtual-path identifiers (VPIs) and virtual-channel identifiers (VCIs) are the labels to identify a particular virtual path (VP) and virtual channel (VC) on the link. The switching node uses these values to identify a particular connection and then uses the routing table established at connection setup to route the cells to the appropriate output port. The switch changes the value of the VPI and VCI fields to the new values that are used on the output link.

From the transmitter, the cell-multiplexing function combines cells from the individual VP and VC into one cell flow. On the other hand, in the receive direction the cell demultiplexing function directs individual cells to the appropriate VP or VC. Figure 1–4 depicts the relationship of VPs and VCs.

Both VPIs and VCIs are used to route cells through the network. Note that VPI and VCI values must be unique on a specific transmission path. Each transmission path between network devices such as ATM switches uses VPIs and VCIs independently. This is illustrated in Figure 1–5. Each switch maps an incoming VPI and VCI to an outgoing VPI and VCI.

The ATM AAL performs the necessary mapping between the ATM layer and the next higher layer. The process is done at the terminal equipment or at the terminal adapter (i.e., at the edge of the ATM network).

The ATM network, the part of the network which processes the functions of the ATM layer, is independent from the telecommunications services it carries, which also means that the user payload is carried transparently by the ATM network. The ATM network does not process the user payload nor does it know the structure of the data unit. There is no timing relationship

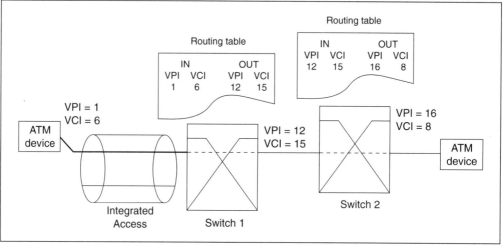

**Figure 1–5** Illustration of VPI/VCI Usage

between the clock of application and the clock of network. Therefore, the network is also time independent. The consequence of this time independence infers that all time-dependence functions required by an application are only provided by services within the AAL.

The function of the AAL is to provide the data flow sent by the user to the upper layers at the receiving end by taking into account any effects introduced by the ATM layer. Within the ATM layer, the data flow can be corrupted by errors during the transmission, or it can suffer cell-delay variation as the result of variable delay in buffers, or through congestion in the network. The consequence is the loss of cells or the incorrect delivery of cells, which may have an impact on the application whether it is in data transfer, video, or voice communication. The AAL protocols must cope with these effects. For each telecommunication service, a separate AAL should be developed. However, considering the common factors within possible telecommunication services, it is possible to have a small set of AAL protocols to support these services.

The functions performed in the AAL depend upon the higher layer requirements. Since the AAL supports multiple protocols to fit the needs of different AAL service users, it is, therefore, service dependent. To minimize the number of different AAL protocols required, a telecommunication service classification is defined based on the following parameters: 1) timing relationship between source and destination, 2) bit-rate (constant or variable), and 3) connection mode (connection oriented or connectionless). Using these parameters, five classes of service have been defined. Figure 1–6 depicts different ATM service classes and AALs.

- Class A: Timing required, bit rate constant, connection oriented
- Class B: Timing required, bit rate variable, connection oriented
- Class C: Timing not required, bit rate variable, connection oriented
- Class D: Timing not required, bit rate variable, connectionless

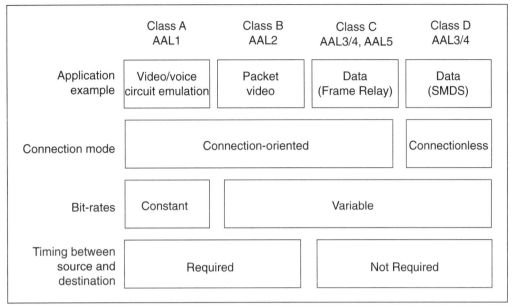

**Figure 1-6**  ATM Service Classes and AALs

• Class X: Unrestricted (bit rate variable, connection oriented or connectionless)

Class A service is an on-demand connection-oriented, constant bit-rate ATM transport service and has end-to-end timing requirements. Class A may require stringent cell loss, cell delay, and cell-delay variation performance. The user chooses the desired bandwidth and the appropriate QoS during the signaling phase of an SVC (simulated virtual connection) call to establish a Class A connection. (In the PVC (permanent virtual connection) case, this is negotiated in advance.) This service can provide the equivalent of a traditional, dedicated line and may be used for video distribution, video conferencing, multimedia, etc.

Class B service is not currently defined by formal agreements. Class C service is an on-demand, connection-oriented, variable-bit-rate ATM transport service and has no end-to-end timing requirements. The user chooses the desired bandwidth and QoS during the signaling phase of a call to establish a Class C connection. Class D service is a connectionless service and has no end-to-end timing requirements. The user supplies independent data units that are delivered by the network to the destination specified in the data unit. Switched multimegabit data service (SMDS) is an example of a Class D service. Class X service is an on-demand, connection-oriented ATM transport service where the traffic type can be either variable bit rate (VBR) or constant bit rate (CBR). Its timing requirements are user defined (i.e., transparent to the network). The user chooses only the desired bandwidth and QoS during the signaling phase of an SVC call to establish a Class X connection. (In the PVC case, it is negotiated in advance.)

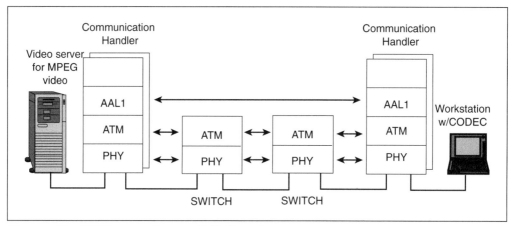

**Figure 1–7**  AAL Functionality over ATM Network

Figure 1–7 depicts the role played by AAL in end-to-end video transport. A communication handler can be implemented on top of AAL1 to support various video applications such as distribution of a motion picture experts group (MPEG) video from a video server to workstation clients over an ATM network. This communication handler provides video-application interface to establish video sessions according to a desired specification.

### 1.1.2  Leaky Bucket Model

The leaky bucket (LB) model is defined in the ATM standard to provide QoS to the user and also enforce the reserved rate from the user to the network. The LB model offers this QoS by guaranteeing each admitted call a constant reserved rate. A source is defined as conformance when it does not transmit beyond its reserved rate. The LB algorithm is the key in defining the meaning of conformance checking for an arriving cell stream against traffic parameters in the traffic contract. (A formal definition of the LB algorithm can be found in the ATM Forum UNI specification or International Telecommunications Union–Telecommunication (ITU-T) I.371.) According to the traffic contract, the LB is analogous to a bucket with a certain depth and a hole on the bottom which causes it to leak at a certain rate. Each cell arrival creates a cup of fluid flow that is poured into one or more buckets for use in conformance checking. The funneling of cell arrival fluid into buckets is controlled by the cell loss priority (CLP) bit in the ATM cell header.

Figures 1–8 and 1–9 show examples of a conforming and nonconforming cell flow, respectively, and also describe the LB's operation. In both examples, the nominal cell interarrival time is 4 cell times, which is also the bucket increment, with the bucket depth being 6 cell times.

Upon arrival of a cell, an agent checks whether the entire bucket increment for a cell can be added to the current bucket contents without overflowing. If the bucket does not overflow, then the cell is conforming; otherwise it is nonconforming. The agent discards the fluid for nonconforming cells to the floor. Fluid from a cell arrival is added to the bucket only if the cell is conforming. At each cell time, the bucket drains one unit and each cell arrival adds a number of

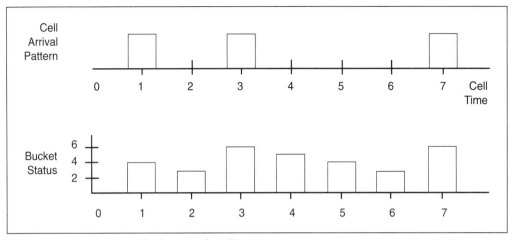

**Figure 1–8** Example of Conforming Cell Flow

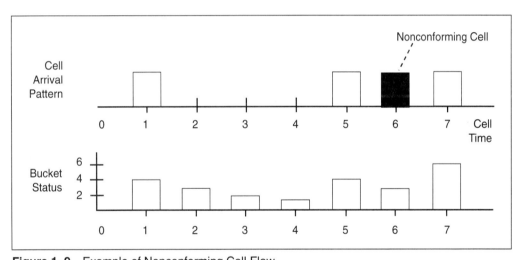

**Figure 1–9** Example of Nonconforming Cell Flow

units to the bucket specified by the increment. A cell's fluid is completely drained out after a number of cell times. This depends on both the LB increment and the leak rate of the bucket.

For the conforming cell flow, an example is shown in Figure 1–8. The first cell arrival finds an empty bucket and fills it to a depth of four units. At the third cell time, two units have drained from the bucket, and a new cell arrives. The agent determines whether the fluid from this cell fills the bucket to the rim or to a depth of six units so that it is conforming and then added to the bucket. The earliest conforming cell arrival time is in the next four cell times or at the seven-cell time. The four increments must first be drained from the bucket for a cell arrival not to cause the depth of the bucket (equal to six units) to overflow.

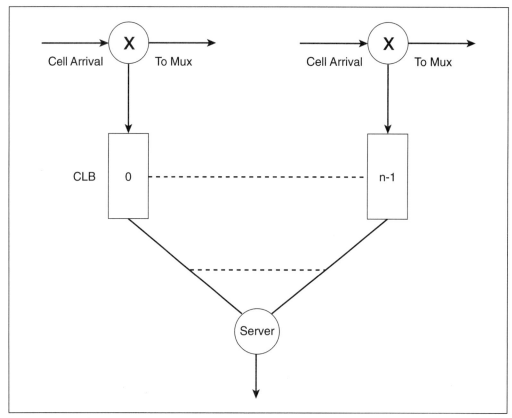

**Figure 1–10** Cooperating Leaky Bucket Model

In the nonconforming cell flow example shown in Figure 1–9, the first cell arrival finds an empty bucket and fills it to a depth of four units. On the fifth-cell time, another cell arrives and fills the empty bucket with four increments of fluid. At the sixth-cell time, a cell arrives, and the agent determines that the bucket would overflow if the new arrival's fluid were to be added. Therefore, this definition determines that this cell is nonconforming. The agent then discards the fluid for this cell and this cell is considered to be lost. Since the fluid for the nonconforming cell was not added to the bucket, the next conforming cell can arrive at cell time 7, completely filling the bucket.

### 1.1.3 Cooperating Leaky Bucket Model

The LB mechanism ensures that the traffic rate entering the network cannot exceed the leak rate of the bucket. Nonbursty traffic is a good candidate for a mean-rate policing scheme such as the LB mechanism. However, as the burstiness of the traffic increases, the LB counter limit has to increase to accommodate the arrival of traffic bursts. There is no consensus on the relationship between the LB counter limit and the burstiness of the traffic source.

The cooperating leaky bucket (CLB) mechanism takes advantage of the fact that multiplexed traffic is much less bursty, but at the same time it makes certain of the fairness in the allocation of bandwidth. It provides fairness to all traffic sources that need additional rate by offering those traffic sources an equal amount of unused bandwidth. The implementation of CLB is similar to that of LB. An LB is located at the entrance to the network. The leak rate is set to the negotiated mean traffic rate and the counter limit is set based on the burstiness of the traffic. There are no standards on the counter limit or the burstiness of the traffic source. The major difference that distinguishes CLB from LB is that once a bucket becomes empty, its leak rate is distributed to other buckets of sources that require an additional rate. When there is no empty bucket, CLB works exactly the same as LB. However, if one or more empty buckets exist, CLB distributes the unused leak rate to other buckets. In the case of LB, the leak rate is wasted if the bucket becomes empty before the arrival of the next burst of cells.

Figure 1–10 depicts the CLB model. The service rate of the server is deterministic and it is set to $n$ times the policed mean rate for identical sources, where $n$ is the number of traffic sources. The traffic sources in this model do not necessarily need to have the same mean rate. The service discipline is round robin with one-limited service. According to this discipline, after serving queue $k$, the server will then select queue $(k+1)$ modulo $n$ for service. If the bucket selected is empty, the server immediately switches to the next bucket unit in which a nonempty bucket is selected. When a leaky bucket is empty, its allocated leak rate is evenly distributed to the other nonempty buckets.

### 1.1.4 Compression Technology

MPEG sources generate different traffic rates that stress different network requirements in terms of the expected throughput, delay, and frame loss. A careful analysis of these requirements is necessary in the development of transport models for supporting MPEG services. An MVC model for MPEG transmission demonstrates an efficient method to accommodate the bandwidth requirements of multiple sources. It is beneficial for the client to reserve just the average rate and to require higher rates from other mechanisms. At the same time, the service provider can accommodate more customers and maximize utilization of system resources.

The majority of traffic is contributed by video transmission due to its huge bandwidth demand. Video compression is essential to reduce the bandwidth needed. The compression standards are critically important to the widespread usage of digital video, digital–video distribution, video-on-demand, and other multimedia services. A number of different video compression standards such as Joint Picture Experts Group (JPEG), Motion Joint Picture Experts Group (MJPEG), Motion Picture Experts Group Standard 1 (MPEG-1), Motion Picture Experts Group Standard 2 (MPEG-2), digital video interactive (DVI/Indeo), and compact disk interactive (CD-I) are available. These compression standards have different characteristics and serve different applications. The following are examples of different applications for the JPEG: color facsimile, quality newspaper wirephoto transmission, desktop publishing, and medical imaging. As for DVI/Indeo and CD-I, they are primarily developed to work with CD-ROMs applications. ITU

H.261 is the widely used international video compression standard for video–conferencing markets. The video portion of Recommendation H.320 defines the technical requirements for narrowband visual telephone services for line transmission of nontelephone signals. Compared to other video compression standards, MPEG has a higher compression ratio, better image quality, better ability to handle fast motion pictures, and is more suitable for long-haul network transmission.

A number of proposed and tested ATM systems for video transmission primarily emphasize throughput and cell-loss rate. There are two shortcomings in those systems. First, the proposed and tested systems can only accommodate CBR sources or they require the subscribing sources to adhere to the initial reserved rate throughout the entire session. As a result, the quality of video transmission changes over time since the bandwidth requirement of the video source changes over time. Second, the proposed systems are not designed to understand that not all cells from MPEG transmission are of equal importance. This lack of ability to make a distinction between cells may cause the system to drop certain cells that are more important than others.

Hać and On (1998) proposed a system to handle the bandwidth requirements needed by the VBR MPEG sources. It is sensitive to the relative importance of data from the MPEG sources. The system performs compression and selective multiplexing of transmission from MPEG sources. The concatenated traffic is then packetized into cells to a CBR channel. A CBR channel is used to transport the VBR streams as a concatenated group.

There are two basic advantages of this approach. First, compression and selective multiplexing of the MPEG sources are introduced to the system or to the transmitter side so that the decoders of the MPEG streams do not need modifications. Second, the contract between the system and the network is simple. The network is required to guarantee the delivery of all cells as long as the data rate of the combined streams does not exceed the CBR channel bandwidth such as optical carrier signal level 1, 3, and 12 (OC-1, OC-3, and OC-12).

The MPEG compression standard, ISO/IEC 11172, provides video coding for digital storage media with a rate of 2 Mbps or less. H.261 and MPEG-1 standards provide picture quality similar to that obtained with a VCR. Both standards are characterized by low-bit-rate coding and low spatial resolution. H.261 supports 352 pixels per line, 288 lines per frame, and 29.97 noninterlaced frames per second. MPEG-1 typically supports 352 pixels per line, 288 lines per frame, and 29.97 noninterlaced frames per second. The MPEG-1 compression ratio is about 100:1 and the data rate for the MPEG stream is in the range of 1.5 to 2 Mbps.

The International Standards Organization (ISO) MPEG working group has produced a specification to code combined video and audio information. The specification is directed to motion video display as compared to still image. MPEG specifies a decoder and data representation for retrieval of full-motion video information from digital storage media in the 1.5 to 2 Mbps range. The specification is composed of three parts: systems, video, and audio. The system part specifies a system coding layer for combining coded video and audio, and also provides the capability of combining private data streams and other streams that may be defined at a later date. The video part specifies the coded representation of video for digital storage media and

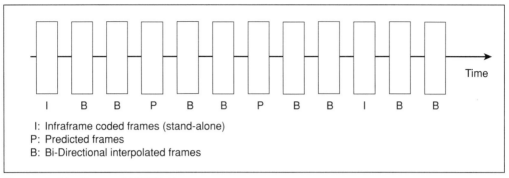

**Figure 1-11** Group of MPEG Frames

specifies the decoding process. The audio part specifies the coded representation of high-quality audio for storage media and the method for decoding of high-quality audio signals.

The MPEG standard embodies the concepts of a group of frames and interpolated frames. Each MPEG stream contains frames that are intraframe coded to facilitate random access to different video scenes. Figure 1–11 shows an MPEG stream that consists of key intraframe coded frames or I frames, predicted frames or P frames, and interpolated frames or B frames. The encoding and decoding process falls into two main categories: intraframe and interframe coding. P and B frames use combinations of key motion-predicted and interpolated frames to achieve a high-compression ratio to accommodate the data rate of the transmission channel. This method compresses every frame of video individually. The intraframe or I frame coding offers the advantage of direct editing of key I frames; however, it produces 2 to 10 times more data than the interframe coding.

MPEG's system-coding layer specifies a multiplexing scheme for elementary streams of audio and video, with a syntax that includes data fields directly supporting synchronization of the elementary streams. Figure 1–12 depicts an MPEG encoder at the functional level. The video encoder receives uncoded digitized pictures called video presentation units (VPUs) at discrete time intervals. Similarly, at discrete time intervals, the audio digitizer receives uncoded digitized blocks of audio samples called audio presentation units (APUs). Note that the times of arrival of the VPUs do not necessarily align with the arrival time of the APUs.

The video and audio encoders produce coded pictures called video access units (VAUs) and coded audio called audio access units (AAUs). The outputs are referred to as elementary streams. The system encoder and multiplexer produce a multiplex stream, M(I), containing the elementary streams as well as system-layer coding.

MPEG-1 uses three types of frames: intrapicture (I) frames; predicted (P) frames; and bidirectional (B) frames. I-type frames are compressed using only the information within the frame by the Discrete Cosine Transform (DCT) algorithm. P frames are derived from the preceding I frames or P frames by predicting forward motion in time. B-interpolated frames are derived from the previous or next I or P frames within the stream. The I, P, and B frames are coded according the MPEG specification. The B frames are required to achieve a low average data rate.

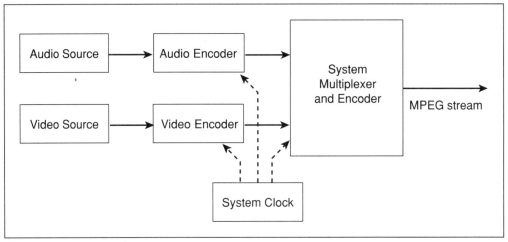

**Figure 1–12** MPEG Encoder

The bandwidth allocated to each type of frame typically conforms to the ratio of 5:3:1 for I, P, and interpolated frames, respectively.

An important aspect of MPEG is synchronization, which is a fundamental issue for multimedia communication. In multimedia, synchronization means that various signal objects comprising the combined signal must be stored, retrieved, and transmitted with precise timing relationships. To achieve synchronization in multimedia systems that decode multiple video and audio signals originating from a storage or transmission medium, there must be a "time master" in the decoding process. MPEG does not specify which entity is the time master. The time master can be (1) any of the decoders, (2) the source stream, or (3) an external time base. All other entities in the system such as decoders and information sources must slave their timing to the master. If a decoder unit is taken as the time master, the time it shows a presentation unit is considered to be the correct time for the use of the other entities. Decoders can implement phase-locked loops or other timing means to ensure proper slaving of their operation to the time master. If the time base is an external entity, all of the decoders and the information sources must slave the timing to the external timing source.

The method of multiplexing the elementary streams of VAUs and AAUs is not directly specified in MPEG. However, there are some constraints that must be followed by an encoder and a multiplexer to produce a valid MPEG data stream. For example, it is required that the individual stream buffers must not overflow or underflow.

### 1.1.5 Multimedia Virtual Circuit Model

The Multimedia Virtual Circuit Model (MVC) consists of four components: virtual circuit negotiator (VCN), multimedia virtual circuit (MVC) queue, shared LB, and video multiplexing server (VMS) with available bit rate (ABR) pool.

# Broadband Networks

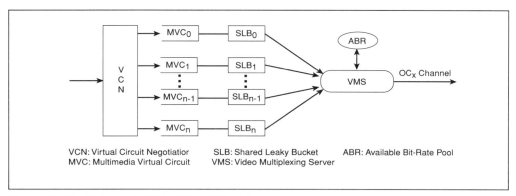

**Figure 1–13** Multimedia Virtual Circuit Model

The VCN performs the call admission in the MVC model. The client establishes an MVC call with the VCN. Each client reserves an average rate with the VCN for each call. The VCN can admit a call into the system until the communication link capacity is exhausted. The reservation of an average rate for each MVC call enables this model to accommodate more clients compared to requiring each call to reserve the maximum rate. Figure 1–13 shows four components of the MVC model.

At discrete time intervals, each call submits its frames to the MVC queue waiting to be serviced and transmitted. The MVC stage receives digitized VPUs and digitized APUs from each call. It then performs the video and audio encoding for VPUs and APUs to VAUs and AAUs. The MVC stage performs MPEG video and audio encoding of each call according to the MPEG specification.

The shared leaky bucket (SLB) model is part of the MVC model. The implementation of SLB is similar to that of the LB where the SLB is set to the negotiated average rate. A major difference that distinguishes SLB from LB is that when a bucket becomes empty its leak rate is distributed to the ABR pool. The SLB principle shares the unused leak rate among each other and takes advantage of statistical characteristics of the traffic sources by sharing the unused leak rate with the needed sources. In LB, the leak rate is wasted when the bucket is empty since it does not share the unused leak rate with other LB. SLB works similarly to CLB except for the method it uses to distribute the unused leak rate. The CLB model distributes the unused leak rate evenly among other CLBs that require additional rates. The SLB model distributes the unused leak rate according to the first-come-first-served principle.

The use of the SLB can enforce the reserved rate for each MVC call. There are two options available when the MVC call requires additional rates. The first is to request an extra capacity from the VCN if unallocated capacities are available. The second is to borrow from the ABR pool resulting from the unused capacity from each MVC call. The ABR pool is a storage of the unused leak rate from each MVC call. The ABR pool size changes dynamically during each cycle or frame period and the unused capacity from the previous frame period does not accumulate to the next period. When an MVC call requires capacity beyond the average rate during which no extra capacity is available from either the VCN or ABR pool, the service provider or telecommunication operator has to perform the frame quality scaling (FQS) method. The FQS method approach reduces the information content of a frame to fit into a particular transmission bandwidth.

The VMS can be represented by a polling system with $x$ SLBs, where $x$ is the number of sessions. The service rate of the VMS is deterministic and the service discipline is round robin. According to this discipline, each time after VMS served SLB $k$, the VMS will select SLB $(k+1)$ modulo $x$ for service. If the SLB selected is empty, the VMS will immediately switch to the next SLB until a nonempty SLB is selected. This polling model guarantees that the leak rate of each SLB is at least the reserved average rate when it is not empty. When an SLB is empty, its allocated leak rate is distributed to the ABR pool. This ABR pool is managed by the VMS unit. The SLB for a call negotiates with VMS for additional rates from the ABR pool when an MVC call requires an additional rate. When a frame from an MVC call requires more than the reserved rate and there is no rate available, the service provider uses the FQS method.

Data in an MPEG coded video stream are of different importance. In the MPEG compression scheme, the output data stream can be divided into macroblocks. Each macroblock has a certain number of bits or code words that can be dropped if necessary. Associated with each macroblock is a function relating to the number of code words or bits retained and the corresponding image quality. With each macroblock, code words or bits are ordered according to their significance, those of less significance are dropped first when necessary. The header information, Motion Vector (MV) and DC components of an MPEG frame are the most important. Among the DCT AC components, those of lower frequencies are more important than those of higher frequencies for two reasons. First, the amplitude square of the DCT AC components tends to decrease along the zig-zag scanning order. Second, human vision is less sensitive to the high-frequency signals. Figure 1–14 depicts the above ordering concept.

In the FQS method, let $N$ be the total number of macroblocks collected from a frame. Let macroblock $MB_i(D)$ be the number of code words or bits in macroblock $MB_i$ that must be retained in order to maintain a distortion level of $D$. The goal is to find a distortion level $D'$ such that

$$MB_1(D') + MB_2(D') + \ldots + MB_N(D') = MB_t \qquad (1-1)$$

Macroblock $MB_t$ is defined as the transmission bandwidth available to transmit the $N$ macroblocks within the frame. As a result, when $MB_t$ is insufficient to transport all the code words or bits from all macroblocks, code words are dropped until the above equality can be achieved. The objective is for all macroblocks to achieve the same distortion level $D$.

When a frame from an MVC call requires an additional rate that is not available from unallocated capacity or ABR, the service provider reduces the number of code words from each macroblock. Each macroblock of the frame has equal quality relative to other macroblocks and at the same time scales down the information contents of the frame to fit into the available rate. This method avoids the dropping of frames when the available rate is insufficient to transmit the

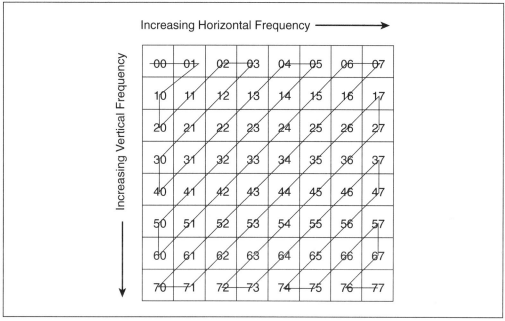

**Figure 1–14** Zigzag Scanning Order of DCT Components

frame. It also smooths out the quality of the picture by extracting away only higher frequency code words from each macroblock.

The MVC model provides benefits to both the service provider and clients. The model is beneficial to the client since it is able to reserve just the average rate as well as acquiring more rates from other mechanisms. If the clients reserve the maximum rate, they have to pay for an idle reserved rate during the call session. The service provider can accommodate more customers and enhance the utilization of system resources.

Two key benefits of the MVC model for MPEG are statistical multiplexing and network simplicity. The MVC model employs statistical multiplexing of video sources since VBR sources have different degrees of activity during a connection. Many future ATM applications exhibit such behavior. Statistical multiplexing allows more sources to be admitted when not all VBR sources are expected to generate cells at their peak rates during the entire connection.

## 1.2 Rate-Based Congestion Control Schemes for ABR Service in ATM Networks

In ATM networks, traffic can be divided into two classes: guaranteed and best effort. Guaranteed traffic requires an explicit guarantee of service given by the network, as in CBR and VBR services. The limit on each connection's usable bandwidth is based on the notion of traffic contract. For these services, congestion control is administered through admission control and bandwidth allocation. However, the bandwidth requirements for data traffic are not likely to be known at connection set-up time. Dynamically sharing the available bandwidth among all active users is required for this service, referred to as best-effort or ABR service, which has a highly bursty nature of data traffic.

### 1.2.1 Services in the ATM Network

The capability of ATM networks to provide a large bandwidth and to handle multiple QoS guarantees can be realized by preparing effective traffic management mechanisms. Traffic management includes congestion control, call admission control, and VP / VC routing. Essential for the stable and efficient operation of ATM networks is congestion control, which is performed between ATM end systems. The ATM end system is the point where an ATM connection is terminated, and the connection goes up to the ATM AAL. It is also defined as the point where VC connections are multiplexed, demultiplexed, or both. Therefore, each ATM segment can, depending on its characteristics, adopt a different congestion-control scheme.

An important issue in the selection of a congestion control scheme is the traffic pattern. In CBR and VBR services the traffic parameters are described, for example, in terms of such measures as peak-cell rate, cell-delay variation, sustainable-cell rate and burst-length tolerance. Once the connection request is admitted, the QoS is guaranteed throughout the session. The congestion control for CBR and VBR services is administered through admission control and bandwidth allocation. If the ATM network cannot deliver the resources demanded by the connection request, the request will be rejected at call set-up time. Voice and video are examples of sources that require guaranteed traffic service.

The remaining bandwidth not used by guaranteed bandwidth services must be shared fairly among all active users by using ABR services, or best-effort services. An example is data communication.

### 1.2.2 Congestion Control for ABR Service

To support ABR traffic, the network requires a feedback mechanism to inform each source how much data to send. The main feedback mechanisms are credit-based and rate-based flow control.

Credit-based flow control schemes make use of hop-by-hop feedback loops. Each link maintains its own independent control loop. The traffic moves across the network through a series of hop-by-hop feedback loops. The receiving end of each link issues "credits" to the transmitting end indicating the number of cells the transmitting station is allowed to send. Source end stations transmit only when they have permission from the network, as shown in Figure 1–15. In this approach, each link in the network runs the flow-control mechanism independently for each virtual circuit. A number of cell buffers are reserved for each virtual circuit at the receiving end of each link. One round-trip worth of cell buffers must be reserved for each connection. The amount of buffering required per connection depends on the propagation delay of the link and the required transmission rate of the virtual connection.

In high-speed Wide Area Network (WAN) based on ATM, the propagation delay is greater then the queuing delay. That is, it takes longer for the data to cross the link than for a switch or

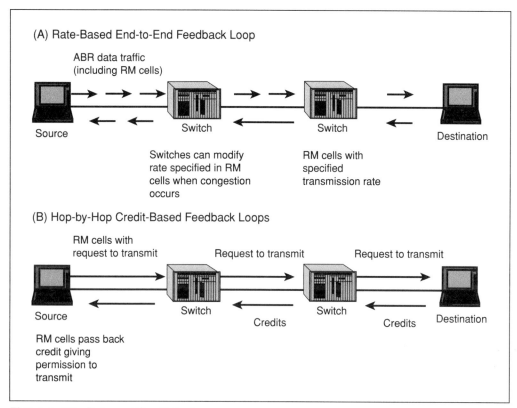

**Figure 1–15** Rate and Credit Basics

the end station to process the transmission. Buffer queues fill up more quickly than the network can accommodate the traffic. The buffer sizes required to support a hop-by-hop credit scheme are impractical.

Rate-based flow-control schemes are end-to-end feedback mechanisms. They have one source and one destination station for each feedback loop, as shown in Figure 1–15. Within the feedback loop, the destination end alerts the source end to slow transmission when congestion occurs. If there are ATM switches between the loop's source and destination, these devices simply forward and augment the flow-control information moving between the destination and the source.

In this end-to-end rate-based scheme, a resource management (RM) cell is used, which is a standard 53-byte ATM cell used to transmit flow-control information. This RM cell carries information over the virtual circuit and is therefore allowed to flow all the way to the destination/source end station. The destination reflects or issues the RM cell, with an indicator to show the status of the traffic. The intermediate switches then simply forward or mark down the rate in the RM cell if needed. The source-end system then uses the information in the RM cells for subsequent transmissions until a new RM cell is received.

### 1.2.3 Rate-Based Control Schemes

There are rate-based congestion control schemes such as the forward explicit congestion notification (FECN) and backward explicit congestion notification (BECN) schemes, proportional rate control algorithm (PRCA), and intelligent congestion control schemes.

**FECN and BECN Schemes.** FECN schemes make use of the explicit forward congestion indication (EFCI) state carried in the payload type identifier (PTI) field to convey congestion information in the forward direction. When the switch becomes congested, it will mark in each VC the EFCI state of all cells being forwarded to the destination. Upon receiving marked cells, the destination sends an RM cell back to the source along the backward path. Then the source-end system must decrease its cell transmission rate accordingly on each VC. A time interval is defined at the destination-end system, and only one RM cell is allowed to be sent. The source-end system is also provided with an interval timer update interval (UI). When the timer expires without an RM cell received, the source recognizes no congestion in the network and increases the transmission rate.

BECN schemes use similar mechanisms except that the congestion notification is returned directly from the point of congestion to the source by the marked EFCI state of the cells. Thus, the response to congestion is faster.

These approaches require interval timers at the end systems and increase the complexity of the implementation. Furthermore, they use the negative feedback mechanism and could cause a collapse of the network in heavily congested conditions due to RM cells that are delayed or lost and the source increasing its cell emission rate because of the absence of RM cells.

**PRCA Scheme.** PRCA is based on a positive feedback rate-control paradigm. This shift in the rate-control paradigm is intended to remedy the problem of network congestion collapse encountered in FECN and BECN schemes. In PRCA, the source-end system marks the EFCI bit in all data cells except for the first cell of every NRM cells, where NRM is the number of data cells issued between two RM cells emitted. The destination-end system instantly sends an RM cell back to the source when it receives a cell with the EFCI bit cleared. If the EFCI bit is set by an intermediate switch because of congestion, the destination takes no action. By using this mechanism, receiving the RM cell implies that there is no congestion in the network, and therefore, the source-end system is given the opportunity to increase its rate. Otherwise, the source continuously decreases its transmission rate.

There are certain problems with PRCA. When VC is set up through more congested links, the EFCI bit of its data cells will be marked by the congested switches more often than those of other VCs set up through fewer congested links. Consequently, such VC will have a lower allowed cell rate (ACR) than other VCs. This undesirable effect of VC starvation is proportional to the number of congested links on which VC transmits its cells, and is referred to as the ACR beat-down problem.

**Intelligent Control Scheme.** In intelligent congestion control schemes, each queue of a switch maintains a variable modified ACR (MACR), which is the value of the estimated optimal cell rate. The RM cells need to contain Current Cell Rate (CCR) and Explicit Rate (ER). Both CCR and ER are modified constantly by the intermediate switches according to the traffic in the network. Although this intelligent scheme shares resources more fairly than the other schemes, there is the expense of having interval timers and VC tables in every switch to detect the congestion status. This intelligent scheme also needs to maintain an MACR for each port in every switch, with the computation to estimate the best possible cell rate on each VC, which can increase the complexity of all switches. In addition, at a switch the effective rate of some connections that experience congestion at another switch can be quite different from the CCR values contained in RM cells; this can lead to misbehavior of the cell-rate control.

A rate-based control scheme proposed by Hać and Ma uses bandwidth fairly and efficiently for all switch connections under the proper congestion control and makes the complexity of switches low. In this scheme, all ATM switches in the network need to keep only one value shared ABR bandwidth (SB) for each output port. When an RM cell is received, the value of SB is modified, no matter to what path the RM cell belongs. Each end-system source modifies the issued cell rate, or ACR, according to the SB value carried in the backward RM cells.

## 1.3 Multiple Access Protocols for Wireless ATM Networks

Wireless ATM (WATM) is considered the framework for the next generation of wireless communication networks. ATM in the wireless environment has to cope with the low-speed and noisy, wireless medium. An appropriate medium access protocol that can efficiently utilize and share this limited frequency spectrum is essential.

ATM supports multimedia services at any speed from time-bounded voice and multimedia communications to bursty data traffic. Broadband wireless networks support traditional voice service as well as mobile communications with multimedia applications.

ATM networks are fixed (optical) point-to-point networks with high bandwidth and low error rates. These attributes are not associated with the limited bandwidth and error-prone radio medium. While increasing the number of cables (e.g., copper or fiber optics) can increase the bandwidth of wired networks, wireless telecommunications networks experience a more difficult task. Due to limited usable radio frequency, a wireless channel is an expensive resource in terms of bandwidth. For wireless networks to support high-speed networks like ATM, we need a new multiple-access approach for sharing this limited medium in a manner different from the narrowband, along with the means of supporting mobility and maintaining QoS guarantees.

Media Access Control (MAC) is a set of rules that attempt to efficiently share a communication channel among independent competing users. Each MAC uses a different media or multiple access scheme to allocate the limited bandwidth among multiple users. Many multiple-access protocols have been designed and analyzed both for wired and wireless networks. Each has its advantages and limitations based on the network environment and traffic. These schemes can be classified into three categories: fixed assignments, random access, and demand assignment. The demand assignment scheme is the most efficient access protocol for traffic of varying bit rate in the wireless environment.

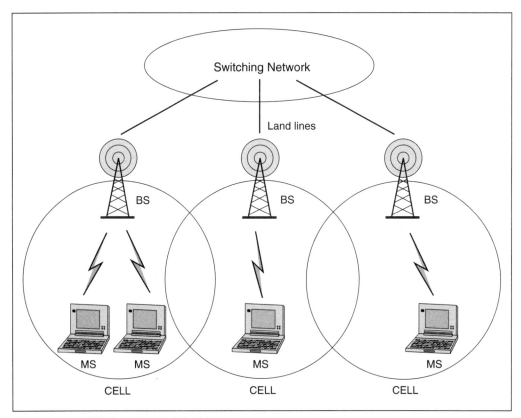

**Figure 1–16** Wireless Network Architecture

### 1.3.1 Wireless Networks

The most widely employed wireless network topology is the cellular network. This network architecture is used in cellular telephone networks, personal communication networks, mobile data networks, and wireless local area networks (WLAN). In this network configuration, a service area, usually over a wide geographic area, is partitioned into smaller areas called cells (Figure 1–16). Each cell, in effect, is a centralized network, with a base station (BS) controlling all the communications to and from each mobile user in the cell. Each cell is assigned a group of discrete channels from the available frequency spectrum, usually a radio frequency. These channels are, in turn, assigned to each mobile user when needed.

Typically, BSs are connected to their switching networks using landlines through switches. The BS is the termination point of the user-to-network interface of a wireless cellular network. In addition, the BS also provides call set-ups, cell handoffs and various network-management tasks, depending on the type of network.

Due to the limited radio frequencies available for wireless communication, wireless networks have to maximize the overall capacity attainable within a given set of frequency channels.

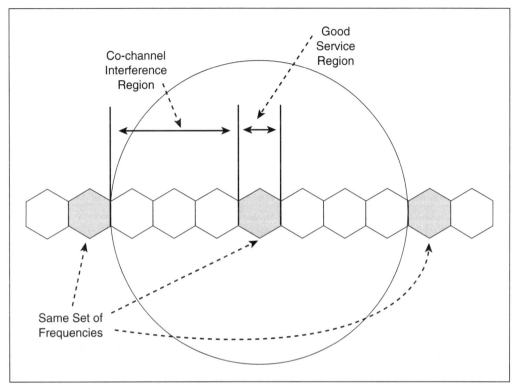

**Figure 1–17**  Illustration of Channel Reuse

Spectral efficiency describes the maximum number of calls that can be served in a given service area. To achieve high spectral efficiency, cellular networks are designed with frequency reuse, initially proposed by the Bell Telephone Laboratories™. If a channel with a specific frequency covers an area of a radius R, the same frequency can be reused to cover another area (Figure 1–17). A typical cellular service area using frequency reuse is shown in Figure 1–18. A service area is divided into 7-cell clusters. Each cell in the cluster, designated one through seven, uses a different set of frequencies. The same set of frequencies in each cell can be reused in the same service area if it is sufficiently apart from the current cell. Cells using the same frequency channels are called cocells. In principle, by using this layout scheme, the overall system capacity can be increased as large as desired by reducing the cell size, while controlling power levels to avoid cochannel interference. Cochannel interference is defined as the interference experienced by users operating in different cells using the same frequency channel. Smaller size cells called microcells are implemented to cover areas about the size of a city block. Research has been done on even smaller cells called picocells.

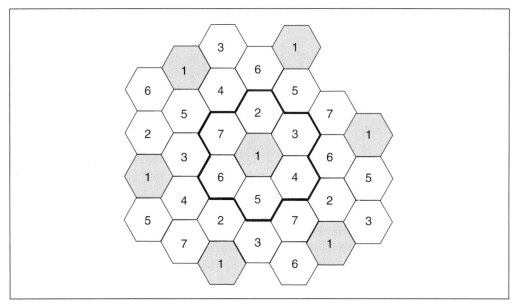

**Figure 1–18** Typical Cellular Frequency Reuse Pattern, with Seven-Cell Clusters

### 1.3.2 Wireless ATM

#### 1.3.2.1 ATM Services

Users request services from the ATM switch in terms of destination(s), traffic type(s), bit rate(s), and QoS. These requirements are usually grouped together and categorized in different ATM traffic classifications. The prototypical ATM services are categorized as follows.

- constant bit rate (CBR): Connection-oriented constant bit-rate service such as digital voice and video traffic.
- real-time variable bit rate (rt-VBR): Intended for real-time traffic from bursty sources such as compressed voice or video transmission.
- non-real-time variable bit rate (nrt-VBR): Intended for applications that have bursty traffic but do not require tight delay guarantee. This type of service is appropriate for connectionless data traffic.
- available bit rate (ABR): Intended for sources that accept time-varying available bandwidth. Users are only guaranteed a minimum cell rate (MCR). An example of such traffic is LAN emulation traffic.
- unspecified bit rate (UBR): Best-effort service that is intended for noncritical applications. It does not provide traffic-related service guarantees.

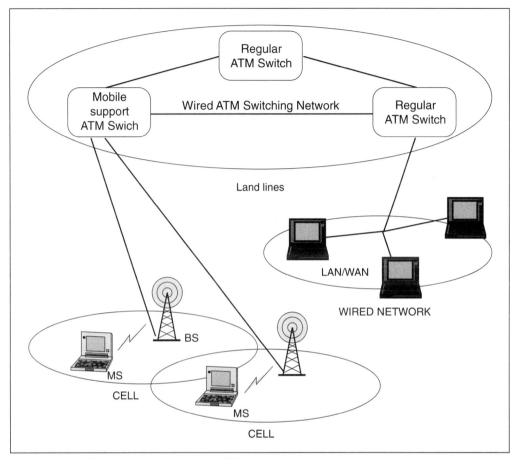

**Figure 1–19** Wireless Integration in an ATM Network

### 1.3.2.2 Wireless Integration

The future of integrated multimedia networks will be dominated by broadband ATM. In addition to providing mobility, wireless ATM networks also allow flexible bandwidth allocation and QoS guarantees that existing wireless LAN is unable to provide. The wired/wireless integration is illustrated in Figure 1–19. Wireless networks are connected to wired networks using high-speed landlines through ATM switches.

This integration raises many serious compatibility issues. First, there is the issue of bandwidth. Wireless medium has a limited (e.g., maximum rate of about 34 Mb/s) and expensive bandwidth, while the ATM was designed for a bandwidth-rich environment. In addition, a wired ATM operates at a very low bit-error rate (BER), whereas wireless medium experiences a noisy and time-varying environment.

### 1.3.3 Multiple-Access Protocol

A multiple-access protocol is a scheme to control the access to a shared communication medium (a radio frequency in this case) among various users. Although many access protocols have been proposed and each has its advantages and limitations, very few are suitable for integrated wireless communications. Access protocols can be grouped according to the bandwidth allocation mechanism, which can be static or dynamic, according to the type of control mechanism implemented. Multiple-access protocols can be categorized into three classes: fixed assignment, random assignment, and demand assignment.

#### 1.3.3.1 Fixed Assignment

Time-division multiple access (TDMA) and frequency-division multiple access are fixed assignment techniques that incorporate permanent subchannel assignments to each user. These 'traditional' schemes perform well with stream-type traffic such as voice, but are inappropriate for integrated multimedia traffic because of the radio channel spectrum utilization. In a fixed-assignment environment, a subchannel is wasted whenever the user has nothing to transmit. It is widely accepted that most services in the broadband environment are VBR service (e.g., bursty traffic). Such traffic wastes a lot of bandwidth in a fixed-assignment scheme.

#### 1.3.3.2 Random Assignment

Typical random assignment protocols like ALOHA and carrier sense multiple access with collision detection (CSMA/CD) schemes are more efficient in servicing bursty traffic. These techniques allocate the full-channel capacity to a user for short periods on a random basis. These packet-oriented techniques dynamically allocate the channel to a user on a per packet basis.

Although there are a few versions of the ALOHA protocol, in its simplest form it allows users to transmit at will. Whenever two or more user transmissions overlap, a collision occurs and users have to retransmit after a random delay. The ALOHA protocol is inherently unstable due to the random delay. That is, there is a possibility that a transmission may be delayed for an infinite time. Various collision resolution algorithms were designed to stabilize and reduce contention in this scheme.

Slotted ALOHA is a simple modification of the ALOHA protocol. After a collision, instead of retransmitting at a random time, slotted ALOHA retransmits at a random time slot. Transmission can only be made at the beginning of a time slot. Obviously, this protocol is implemented in time-slotted systems. Slotted ALOHA is proven to be twice as efficient as a regular or pure ALOHA protocol.

CSMA/CD takes advantage of the short propagation delays between users in a typical LAN and provides a very high throughput protocol. In a plain CSMA protocol, users will not transmit unless the protocol senses that the transmission channel is idle. In CSMA/CD, the user also detects any collision that happens during a transmission. The combination provides a protocol that has high throughput and low delay. However, carrier sensing is a major problem for radio networks. The signal from the local transmitter will overload the receiver, disabling any attempts to sense remote transmission efficiently. Despite some advances in this area, sensing still poses a problem due to severe channel fading in indoor environments. Similarly, collision detection proves to be a difficult task in wireless networks. While it can be easily done on a wired network by measuring the voltage level on a cable, sophisticated devices are required in wireless networks. Radio signals are dominated by the terminal's own signal over all other signals in the vicinity, preventing any efficient collision detection. To avoid this situation, a terminal transmitting antenna pattern has to be different from its receiving pattern. This requires sophisticated directional antennas and expensive amplifiers for both the BS and the mobile station (MS). Such requirements are not feasible for the low-powered mobile terminal end.

Code-division multiple access (CDMA) is a combination of both fixed and random assignment. CDMA has many advantages such as near zero channel-access delay, bandwidth efficiency, and excellent statistical multiplexing. However, it suffers from significant limitations such as limited transmission rate, complex BS, and problems related to the power of its transmission signal. The limitation in transmission rate is a significant drawback to using CDMA for integrated wireless networks.

#### 1.3.3.3 Demand Assignment

In this protocol, channel capacity is assigned to users on a demand basis as needed. Demand assignment protocols typically involve two stages: a reservation stage where the user requests access, and a transmission stage where the actual data is transmitted. A small portion of the transmission channel, called the reservation subchannel, is used solely for users requesting permission to transmit data. Short reservation packets are sent to request channel time by using some simple multiple-access schemes, typically, TDMA or slotted ALOHA. Once channel time is reserved, data can be transmitted through the second subchannel contention free. Unlike a random-access protocol where collisions occur in the data transmission channel, in demand assignment protocols, collisions occur only in the small-capacity reservation subchannel.

This reservation technique allows demand-assignment protocols to avoid bandwidth waste due to collisions. In addition, unlike fixed-assignment schemes no channels are wasted whenever a VBR user enters an idle period. The assigned bandwidth will simply be allocated to another user requesting access. Due to these features, protocols based on demand-assignment techniques are most suitable for integrated-wireless networks.

Demand-assignment protocols can be classified into two categories based on the control scheme of the reservation and transmission stages. They can be either centralized or distributed. An example of a centralized controlled technique in demand assignment is polling. Each user is sequentially queried by the BS for transmission privileges. This scheme, however, relies heavily on the reliability of the centralized controller.

An alternative approach is to use distributed control, where MSs transmit based on information received from all the other MSs. Network information is transmitted through broadcast channels. Every user listens for reservation packets and performs the same distributed scheduling algorithm based on the information provided by the MS in the network. Requests for reservation are typically made using contention or fixed-assignment schemes.

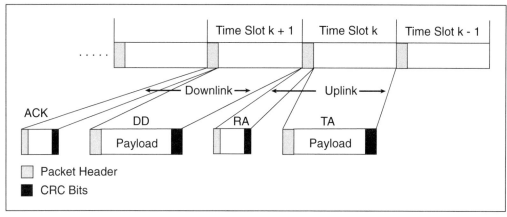

**Figure 1–20** Radio Channel Classification: Slot-by-Slot

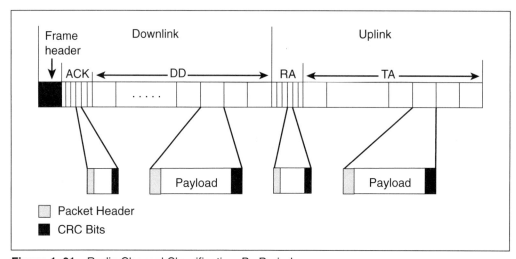

**Figure 1–21** Radio Channel Classification: By Period

### 1.3.4 Demand Assignment Multiple-Access (DAMA) Protocols

Most DAMA protocols use time-slotted channels that are divided into frames. Depending on the transmission rate and the type of services, the channel bandwidth can be represented by a single or multiple frame(s). Each frame is divided into an uplink and a downlink period (i.e., channel). These periods are further divided into two subperiods or slots. They can be partitioned on a slot-by-slot or period basis (Figures 1–20 and 1–21, respectively). In the slot-by-slot method, each uplink and downlink period consist of a single time slot. In the method by period, the uplink and downlink period contains multiple time slots, encapsulated as a frame. The uplink and downlink communications can be physically separated using different frequency channels or dynamically shared using the time-division duplex (TDD) system.

The uplink channel (e.g., mobile-to-base) is divided into the request access (RA) and data transmission access (TA) subperiods. On the other hand, the downlink channel is divided into the acknowledgment (ACK) and the data downstream (DD) subperiods. A user requests bandwidth using the RA subperiods (i.e., uplink). When the BS hears a successful request such as no collision, it will notify the corresponding user through the ACK subperiods (i.e., downlink). Successful users are then assigned bandwidth, if available, in the TA subperiods. The DD subperiods are used by the BS to transmit downstream data to mobiles. These subperiods, also known as slots, vary in length depending on the type and amount of information they carry as determined by the protocol designer. The RA and ACK slots are much smaller than the data slots; hence, their time intervals are called minislots. Depending on the protocol, they may not have equal lengths.

DD transmissions are controlled by the BS and are performed contention free; typically using a time-division multiplexing (TDM) broadcast mode. These transmissions are performed with little delay and are not a crucial performance driver of the system.

In a wireless service area, the number of mobiles a BS covers is much larger than the available channel bandwidth. However, not all the mobiles are active simultaneously. Therefore, mobiles request access using random-access schemes in the RA subperiods. Access methods like ALOHA and its variations are usually used. On the other hand, variations of fixed-assignment schemes are typically used in the TA channel. The methods are the TDMA and CDMA schemes, of which TDMA is easier to implement.

**Resource Auction Multiple Access (RAMA).** The RAMA protocol was proposed as a fast resource or call-level assignment and handoff mechanism. RAMA is a deterministic DAMA protocol in which mobile users request access by transmitting their b-bit IDs, assigned during call set-up, on a symbol-by-symbol basis. For example, the b-bit ID could be a 10-digit number. Each digit represents a symbol and is transmitted one at a time. After each transmission, the BS acknowledges the symbol with the largest value and ignores the rest of the symbols. Mobiles that do not hear their symbol drop out of the auction. After 10 rounds of these transmission and acknowledgement processes, a single winning mobile remains (e.g., the mobile with the highest ID). The BS then assigns an available communication channel to the winner.

The RAMA protocol provides assignment at a rate that is suitable for many applications, including some statistical multiplexing of voice. For this protocol to be employed in ATM networks, packets need to be used. In that case, RAMA can be implemented on a slot-by-slot basis (i.e., each time slot identifying a mobile to transmit a packet). Like most DAMA, the transmission time slots are never wasted when any mobiles have packets to transmit. However, the overhead experienced during the auction phase is substantial. The total time required for a contention period (i.e., auction) is given by $t = K_m(2t_d + 2T_s)$, where $K_m$ is the number of $M$-ary digits representing the mobile's ID, $T_s$ is the symbol duration, and $t_d$ is the guard interval after each symbol. The guard interval $t_d$ must be sufficiently long to accommodate the on/off switching of transmitters and to allow for the processing and propagation delays between consecutive uplink and downlink transmissions.

**Packet Reservation Multiple Access (PRMA).** PRMA was proposed for packet-based voice transmission over wireless networks. PRMA is designed to increase the bandwidth efficiency over fixed assignment TDMA. In this protocol, time slots are grouped into frames and each timeslot is labeled as either reserved or available according to an acknowledgement from the BS at the end of each slot. When a mobile that is generating periodic traffic successfully transmits a packet in an available slot (using the slotted ALOHA protocol), the mobile also reserves that slot in the future frames. There is no subsequent contention with other mobiles in that slot until it is released. A mobile releases a time slot at the end of a burst by leaving the reserved slot empty. Packets from random traffic also contend for available time slots using slotted ALOHA. However, when a random packet is successfully transmitted, the time slot is not reserved in subsequent frames.

Although PRMA can be used in most packet-oriented networks, it proves to be inefficient for wireless ATM. PRMA has a variable-channel access delay (packet or burst level), which does not suit the requirements of the ATM services.

**Distributed Queuing Request Update Multiple Access (DQRUMA).** The DQRUMA protocol is designed specifically for data-packet (e.g., ATM) networks. It attempts to provide an efficient bandwidth-sharing scheme that can satisfy QoS parameters and support various types of ATM services. DQRUMA is designed for fixed-length packets (e.g., ATM cells) arriving at the mobile at some bursty random rate. The uplink and downlink periods are configured on a slot-by-slot basis. The uplink slot comprises a single data transmission slot (i.e., TA slot) and one or more RA minislots.

DQRUMA introduces the concept of a dynamic uplink slot where the uplink slot may be converted into a whole RA channel filled only with RA minislots. This conversion occurs when the BS senses that there is a significant amount of collisions in the RA channel. It allows as many as 25 RA minislots in a single time slot where mobiles can send their requests. This method drastically reduces request contention in the RA channel. Requests in the RA channel use a random access protocol like slotted ALOHA. The downlink slot contains the typical DD time slot, one or more ACK minislots, and a transmission permission, or perm slot. In situations where the uplink slot is converted into a series of RA minislots, the subsequent downlink slot is converted into a series of corresponding ACK minislots.

When a mobile terminal transmits its RA packet, it listens to the downlink slot for its ACK. ACK only indicates that the request has been received by the BS. Mobile users may not transmit their data until they hear their b-bit access ID in the perm slot. Upon hearing the transmission permission (b-bit ID), users may transmit their data in the next uplink time slot. This is the distributed queuing aspect of the protocol, where packets are queued at the mobile's buffer until the BS services them according to a scheduling policy (e.g., in a round-robin fashion).

DQRUMA also introduces an extra bit called the piggyback (PGBK) bit in the uplink channel. Each time a mobile transmits a packet (uplink), it also includes this PGBK bit to indicate whether it has more packets in the buffer. This bit serves as a contention-free transmission request for a mobile transmitting a packet. The BS checks this bit and updates the appropriate entry in its request table in the BS, accordingly. When this bit is included, a mobile does not need a request for channel access in the following time slot. The BS knows that the mobile has more data to transmit and will assign a time slot to the mobile accordingly. This is the update portion of the protocol. The PGBK bit drastically reduces contention in the RA channel and greatly improves the overall protocol performance, especially for bursty traffic. Nevertheless, DQRUMA suffers from some channel-access delay problems due to its inability to distinguish different types of traffic.

RAMA performs like an ideal system where transmission time slots are never wasted when mobiles are transmitting. Unfortunately, it produces very significant overhead due to the auction process. Under certain reasonable ATM conditions, RAMA may produce an overhead of approximately 10%. Such substantial overhead makes it very impractical for packet-level access in the limited-bandwidth wireless environment. In addition, unlike DQRUMA where the BS is capable of controlling the order in which mobiles transmit their data based on some scheduling policy, RAMA is solely based on the value of their IDs (i.e., the largest value wins). This is not desirable in the ATM environment where there is an integrated mix of multimedia traffic with different priorities due to different QoS requirements.

PRMA offers a simpler strategy with little overhead and can be easily implemented in any packet network. However, a more sophisticated protocol like DQRUMA provides superior performance in several ways. PRMA wastes a single time slot at the end of each transmission burst by leaving the reserved slot empty (i.e., available slot). In addition, PRMA uses the entire uplink transmission channel for requests (slotted ALOHA), whereas DQRUMA uses only the RA channel. Therefore, when collisions occur, DQRUMA wastes less bandwidth (only a portion of the uplink channel). Finally, as in the comparison with RAMA, DQRUMA allows the BS to control the packet-transmission policy, allowing it to accommodate different types of traffic with various transmission rates. PRMA provides only a limited support for periodic traffic using a reserved packet.

DQRUMA's ability to reduce contention in the request channel with the PGBK bit and the dynamic RA channel conversion significantly improves its performance as a general multiple-access protocol. The flexibility of its BS control channel assignment offers a potential to support different traffic types like those in the ATM network.

Like most protocols, DQRUMA has its drawbacks. Although it is capable of supporting various type of services (e.g., CBR, VBR, etc.), it does not offer any distinction between different types of traffic during the reservation phase. Therefore, time-sensitive application that produces real-time CBR traffic is not given priority over ABR or UBR traffic. A more discriminating request scheme that could assign priority to requests from different traffic types is desired. While DQRUMA claims to support various service types, no detailed procedures for handling different services have been proposed. In addition, admitted but idle users (e.g., VBR) are required to perform a new request when resuming transmission. There are no distinctions between them and other newly arrived services. In order to distinguish between new and recurring VBR traffic, data packets have to carry additional header bits while increasing the complexity of the scheduling policy in the BS. This is a difficult trade-off problem between the channel access delay and the protocol overhead (bandwidth and complexity). The traffic type in a specific network environment plays a significant role in such protocol design.

**An Adaptive Request Channel Multiple Access (ARCMA) Protocol.**
ARCMA is a demand assignment multiple-access protocol with dynamic bandwidth allocation. Its basic architecture is modeled after the DQRUMA protocol. This scheme is designed to function in a cell-based wireless network with many MSs communicating with the BS of their particular cell. Transmissions are done on a slot-by-slot basis without any frames. As with DQRUMA, each slot is divided into a TA slot and a RA minislot. However, the RA channel in ARCMA is capable of carrying additional information for different classes of ATM service (e.g., CBR, VBR, etc.). This additional information is used by the BS to provide better QoS support for different classes of traffic. As in PRMA, transmission from CBR traffic may reserve an incremental series of slots in the duration of their transmission. No further request is needed until the CBR transmission finishes.

The BS maintains a request table to keep track of all successful requests, and assigns permission to mobiles for transmission at different time slots. In ARCMA protocol, the BS inspects the service class of a request and gives transmission priority to delay sensitive data (e.g., CBR). A piggyback bit is used in the uplink channel to reduce contention in the RA channel. This is especially beneficial for bursty traffic.

A dynamic RA channel similar to that of DQRUMA is used where an entire uplink channel can be converted into multiple RA channels. This conversion is done when the request table is empty, which in most cases indicates heavy collisions in the request channel. ARCMA uses a more complex algorithm that takes advantage of the random-access scheme in the RA channel. The slotted ALOHA with binary exponential backoff (BEB) is used as the random access protocol for ARCMA.

ARCMA improves the spectral efficiency by reducing collisions in the RA channel while improving support for the various classes of ATM services.

CHAPTER 2

# Architecture of Telecommunications and Wireless Networks

## 2.1 Multimedia Network Architecture

Multimedia communications is the field referring to the representation, storage, retrieval and dissemination of machine-processable information expressed in multimedia, such as voice, image, text, graphics, and video. With the advent of high-capacity storage devices, powerful and yet economical computer workstations, and high-speed integrated-services digital networks, providing a variety of multimedia communication services is becoming not only technically but also economically feasible. Multimedia conference systems can help people interact with each other from their homes or offices while they work as teams by exchanging information in several media, such as voice, text, graphics, and video. A multimedia conference system allows a group of users to conduct a meeting in real time. The participants can jointly view and edit relevant multimedia information, including text, graphics, and still images distributed throughout the LAN. Participants can also communicate simultaneously by voice to discuss the information they are sharing. This multimedia conference system can be used in a wide variety of cooperative work environments, such as distributed-software development, joint authoring and group-decision support.

Multimedia is the integration of information that may be represented by several media types, such as audio, video, text, and still images. The diversity of media involved in a multimedia communication system imposes strong requirements on the communication system. The media used in the multimedia communications can be classified into two categories: discrete media and continuous media.

Discrete media are those media that have time-independent values, such as text, graphics, or numerical data, bit-mapped images, geometric drawings, or any other nontime-dependent data format. Capture, storage, transmission, and display of nonreal-time media data does not require that it happen at some predictable and fixed time or within some fixed time period.

Continuous media data may include sound clips, video segments, animation, or timed events. Real-time data requires that any system that is recording or displaying be able to process the appropriate data within a predictable and specified time period. In addition, the display of real-time data may need to be synchronized with other data or some external (real-world) event.

Multimedia data can be accessed by the user either locally or remotely during multimedia communication.

Locally stored data typically resides in conventional mass storage systems such as hard disks, CD-ROMs, optical disks, or high-density magnetic tape. It can also be stored to and recalled from analog devices that are under the control of the system, video tape disks, video-disk players, CD-audio disks, image scanners and printers. In addition, media data can be synthesized locally by the systems or its peripherals. Multimedia data is typically recorded and edited on local systems for distribution on some physical media and is later played back using local devices.

Remotely stored multimedia data is accessed via a network connection to a remote system. The data is stored on that remote server and recalled over the network for viewing, editing, or storage on the user's system.

Multimedia communications cover a large set of domains including office, electronic publishing, medicine, and industry. Multimedia communication can be classified into real-time applications and nonreal-time applications.

Multimedia conferencing represents a typical real-time multimedia communication. In general, high conductivity is needed for real-time multimedia communications. A guaranteed bandwidth is required to ensure real-time consistency, and to offer the throughput required by the different media. This bandwidth varies depending on the media involved in the application. There is also a need for synchronization between different users, and between different flows of data at a user workstation.

Nonreal-time multimedia communications, such as multimedia mail, are less demanding than real-time applications in terms of throughput and delay. However, edition tools, exchange formats, and exchange protocols are essential. Multicast service and synchronization has to be offered at presentation time.

Local nonreal-time multimedia characterizes most typical personal-computer applications such as word processing and still image editing. Typical text-based telecommunications can be described as remote nonreal-time processes. Database of text and still images may be interactively viewed and searched, and audio or video data (perhaps included in mail messages) can be downloaded for display locally.

Multimedia workstations are generally characterized by local real-time applications. Data from video and audio editing and annotations, interactive-animated presentations, and music recording are stored on local devices and are distributed on physical media for use locally.

Networks that can provide real-time multimedia communication via a high speed network connection will enable the new generation of multimedia applications. Real-time remote workstation-based multimedia conferencing, video and audio remote database browsing, and viewing of movies or other video resources on demand are typical for these systems.

In multimedia conferencing, each participant has a workstation that may include a high-resolution screen for computer output, keyboard and pointing device, microphone and speaker, and possibly camera and video monitor. Parts of each participant's screen can be dedicated to display shared space where everyone sees the same information. The voice communication equipment can be used by the conference participants for discussion and negotiation. The video communication can add the illusion of physical presence by simulating a face-to-face meeting. The conversational references (such as "this number" or "that sentence") can be clarified by pointing at displayed information. The displayed information can be dynamically edited and processed, permanent records can be saved, and new information that is relevant to the discussion can be retrieved for display at any time. Participants can, in addition, have private spaces on their screens that allow them to view relevant private data or to compose and review information before submitting it to the shared space.

The emphasis in multimedia conferencing is somewhat different from teleconferencing in that it uses video communication to provide simulation of face-to-face contact among participants. Video teleconferencing is valuable when nonverbal communication, in the form of gestures and facial expressions, is an important part of the discussion and negotiation that takes place. Video communication can be a useful enhancement to multimedia conferencing, but it is less critical than voice communication and the ability to share information and software.

Computer-based information-sharing tools have the advantage over informational aids typically included in teleconferencing systems, such as facsimile transmission and telewriting: such aids do not allow information that originated from one location to be modified by participants at other locations. Electronic blackboards allow joint manipulation of shared images, but are subject to the same limitations as the physical blackboard. That is, the only manipulations possible on a real or simulated blackboard are writing, drawing, and erasing. It is not possible to move and rearrange text or the components of a drawing easily, or to change an entry in a spreadsheet and have the totals automatically recomputed, or to run a complex simulation model in order to evaluate the proposed plan. Furthermore, the only information that can be shared by means of these aids at face-to-face meetings or teleconferencing is that which was brought to the meeting or composed during the meeting. Participants cannot access the increasing multitude of private, corporate, and public databases to satisfy unanticipated information needs. The participant's ability to access and manipulate information dynamically by making use of powerful computer-based tools is the distinguishing characteristic of the multimedia conference.

A number of multimedia-conference systems have been designed and implemented. There are basically two platforms involved in these multimedia-conference systems: general-purpose workstations and personal computers. The common goal of these multimedia-conference systems is to emulate important characteristics of face-to-face meetings. These systems can allow groups of users to conduct meetings in real time and the conference participants can jointly view, edit, and discuss relevant multimedia documents.

### 2.1.1 Design of the Multimedia Conference System

A software architecture supporting multimedia conference services is presented by Hać and Lu (1997). The architecture is based on the concepts of partitioning and representation and allows for great extendibility—fast and easy design, and integration of new capabilities and features. The architecture includes two components: end user agent (EUA) and multimedia conference bridge (MCB).

The idea of partitioning is to split the software up into components that have only limited understanding of each other. This provides a mechanism to limit, clarify, and standardize interactions, and thus, supports the goal of extendibility and easy implementation. Architecture partitioning is used here to split the software system up into two components: EUA and MCB.

The components in the software architecture represent the end user, conference connection, and conference-access control. The software architecture uses the concepts of EUA to represent the end user, and MCB to represent conference connection and access control.

Based on the concepts of partitioning and representation, MCB is designed to store, buffer, and distribute information relating to a specific conference relationship. MCB abstracts the connection topology of a conference relationship without regard for the specific protocols or intents of the end users. Since MCB works with relationships, the major part of its function is the negotiation of protocols between multimedia conference participants so that the communication relationship can be established. A connection uses both transport paths which carry the information between the EUAs, and the signaling paths necessary to establish and support that transport. MCB does not care why the connection exists; this is the EUA's job. MCB provides the connection at the request of the end user. It is the data in the EUA component that drives the protocol negotiation process when the MCB attempts to create or change the connection.

Protocol negotiation is needed to enter into a conference relationship. For example, in a voice-conference environment, call waiting is a protocol for dealing with the situation where one party wants to talk to another party who is currently talking with a third party. There are also call-suspending protocols and call-terminating protocols. MCB does the protocol matching and negotiation, based on the information about preferred and acceptable protocols provided by the EUA. MCB plays the most important role in the multimedia conference system. It provides the total conference management for both two party and multiparty conferences including: multiple-call setup, conference status transmission, real-time control of audio, dynamic allocation of network resources, and multipoint voice and data transfer.

EUA is the projection of the customer's identity, language protocols, capabilities, and limitations into the switch, as seen through the customer premises equipment. In keeping with the pertaining rule, the EUA is not concerned with the overall connection topology or specific hardware-signaling protocols. EUA collects users' requests for service and sees to it that those requests are satisfied by the MCB and other central network software entities to reflect the users

# Multimedia Network Architecture

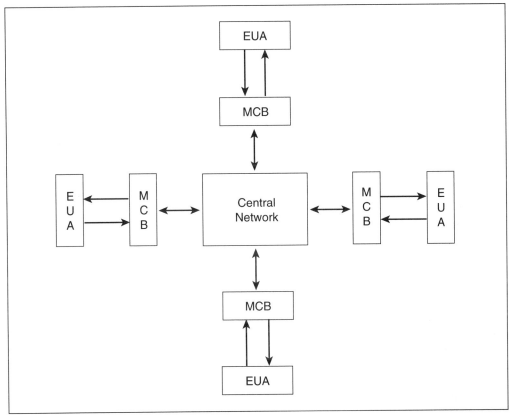

**Figure 2–1**  Software Architecture of the Multimedia Conference System

state, pass users' requests to the MCB, and filter the features, capabilities, and limitations of the user. EUA is not concerned with the details of the signaling protocol or how the MCB will satisfy the request for conference. Once the request is made, the EUA only wants to know whether or not the request will be satisfied. Both signaling protocol and connection protocol are abstract things to the EUA. Figure 2–1 shows the software architecture of the multimedia-conference system.

The multimedia-conference system communication model is designed based on the OSI model. According to the actual requirements of the multimedia-conference implementation, three top layers of the OSI model are combined into one called the multimedia conference layer. The bottom two layers of the OSI model are also combined into a single layer called data-link layer. The communication model for the multimedia conference is shown in Figure 2–2.

The data-link layer of the communication model defines the network hardware characteristics (i.e., Ethernet, token ring, etc.). Design and implementation of the data-link layer mainly depends on the kind of network used. The LAN Ethernet is used, the data-link layer is the Ethernet and uses Ethernet protocols.

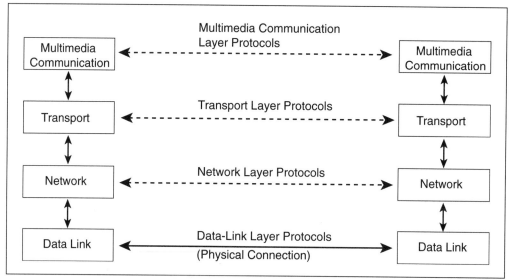

**Figure 2–2**  Multimedia Conference Communication Model

    The transmission control protocol/internet protocol (TCP/IP) suite is used for the network and transport layers. TCP/IP is a set of protocols allowing computers to share resources across the network. Although the protocol family is referred to as TCP/IP, User Datagram Protocol (UDP) is also a member of this protocol suite.

    The TCP/IP suite used as network and transport layers has the following advantages: (1) it is not vendor specific, (2) it has been implemented on most systems from personal computers to the largest supercomputers, and (3) it is used for both LANs and WANs. Using TCP/IP also makes the multimedia conference system portable. Program portability is one of the system-design goals.

    The network layer uses IP. IP is responsible for routing individual datagrams and getting datagrams to their destination. The IP layer provides a connectionless and unreliable delivery system. It is connectionless because it considers each IP datagram independent of all others and any association between datagrams must be provided by the upper layers. Every IP datagram contains the source address and the destination address so that each datagram can be delivered and routed independently. The IP layer is unreliable because it does not guarantee that IP datagrams ever get delivered or that they are delivered correctly. Reliability must be provided by the upper layers.

    The transport layer uses user datagram protocol (UDP) and TCP. Multimedia conference processes interact with UDP or TCP to send or receive data. TCP is a connection-oriented protocol that provides a reliable, full-duplex, byte stream for the multimedia-communication process. TCP is responsible for breaking up the message into datagrams, reassembling them at the other end, resending anything that got lost, and putting everything back in the right order. TCP handles the establishment and termination of connections between processes, the sequencing of data that

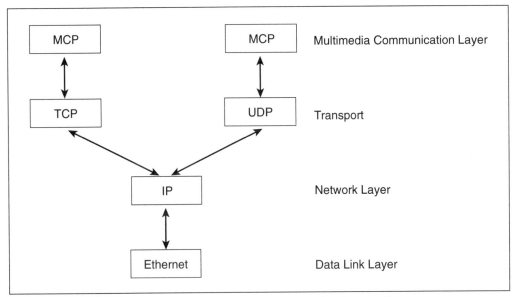

**Figure 2-3** The Communication Model for the Multimedia Conference System

might be received out of order, the end-to-end reliability (checksums, positive acknowledgments, timeouts), and the end-to-end flow control. UDP is a connectionless protocol for user processes. Unlike TCP, which is a reliable protocol, there is no guarantee that UDP datagrams ever reach their intended destination.

The voice communication in the conference requires real-time delivery of speech information, even at the expense of an occasional loss of a voice packet. UDP is used for voice communication in the conference. UDP is less reliable but transfers data faster because the data are not held up by earlier messages awaiting retransmission. On the other hand, the document communication (text, graphics, and still images) needs reliable, sequenced delivery; real-time delivery may not be of utmost importance. TCP protocol is used for this kind of file transfer. The actual communication model for the multimedia conference system is shown in Figure 2-3.

The top layer is the Multimedia Conference Protocol (MCP). MCP is an application protocol similar to TELNET and file transfer protocol (FTP). Multimedia data are transmitted between hosts using services provided by transport and network layer protocols, that is, UDP/IP and TCP/IP.

Because the software architecture of the multimedia conference is designed by using the EUA and MCB model, the multimedia conference protocol is also designed as an EUA protocol and MCB protocol. EUA protocol is responsible for the conferences between the EUA and the MCB. MCB protocol is responsible for the communications between MCB. MCPs are designed as internal layers. Figure 2-4 shows the multimedia conference protocols.

The MCB plays the most important role in the multimedia conference system. It provides conference management for both two-party and multiparty conferences including: conference

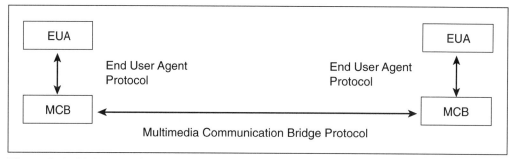

**Figure 2–4** Multimedia Conference Protocols

set-up, conference status transmission, real-time control of audio, dynamic allocation of network resources, and multipoint voice and data transfer.

In the multimedia conference system, depending on the type of information transmitted, data delivery, processing and presentation have different requirements. Voice data is time-constrained, thus, requiring each participant's voice data to be delivered to and processed by all other conference participants in real time. In addition, for natural-voice conferencing, the voice-delivery subsystem should accommodate the possibility of multiple, simultaneously active speakers. When several speakers are active, speech samples from the individual speakers must be combined before playback.

File data (i.e., text, graphics, and still images) transmission during the conference requires acknowledged, error-free transfer of a possibly large amount of data to all participants. In multimedia conferencing, transferring a large file takes time, and the participants need to use voice communication to coordinate their work. Therefore, simultaneously transferring voice data and file data in the multimedia conferencing system is needed.

According to the above-stated data delivery requirements of the multimedia conference system, two communication channels are used in the conference system. The UDP channel is used in the conference system for voice communication and conference control. The TCP channel is used for file transfer in the conference system.

The multimedia conference protocols are designed as internally layered protocols. The MCB protocols are at the bottom of the layer and are responsible for the communication with the transport-layer protocols (e.g., TCP and UDP).

Because two types of communication in the multimedia conference system (voice communication and conference control) use UDP protocols, and the file transfer uses TCP protocols, the MCB protocols are also designed with two types, MCB/UDP protocols and MCB/TCP protocols.

MCB/UDP protocols are designed and implemented for voice communication and conference control. Each MCB/UDP packet consists of two parts: the MCB/UDP header and MCB/UDP data. The MCB/UDP header is divided into two 2-byte fields: the packet-type field used to set up the MCB/UDP packet types, and the length field used to give the packets length. The types of MCB/UDP packets are defined and used in the conference system for the conference control, file transfer, and voice communication.

MCB/TCP protocols are designed and implemented for file transfer in the conference system. Each MCB/TCP packet consists of two parts: an MCB/TCP header and MCB/TCP data. The MCB/TCP header is divided into two 2-byte fields (packet type and length field) and one 20-byte field. A packet-type field is used to set up the MCB/TCP packet type, the file name field is used to give the file name to be transferred, and the length field is used to provide packet length.

## 2.2 Congestion Control for ABR Traffic in ATM Networks

ATM is a high-bandwidth, low-delay switching and multiplexing technology becoming available for both public and private networks. This technology is very suitable for a multimedia communication environment. Multimedia services require a combination of text, image, voice, video storage, and communication, and have potential applications in many areas such as education, banking, and medical imaging. ATM offers a great flexibility in bandwidth allocation.

There are four service categories defined by the ATM Forum. They are based on the kinds of traffic the ATM networks carry and what services their customers want. These are CBR, VBR, ABR, and UBR.

In CBR cell transmission, the rate will never change. It is suited to all interactive audio and video streams.

VBR is divided into two subclasses, real time and nonreal time. Real-time VBR is intended for services that have variable-bit rates combined with stringent real-time requirements, such as interactive compressed video and video conferencing. For this service, both the average cell delay and the variation in cell delay must be tightly controlled. However, an occasional lost bit or cell is tolerable and is best just ignored. The other VBR subclass is for traffic where timely delivery is important, but a certain amount of jitter can be tolerated by the application.

ABR is designed for bursty traffic whose bandwidth range is roughly known. The main motivation for its development is for supporting data traffic. Because data traffic applications usually require a service that dynamically shares the available bandwidth among all active sources, such traffic is referred to as a best-effort traffic class. A typical example might be for use in a company that currently connects its offices by a collection of leased lines. Typically, the company has a choice of putting in enough capacity to handle the peak load, which means that some lines are idle part of the day, or putting in just enough capacity for minimum load, which leads to congestion during the busiest part of the day. Cell loss for ABR traffic is expected to be low. Traveling ABR is like flying standby: if seats are left over (excess capacity), standby passengers are transported without delay. If there is insufficient capacity, the cells have to wait unless some of the minimum bandwidth is available. The ABR service is potentially useful for a wide variety of applications.

UBR gives no feedback about congestion. All UBR cells are accepted, and if there is capacity left over, they will also be delivered. If congestion occurs, UBR cells will be discarded.

This class is for applications that have no delivery constraints and want to do their own error control and flow control.

QoS is an important issue for ATM networks. When a virtual channel is established, both the customer and the network must agree on a contract defining the service. To make it possible to have reasonable traffic contracts, the ATM standard defines a number of quality parameters whose values the customer and carrier can negotiate:

- Peak Cell Rate (PCR) is the maximum rate at which the sender is planning to send cells. It may be lower than the bandwidth of the line.
- Allowed Cell Rate (ACR): the required cell rate averaged over a long time interval. For CBR traffic, ACR will be equal to PCR. For all the other service types, it may be lower.
- Minimum Cell Rate (MCR): the minimum cell rate that the customer can accept. For ABR service, the actual bandwidth used must lie between MCR and PCR.
- Cell Loss Ratio (CLR) measures the fraction of the transmitted cells that are not delivered at all or are delivered too late.
- Cell Transfer Delay (CTD) is the average transit time from source to destination.
- Cell Delay Variation (CDV) shows how uniformly the cells are delivered.
- Cell Error Ratio (CER) is the fraction of cells that are delivered with one or more bit incorrect.
- Severely-Errored Cell Block Ratio (SECBR) is the fraction of $N$-cell blocks of which $M$ or more cells contain an error.
- Cell Misinsertion Rate (CMR) is the number of cells per second that are delivered to the incorrect destination due to an undetected error in the header.

ATM with ISDN (Integrated Services Digital Network) offers video on demand, live television from many sources, multimedia communication, and high-speed data transport for science, industry, and many other services. QoS is an important issue for ATM networks, in part because they are used for real-time traffic, such as audio and video. ATM networks do not automatically meet the performance requirements. Congestion at intermediate switches is always a potential problem. The ATM must deal with both long-term congestion, caused by more traffic coming in than the system can handle, and short-term congestion, caused by bursty traffic. Without congestion control, the ATM network is not able to provide the desired high-speed and low-delay service for users. Since the ABR service can be used in very wide application areas, and offer users more flexibility, congestion control is more important for this type of service.

With CBR and VBR traffic, it is generally not possible for the sender to slow down, even in the event of congestion, due to the inherent real-time or semi-real-time nature of the information source. For these two classes of traffic, congestion control is directed through admission control and bandwidth allocation. The limit on each connection's usable bandwidth is based on the notion of a traffic contract. Before the connection is set up, each source system has to declare its traffic parameters to the network. Once the connection request is admitted, its QoS is guaranteed throughout the entire session. CBR and VBR are also called guaranteed traffic classes. If there are not enough network resources, the requested connection will be rejected.

In UBR, if there are too many cells, the extra cells are dropped. In ABR traffic, it is possible and reasonable for the network to signal one or more senders and ask them to slow down temporarily until the network can recover.

Many solutions have been proposed for ABR service during the development of ATM networks. One of them is a resource-reservation scheme. A sender first sends a special cell reserving the necessary bandwidth. Congestion will never occur because the required bandwidth is always there when it is needed. However, there is a potentially long delay before a sender may begin to send, because it takes time to set up the reservation. Another proposal uses packet boundaries that are marked by a bit in the last cell. The idea here is to discard cells to relieve the congestion in a highly selectively manner. A switch scans the incoming cell stream for the end of a packet and then throws out all the cells in the next packet. Each discarded cell may belong to a different packet so the entire packet needs to be resent. Thus, a large portion of the bandwidth available to ABR service is wasted because the portion bandwidth is occupied by cells which belong to packets already corrupted by cell loss. This kind of loss scheme is not efficient and was rejected by ATM standard committee. The discussion about congestion control for ABR focuses on two ideas, a credit-based solution and a rate-based solution. Both the credit-based and rate-based solutions are closed-loop feedback control mechanisms.

The credit-based solution is essentially a dynamic sliding-window protocol. Each link in the network runs the flow-control mechanism. Figure 2–5 shows the basic configuration of a credit-based scheme. A certain number of cell buffers are reserved for each virtual connection at the receiving end of each link. This reservation is given in the form of a credit balance. A credit balance is maintained at the sending end of the link. Each unit of credit represents one empty buffer at the receiving end. A connection is allowed to continue cell transmission as long as it gains credit from the receiving end. When each cell is transmitted by the sending end, the credit balance for that connection is decremented. If a virtual connection runs out of credit on a particular link, it will stop transmitting cells. As each cell is removed from the buffer at the receiving end, a credit is returned to the sending end of each virtual connection. This link-by-link feedback mechanism (virtual connection) can relieve congestion effectively, and no congestion cell loss occurs because the connection has to wait for credit before it can send cells. Some reasons not to adopt the credit-based schemes are that link-by-link flow control requires complex hardware at the switches, and the switch architectures for credit-based schemes also limit flexibility of implementation. Transmission delay may also be too long.

### 2.2.1  FECN Rate-Control Method

FECN is an end-to-end scheme in which most of the control complexity resides in the end systems. An ACR is adjusted at the source according to the congestion status of the network. The congestion is detected at each intermediate switch by monitoring the queue length of the cell

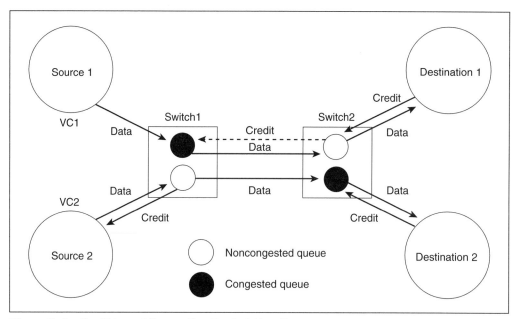

**Figure 2–5** Credit-based Congestion Control Algorithm

buffer. If the queue length exceeds the threshold value of a particular switch, this switch is detected as congested.

To indicate congestion, this switch will mark an EFCI bit in the header of all passing data cells on the path in a forward direction. When the destination end system receives a data cell in which the EFCI bit is marked, it will generate an RM cell, whose PTI is "110". This RM cell is sent back to the source-end system along the backward path. Once the source receives an RM cell, it must decrease its ACR by a rate-decrease factor, but no more than MCR.

$$ACR = \max(ACR*DF, MCR) \qquad (2\text{--}1)$$

where DF is the rate decrease factor, and MCR is the minimum-allowed cell rate. A time interval is defined at the destination end systems. Only one RM cell is allowed to be sent back in an RM time interval. There is also an interval timer in the source-end system. The source recognizes no congestion in the network without receiving an RM cell after an interval time. Then the source will increase ACR.

$$ACR = \min(ACR*IF, PCR) \qquad (2\text{--}2)$$

where IF is the rate increase factor and PCR is the peak cell rate of this connection. Figure 2–6 illustrates a basic configuration of an FECN rate-based congestion control method in the ATM network, whereby the connections are terminated at the source and destination-end systems. There are two source-end systems, two destination-end systems, and two switches between the sources and the destinations. Each VC has its own separate queue. Suppose that in switch 1, the VC1 queue length exceeds the threshold value, meaning that switch 1 is congested, then switch

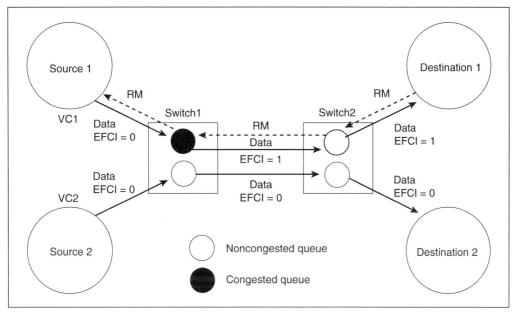

**Figure 2–6** FECN Scheme

1 marks the EFCI state of all passing data cells in a forward direction on VC1 to indicate congestion. When destination 1 receives the marked-data cell, it sends an RM cell back to source 1 notifying it about congestion. According to the RM cells, source 1 starts to reduce its cell-emission rate.

### 2.2.2 BECN Rate-Control Method

Unlike FECN rate-control scheme, in the BECN scheme, the congestion information is returned directly from the point of congestion back to the source for each VC. The source changes its cell emission rate on each VC in a manner similar to the FECN scheme. When the VC queue length exceeds a congestion threshold value in a particular switch, this switch can generate an RM cell and send it back to the source in a backward direction on the VC, which is currently sending traffic through this congested queue.

Once a source receives an RM cell, it will reduce its cell emission rate.

$$ACR = \max(ACR*DF, MCR) \qquad (2\text{–}3)$$

If no RM cells are received within a certain recovery time period by the source, the current cell emission rate is increased once each recovery time period until it reaches the peak cell rate.

$$ACR = \min(ACR*IF, PCR) \qquad (2\text{–}4)$$

In the BECN scheme, the switches not only generate RM cells, but also filter the congestion information to prevent excessive RM cells from being generated. There is no reason to send

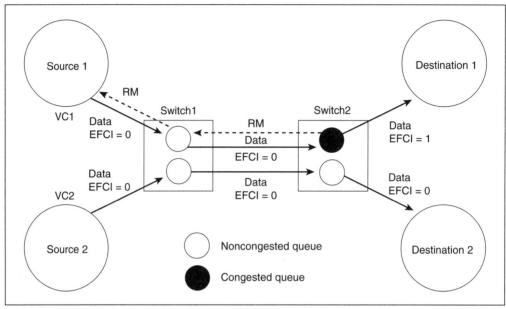

**Figure 2–7** BECN Scheme

another RM cell back to each source until the previous feedback has had time to take effect. When a switch is congested, the filter should transmit no more than one RM cell to each active source during a filter time period. To allow the previous RM cell to take effect, the filter time period should be of the same order of magnitude as the maximum propagation delay for which the ATM network is designed. Figure 2–7 shows the data and control traffic of BECN in a congested network with switches that have an output buffer. If switch 2 is congested, it generates an RM cell and also marks the EFCI state of passing data cells in the forward direction. When source 1 receives the RM cell, it starts to decrease its cell emission rate.

In both FECN and BECN schemes, a switch is considered congested when the queue length exceeds a certain threshold value. According to the congestion notification cell received, a source will reduce its cell emission rate for the corresponding VC. On the other hand, the current cell emission rate will increase if no congestion notification cell is received within a predetermined time period until it reaches the peak cell rate.

An obvious advantage of BECN over FECN is a faster response to congestion because the FECN only sends back notification at the destination, while the BECN sends back congestion information when it occurs. BECN is also more robust against faulty or noncompliant end systems because the network itself generates the congestion notification. The hardware is more complicated in BECN than in FECN, thus, the implementation of BECN is expensive. Both FECN and BECN schemes have network congestion collapse problems. If the congestion notification cells returning to sources in backward traffic experience extreme congestion, the sources cannot receive any congestion notification cells in a certain period, thus, they all attempt to increase their cell emission rate to peak cell rate. At this point, network congestion becomes more serious and causes a collapse.

### 2.2.3  PRCA

Both FECN and BECN schemes adopt the negative feedback approach, because sources receive the RM cells and reduce the cell emission rate only if the network experiences congestion. The PRCA is based on a positive feedback rate-control paradigm. In PRCA, a source only increases its cell emission rate for a VC when it receives an explicit positive indication from the destination; otherwise, the source will continually reduce its cell emission rate. The increments and decrements of the cell emission rate of each VC are proportional to the current cell emission rate, thus eliminating the need for timers and timer-value selection like in previous rate-based proposals. The FECN and BECN are also called timer-based approaches, while PRCA is called a counter-based approach. The object of PRCA is to remedy the problem of network congestion collapse.

In PRCA, the source-end system marks the EFCI bit in all data cells except for the first data cell and once every $N$ cells that follow. For marked cells, EFCI = 1, and for unmarked cells, EFCI = 0. The parameter $N$ is predetermined according to the response time required by the ATM network, and will affect the response time to congestion and backward-link utilization. The destination end system instantly sends an RM cell back to the source when it receives a data cell with EFCI = 0. The source will increase its cell emission rate only when it receives an RM cell; otherwise, it continually decreases its cell rate. If any intermediate switch experiences congestion, it will set the EFCI = 1 instead of EFCI = 0, and the destination takes no action. By using this mechanism, a positive congestion signaling is established. Receiving an RM cell implies that there is no congestion in the network, and therefore, the source-end system is given an opportunity to increase its rate. Figure 2–8 illustrates the data and control traffic of PRCA in a congested network where VC1 is congested at switch 1. Because the cell emission rate is always decreased whenever the source does not receive an RM cell, the network can be restored even if heavy congestion results in all RM cells being discarded. Thus, the network congestion collapse can be remedied because if the source does not receive an RM cell, the source will decrease the rate to send data.

### 2.2.4  Beat Down Problem

There are still unsolved problems in PRCA (described by Chang, Golmie and Su). One of the important issues is referred to as a "Beat down" problem. Each source of active connections that experiences congestion in several switches has less opportunity to receive positive feedback than sources of other connections with fewer switches because of the way PRCA sends an RM cell. When a VC is passing through more congested links, its data cells will be marked more often than those of other VCs going through fewer congested links. Consequently, such a VC will have a lower allowed rate than others and the corresponding source will receive fewer RM cells than the others will. This undesirable effect of VC starvation is proportional to the number of

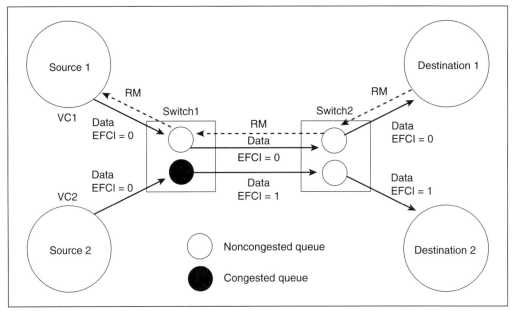

**Figure 2–8** PRCA Scheme

congested links on which a VC transmits its cells, so fairness among connections cannot be achieved. Another problem is that PRCA requires a considerable amount of buffer space when there is a large number of active connections. When the propagation delays are large, the queue length grows temporally because of the control information delays. This is an intrinsic and unavoidable feature of high-speed networks.

There are several ways to solve the Beat down problem. First, to achieve better fairness among all connections, a congestion indicator should be sent to particular sources rather than to all sources. The goal is to decrease the traffic from the sources that cause congestion instead of the traffic on every VC passing through the congested switch. Fairness can be achieved if each connection is maintained separately at the switch, which is called per-VC accounting. But the implementation of this approach increases the hardware complexity and memory requirement of the switches significantly. For example, each switch can have a VC table to record the number of active connections. Each VC entry is marked or unmarked according to the status of its corresponding VC and the number of marked entries represents the number of active connections. The rates of all sources are adjusted through RM cells in one round-trip time when there is a congested switch in the network.

### 2.2.5 An Adaptive Congestion Control Scheme

For most of the proposed rate schemes, the switches are required to monitor their load and set a bit in the data cells or control cells if congestion occurs. The end systems use this one-bit information contained in each cell to adjust their cell rates. The switches make no distinction among different VCs. Lacking a mechanism to distinguish between the cell rates of various sources, a switch experiencing congestion can only attempt to reduce the cell rates among all VCs in the same way. This means that a congested switch will mark all data cells regardless of the individual VCs' cell rates. When a VC goes through more congested links, its data cells will be marked more often than those of other VCs going through fewer congested links. Thus, such a VC will have a lower allowed cell rate than others. This Beat down problem causes extreme unfairness among all VCs.

Hać and Lin proposed an algorithm which computes an ideal rate for each source according to the current ACR and output buffer queue length in the switch. This algorithm does not require per-VC accounting, so it is easier and cheaper to implement. It uses queue length to adjust the convergence parameter, which makes it converge fast to an ideal rate.

All rate-based congestion control schemes are designed to achieve better fairness among connections as well as to increase link utilization and to reduce buffer size at switches; a definition for the fairness measure is needed to evaluate the degree of fairness. The ATM forum has discussed fairness criteria. Fairness for ABR service is defined by a max-min criterion, in which all active connections are served fairly if the following conditions are met:

1. Each connection must pass through at least one bottlenecked switch
2. The available bandwidth should be shared fairly when it is assigned to connections that do not pass through nonbottlenecked switchs.

There are four other definitions of fairness measures proposed by Yin (1994) and Hughes (1994):

1. Minimum usable cell rate (MCR) plus equal share.  In this definition, each active source is first allocated an MCR and then the rest of the available bandwidth is assigned fairly according to the max-min criterion. That is, the $n$-th active connection's rate $R_n$ is given by

$$R_n = \text{MCR}_n + \frac{C - \sum_{i=1}^{N_{vc}} \text{MCR}_i}{N_{vc}} \quad (2\text{--}5)$$

where $1 \leq n \leq N_{vc}$, $C$ is the available bandwidth, and $N_{vc}$ is the number of active connections at the switch, and $\text{MCR}_n$ is the MCR of the $n$-th connection.

2. Maximum of MCR or max–min share.  In this definition, each connection acquires MCR if MCR is larger than the bandwidth equally divided by all connections:

$$R_n = \max\left(\frac{C}{N_{vc}}, \text{MCR}_n\right) \quad (2\text{--}6)$$

where $1 \leq n \leq N_{vc}$.

3. Allocation proportional to MCR. The third definition assigns the available bandwidth to unconstrained connections in a weighted manner.

$$R_n = C \frac{\text{MCR}_n}{\sum_{i=1}^{N_{vc}} \text{MCR}_i} \qquad (2\text{–}7)$$

where $1 \leq n \leq N_{vc}$. This definition cannot be applied to connections with MCR = 0. The bandwidth allocated to each connection is proportional to its MCR.

4. Allocation proportional to PCR. This definition assigns the available bandwidth to unconstrained connections in a weighted manner.

$$R_n = C \frac{\text{PCR}_n}{\sum_{i=1}^{N_{vc}} \text{PCR}_i} \qquad (2\text{–}8)$$

where $1 \leq n \leq N_{vc}$. The bandwidth allocated to each connection is proportional to its PCR.

An appropriate definition of fairness depends on the environment in which it is used. Requirements for charges, for example, may become an important factor in deciding which is appropriate. In the public WAN environment, the third definition is suitable if the tariff is determined in proportion to the MCR. In private networks, on the other hand, the first definition might be more suitable because its implementation is less expensive. In the above definitions, the different values of PCR for connections are not taken into account. Since various speeds of interfaces are currently defined, PCR may also have to be incorporated in the future:

1. At the beginning of transmission, the source issues an RM cell. The RM cell contains the ACR, and the rate at which the source wishes to send the cell (ER). The value of the ER is the same as the PCR. The ACR is set to a value between MCR and PCR. At first, it may be as same as PCR.
2. The source starts to send data cells.
3. For every N data cell transmitted, the source emits an RM (containing ACR and ER parameters) cell to the network, where the value of ER is the same as PCR. The N will determine the response time of rate adjustment.
4. The source continues to decrease its ACR after sending out a cell until it gets an RM cell returned from the destination. The decrease rate is predefined.
5. In the network, each switch periodically estimates the current, optimal estimated allowed cell rate (EACR) according to the changing load situation and the ACR of each VC.

When a noncongested switch receives an RM (ACR, ER) cell from a source, if ACR ≤ EACR, there is no need to adjust the EACR. Once the RM cell is sent back to source, the ACR will be increased. If EACR ≤ ACR, meaning EACR is not large enough, the EACR should be increased using the following formula:

$$EACR = EACR + \beta \times (ACR - EACR) \qquad (2\text{--}9)$$

$\beta$ is the averaging factor and must be less than 1 to make the EACR converge to an optimal ACR. $\beta$ is adjusted according to the output port queue length of the switch, instead of using a fixed constant. The new EACR is calculated using the old EACR and a small percent of error between ACR and the old EACR.

When a congested switch receives an RM (ACR, ER) cell from the source, if EACR ≤ ACR, there is no need to change EACR because the current EACR will result in ACR reduction. If ACR ≤ EACR, there is a need to make EACR smaller. We replace EACR by a new computed allowed cell rate as follows.

$$EACR = EACR + \beta \times (ACR - EACR) \qquad (2\text{--}10)$$

where $\beta$ is the averaging ratio and calculated according to the output port queue length.

6. A destination returns the RM (ACR, ER) cell to the source upon receiving each RM cell. This backward RM cell will be sent back in the same path as a forward RM cell for simplicity.
7. When a congested switch (e.g., when the queue length exceeds the congestion threshold) receives an RM (ACR, ER) cell from a destination, because the current queue length is greater than a threshold, it will replace ER in the RM cell by

$$\min(ER, \gamma \times EACR) \qquad (2\text{--}11)$$

if the CI bit is set to 1 to indicate congestion.

When a noncongested switch receives a backward RM cell, it will replace ER with minimum of EACR and ER in the RM cell by

$$\min(ER, EACR) \qquad (2\text{--}12)$$

if, CI = 0.

$\gamma$ is calculated according to the output-port queue length instead of using a constant.
8. If a source receives an RM cell, it increases its ACR by no more than EACR by an increasing ratio.

An important parameter is the averaging factor $\beta$, which is calculated based on the queue length relative to the congestion threshold. The smaller the difference between the queue length and congestion threshold, the smaller the $\beta$. This means, for a non-congested switch, as the queue length approaches the congestion threshold, the EACR should increase slowly. If the queue length is far from the congestion threshold, the EACR can increase relatively fast. For the congested switch, the queue length exceeds the congestion threshold. If it is close to the threshold, a small reduction of EACR could bring the congestion down to normal. If the queue length far exceeds the congestion threshold, then a larger reduction factor is needed to reduce the estimated rate quickly to restore it back to normal:

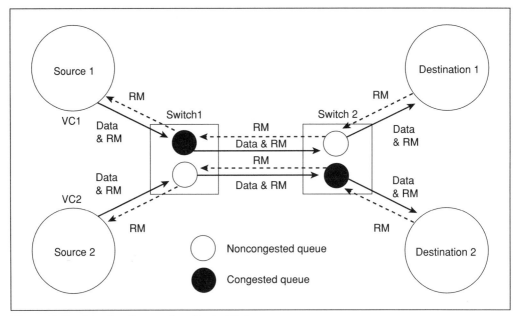

**Figure 2-9** Adaptive Congestion Control Scheme

$$\beta = \frac{|\text{queue length} - \text{congestion threshold}|}{\text{total queue length}} \quad (2\text{-}13)$$

Another important parameter is the fast-down parameter $\gamma$. If a switch is congested, $\gamma$ will be calculated similarly to $\beta$. If the queue length is far from the threshold (too long), a further reduction of the EACR will be used:

$$\gamma = \frac{|\text{queue length} - \text{congestion threshold}|}{\text{total queue length}} \quad (2\text{-}14)$$

The basic configuration of the adaptive scheme is shown in Figure 2–9. In the figure, VC1 is congested at switch 1, and VC2 is congested at switch 2. The forward RM cells contain the current-allowed cell rates and the MACR of the sources, and the backward RM cells contain the new computed cell rates of switches.

## 2.3 Congestion Control Mechanisms in the Intelligent Network

The Intelligent Network (IN) is characterized by service-independent architecture, integrated operations and service management systems, and programmable network entities. Intelligent networks provide infrastructure for rapid creation of telephone service. Their flexible architecture allows for services with a wide variety of features, and options that users can customize. Figure 2–10 shows an IN configuration. The architecture of the IN is built with the following components.

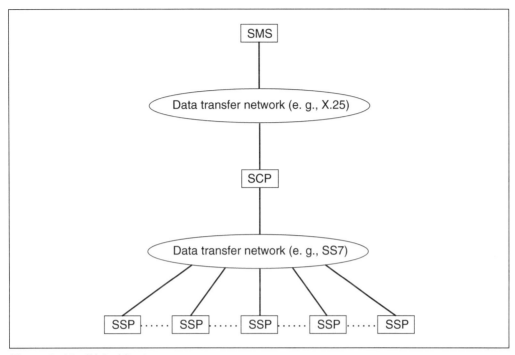

**Figure 2–10** IN Architecture

1. Service Switching Point (SSP), which represents a distributed-switching node that exchanges messages by interacting with Service Control Point (SCP).
2. SCP, which is a centralized node-handling database that contains service-control logic.
3. Service Management System (SMS), which is a centralized operation system for service creation and management.

The intelligent peripheral is often introduced to provide communication between the network and users.

IN supports heavily used calling services. Each call in an IN requires the exchange of several messages between the SCP and SSPs via a signaling system. The aggregated messages in the SCP, the central point of IN architecture, can cause congestion when the SCP is overloaded. Therefore, the SCP node introduces a bottleneck which can lead to unacceptable additional delays in call set-up times, and reduces the network call handling capacity. To avoid degradation of performance caused by overload traffic, a congestion-control mechanism is used to control the flow of input traffic.

Research has been done to control overload mechanisms in SCP. Pham (1994) compares the rate-based control (call gapping) and the window methods. The simulation results of Pham's experiment show that the window method consistently outperforms the method based on flow-rate control. The difference in performance, for experimental results, is the most significant with larger numbers of SSPs, when the offered load in different SSPs varies rapidly and load levels are just above the SCP throughput limit. Based on the simulation results, Pham concludes that the SCP rate-control algorithms are unable to effectively redefine new flow rates to the SSPs with the accuracy and speed necessary to maintain optimal throughput. Pham reasons that the window method shows better performance because it utilizes a tight-feedback loop between SCP and SSP. With each response message from the SCP, the SSP is informed of the current SCP congestion status. The SSP reacts immediately to congestion indication by reducing flow to the SCP. The flow will be gradually increased with the congestion release. Another advantage of the window method is its simplicity: it requires minimal SCP functionality, leading to a simpler implementation. Based on the simulation results and discussion, Pham claims that the window method has a very wide potential range of applications and can be successfully incorporated into most packet protocols as the primary defense against terminal or network node congestion.

Tsolas et al.'s (1992) discussion draws the opposite conclusion. Tsolas believes the slow reaction to the overload indication in the window method can result in severe degradation of network performance. The fairness among SSPs that can be degraded by using the window method is another disadvantage. Tsolas shows that the call gapping method is flexible because it can be used selectively for the originating or terminating number. This method can distinguish different types of abnormalities, such as, general overload, mass calling to a destination or from an origination, and mass calling to a vacant number. The call gapping (CG) method can react quickly to the SCP overload by closely monitoring the traffic at the SCP from various sources. Tsolas also shows that CG can adjust to different network configurations. Therefore, Tsolas concludes that the CG algorithm is a promising overload control method, inherently flexible and able to adjust to various network configurations, service types, and new IN components.

Complex telecommunication networks can perform poorly in uncontrolled-overload conditions. Network congestion is defined by Haas (1992) as the network state, in which some network resource is oversubscribed/overdemanded, and in which the availability of that resource decreases because of the oversubscription/overdemand.

The objective of the congestion control algorithm is to achieve the optimal use of both network and resources in the event of congestion. Hence, the primary goal is efficient network use. The congestion control algorithm should be able to handle a wide variety of traffic patterns (average frequency input, burstiness, etc.), and to meet performance requirements (delay, loss tolerance, etc.). Performance objectives under real-time congestion are to maintain small loss probability, high throughput, provide low delay and jitter to applications which depend on them critically, and to allow available resources to be shared fairly by users. The objective of achieving a degree of fairness is critical for networks.

If the traffic in the network can be predicted, congestion control is simple. In real examples, the traffic is either completely unknown or insufficiently specified. Thus, the congestion control has to be robust enough to perform relatively well in real cases. A single scheme may not be adequate to solve congestion control. A combination of methods is needed, with each emphasizing different aspects of congestion control.

Congestion control schemes can be classified into two categories: proactive and reactive methods. Reactive methods monitor the network performance either directly, by monitoring the state of the resource, or indirectly, by monitoring other correlated parameters. Upon detection of congestion, the reactive methods take an action to drive the network out of a congested state. The reactive methods rely heavily on feedback from the network. Thus, these methods can successfully cope with congestion in the network only when the congestion develops with a time constant on the order of at least a round-trip delay. The windowing schemes are reactive methods.

Proactive methods constantly activate mechanisms that reduce the possibility of network congestion. Proactive methods use bandwidth and admission control (i.e., LB and call gapping).

The disadvantage of reactive methods is the relatively slow reaction to congestion building. These methods rely on an indication from the network about the increasing congestion, which is usually provided only after the actual transmission takes place. Proactive methods, in general, do not have this disadvantage since these schemes are continuously applied. However, most proactive methods that rely on an open-loop control scheme are not robust enough and cannot, therefore, protect the network against all possible traffic patterns.

In the IN, the optimal SCP throughput and minimum blind load on the SCP, for a wide range of input traffic, results in minimal service performance degradation. All SSPs have fair access to SCP resources. The optimal SCP throughput is achieved by adjusting the SSP and SCP logic and operational parameters. When the SCP is overloaded, the congestion algorithm must adjust traffic to match the SCP capacity (or be slightly below it). Overcontrol and undercontrol should be avoided with the application of the congestion-control method. The blind load refers to calls rejected by the SCP. This represents a waste of SCP Central Processor Unit (CPU) time. Every call entering the SCP queue must be processed by the CPU. To maximize the SCP throughput, the rejection or blocking of calls is done close to the source of the traffic (i.e., the SSPs). This unloads the SCP and minimizes the waste of resources at the intermediate nodes (those between the SSP and SCP).

To achieve fairness, the control algorithm should only block calls from certain SSPs that cause congestion. Calls from other SSPs which do not contribute to overload should not be blocked. Different definitions of fairness have been suggested. (1) All calls originating in the network should have the same probability of success. (2) Calls from SSPs that do not generate overload should be guaranteed success. (3) Some SSPs have a higher priority to send more calls than other SSPs.

The SCP congestion-control algorithm should achieve good performance (i.e., optimal SCP throughput, fair control among SSPs, etc.) without prior knowledge of the number of sources (i.e., SSPs) overloading the SCP and the source's traffic intensity. The algorithm receives this information from measurements. The control algorithm should be robust for different overload patterns (i.e., a variety of abnormal situations), call mixes, customer behaviors, and the SCP architecture.

IN services are offered by using a signaling system (i.e., SS7). The exchange messages between the SCP and SSPs go through the signaling system that has its own flow- and congestion-control mechanisms. Tsolas and coworkers (1992) show that the SCP congestion-control problem is an overload control at the user or application level. Therefore, the effect on the overall service performance caused by controls at the lower levels can be ignored. Zeft (1994) concludes that the application level congestion control in the IN does not interact with signaling network-congestion situations.

IN congestion control has to protect the SCP from overload traffic that decreases the SCP throughput, while at the same time allowing enough traffic from the SSPs so that the SCP throughput is optimized. There are two fundamental methods to control SCP congestion. (1) SCP uses flow control commands to the individual SSPs to indicate the maximum rates at which calls should be sent to the SCP. (2) SSP uses a window or quota of requests, which can be outstanding at any given time. In both methods the SSPs contain a flow-control mechanism that regulates the sending of requests to the SCP based on the control feedback from the SCP. The primary difference is that rate-based control requires SCP to observe input traffic, compute a new target flow rate, and send flow rate commands to SSPs. In the window method these functions are not present.

Studies on overload control in IN use the window and rate-based control CG methods. The fundamental feature of the rate control methods is that SCP must allocate its (estimated) processing capacity to different SSPs by using flow rate commands. This allocation usually takes into consideration flow measured from each SSP in the previous observation interval to predict traffic for the next interval.

CG is an example of a rate-control algorithm. The CG method is based on a general overload-control principle that overload detection should be done at the traffic destination (SCP) and load rejection should be done at the traffic sources. SCP determines, based on its capacity and the incoming load (calls) from the SSPs, how much traffic each SSP should reject. This information is sent to the SSPs in a CG message containing a gap ($T$) and a duration time interval ($D$), indicating to the SSPs that one call per $T$ seconds should be accepted for the next $D$ seconds. Parameters $T$ and $D$ are dynamically adjusted according to the incoming traffic. In the CG control algorithm, the other parameters can also be used. For example, a percentage value can be sent indicating the portion of the traffic to be rejected. The CG message can also be sent randomly. The message can be broadcast to all the SSPs, or the SCP can determine where to send the message.

Smith (1993) analyzes performances of two versions of the CG control algorithm in SCP, the table-driven and adaptive CG algorithms. Table-driven CG works as follows. At all times, each source maintains a view of the SCP congestion level, which is a nonnegative integer value. If the source's view equals $n$, then the source uses the $n$th value of the gap interval and the $n$th gap duration value in the table whose values are already predetermined. Instead of reacting to SCP congestion with the table of fixed values, the adaptive version tries to better control SCP congestion in all overload scenarios in real time. The idea of an adaptive CG is that each time the SCP tells the source that its congestion level has increased, the source cuts back its traffic rate to the SCP by a certain amount (e.g., 10%, 15%, etc.). When the congestion level decreases, the

SCP recalculates the gap interval and SSPs can increase their sending rates. Based on simulation results, Smith (1995) concludes that table-driven controls do not perform as well as adaptive controls even when the parameters of table-driven control methods are a good match with the source characteristics. Adaptive control is more robust to traffic patterns and SCP internal operation than table-driven controls.

In the window approach, SCP does not use explicit control messages and there is no observation function to determine actual call rates from SSPs. SCP reacts to congestion (input, queue full) by discarding new requests. The window algorithm in IN works as follows. A new call at the SSP is sent to SCP if the number of outstanding messages (i.e., no ACK has been received from SCP) is smaller than the value of the window size variable (WIN). When the number of outstanding messages (OUT) equals WIN, the transmission window is closed, and the SSP is not allowed to send new requests. Otherwise, SSP can send new requests, incrementing the OUT variable with each transmission. Thus, there are two variables, WIN and OUT, used in SSP to determine when a request can be sent to SCP. WIN indicates the number of requests which are allowed during any given time period. The counter OUT represents the actual number of outstanding requests. The maximum number of outstanding messages (window size) is controlled by timers. The timer is started at the SSP every time a message is sent to SCP. If the timer expires without an ACK, then the message is considered lost (congestion in SCP) and the WIN is decreased by one. When a positive response (ACK) is received, the corresponding timer at SSP is terminated and SSP increments counter C (the counter used by each SSP to count the number of ACKs received by each SSP). The size of WIN is increased by one every time the number of counter C (the number of ACKs) reaches the value $C_{max}$. The minimum value of WIN is one and there is a maximum value $W_{max}$. At SCP, if the queue of incoming messages is full, then the new arrival is discarded. In the window approach, there are no overload messages exchanged between SSP and SCP. The corresponding timer at SSP will expire. In the SCP window algorithm, WIN is variable. If the response is timed out, SSP interprets this as indication of SCP congestion, and reacts by decrementing its WIN.

Pham and Betts (1994), Galletti and Grussini (1992), and Tsolas and coworkers (1992) compare the CG and window control algorithms in IN shown in Figure 2–10.

Window control is easy to implement, reacts quickly to short traffic bursts, and has low overhead. On the other hand, window control reacts slowly to the SCP overload, and cannot control the SSPs individually.

CG is flexible and can prevent an overload by continuously monitoring the SCP queue length. On the other hand, CG has a slow reaction to transient traffic (e.g., burst traffic): its control parameters are sensitive to input traffic, it requires storage for measurement data, and for processing overhead increases with the number of SSPs.

High-speed and fiber-optic networks can have transmission speeds of up to Gigabits per second. The services for multimedia high-speed networks require real-time communications and fast network response. These factors have a significant impact on the design of network protocols and control procedures. Conventional mechanisms for controlling congestion in networks based on end-to-end windowing schemes are not suitable. Window-based mechanisms rely on the end-to-end exchange of control messages, which are used as feedback by the source node to regulate its traffic flow into the network. In high speed networks, the propagation delays across the network dominate the switching and buffering delays. The feedback from the network is outdated and the source action taken is slow to resolve congestion. The propagation delays of the end-to-end control schemes can be much larger than the transmission time of service requests. Preventing congestion in high-speed networks can be done by rate-control algorithm and by using the CG control method.

The CG control method is highly flexible. The CG message can either be randomly sent to some SSPs, broadcasted to all SSPs, or the SCP can determine where to send it. The most successful version of CG message control logic is when the SCP sends flow-rate commands only to SSPs, which are identified as overload sources (i.e., those that generate higher than threshold traffic than in previous intervals). SSPs with lower traffic loads in the previous interval will not be controlled during the next time interval. This can cause serious congestion problems to the SCP if the incoming traffic pattern changes fast (i.e., the heavy load traffic switches to uncontrolled SSPs during the next time interval). Broadcasting messages to all SSPs can avoid the possible congestion problem because all SSPs are constantly controlled. However, it can add a burden to the network because the SCP has to send the CG control message to all SSPs at each time interval.

According to Hać and Gao (1998), the SCP control scheme is proposed to set up a fair value to every SSP according to the current network configuration. The SCP selectively sends the CG control message to those SSPs that have overload traffic. SSPs that do not get CG control messages apply the jumping window scheme to control incoming traffic during the next interval. The fair value is used as the jumping window size. This method can diminish the network overhead caused by the broadcast message between the SCP and SSPs. The jumping window control scheme is desirable because there is no ACK sent between SCP and SSPs.

The SCP has to estimate the SSP offered load before it can make a control decision. In Hać and Gao's algorithm, a generalized CG method is used to estimate traffic in SSPs close to the real traffic pattern by counting the discarded calls outside the SSPs. The generalization of traffic control mechanisms enables the discarded traffic information (i.e., the discarded calls by SSPs in our study) to play a role in the control-decision process. These generalized versions of control mechanisms perform much better than their original counterparts.

The window method can react quickly to short traffic bursts if the window is not full. On the other hand, the CG method has a slower reaction to transient traffic. The CG method is based on the general overload-control principle that overload detection should be done at the node and load rejection should be done at the traffic sources. The SCP determines, based on its capacity and the incoming load, how much traffic each SSP should reject. This information is sent to the SSPs in a CG message containing a gap ($T$) and a duration-time interval ($D$), indicating to SSPs that one call per $T$ time units should be accepted for the next $D$ time period. The parameters $T$ and $D$ are dynamically adjusted according to the incoming traffic. An example of the CG control

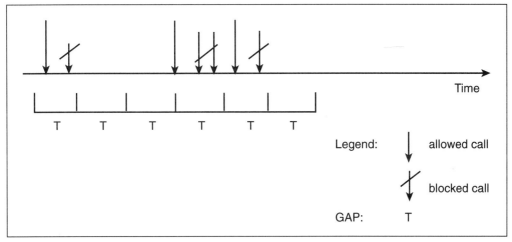

**Figure 2–11** The Original Call Gapping Control Logic

algorithm is shown in Figure 2–11. In Figure 2–11, a call is accepted per gap ($T$) in the CG control method. Three of seven incoming calls are accepted during a five-gap time period. The maximum number of accepted calls is five. The call accepting capacity is not fully utilized. This example shows that the original call gapping method can not handle bursty traffic well. If the unused accepting capacity can be saved for subsequent use, the ability to handle the burst-load pattern can be enhanced.

According to Hać and Gao (1998), the proposed control algorithm uses a modified CG to regulate traffic between the SCP and SSPs. A counter is used to save the unused capacity. The method works as shown in Figure 2–12. The incoming calls can go through the control as long as the value of the counter is larger than zero. The value of the counter is incremented by one at the beginning of each gap T. In Figure 2–11, 6 out of 10 incoming calls are accepted during a six-gap time period. The unused capacity is used in the subsequent gaps when there is more than one call during a gap.

## 2.4 Guaranteed QoS in Wireless ATM Networks

Virtual LAN (VLAN) over ATM networks allows a virtual LAN segment to coexist across the whole ATM network without being bounded by the actual physical location. This has a major advantage over the real, physical LAN where there is overhead involved in manual reconfiguration of station movements. A major difference between conventional LANs and ATM networks is that LANs are connectionless, whereas ATM supports only connection-oriented services (i.e., data may not be sent until a virtual channel connection (VCC) is established). Thus, an important function for a LAN emulation service is to be able to support the shared broadcast nature of LAN over ATM. The process of mapping shared broadcast to point-to-point communication is the ATM LAN emulation (LE) service developed by the ATM Forum that has become a standard.

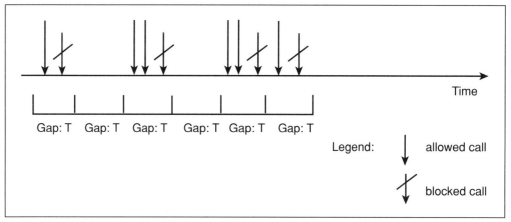

**Figure 2–12** The Modified Call Gapping Control Logic

The ATM Forum's LE service is based on a client-server architecture. The components of an emulated LAN network include LE clients (LECs), an LE server (LES), and a broadcast and unknown server (BUS). (Emulated LAN is the terminology of the ATM Forum specification. It supports only ATM-attached devices, while VLANs are defined for both ATM and nonATM network devices. VLANs can be seen as supersets of emulated LANs.) The LEC is the entity within ATM workstations or ATM bridges, which performs data forwarding, address resolution, and other control functions. The LES provides a facility for registration and resolving MAC addresses to ATM addresses. The BUS handles the data to broadcast addresses and to multicast addresses. It may also handle the initial unicast data, which are sent by a source LEC, before a direct virtual connection to the destination LEC is set up. Stations within the same emulated LAN are connected to the same LES/BUS pair, and stations belonging to different emulated LANs are attached to different LES/BUS pairs. In this respect, a multicast frame is first sent from an LEC to the BUS, which then relays it to all the LECs on the emulated LAN, but does not forward to the other emulated LANs. This result, in a secured broadcast domain, is known as a virtual LAN. For an LE-based virtual LAN, the membership of a station is primarily determined by the LES/BUS pair to which it is connected, not by its physical location. The VLAN is potentially more rewarding in a wireless environment where constant moving of mobile stations is highly anticipated.

Huang et al. show how the mobility issues such as location management, connection management, membership management, etc., can be done efficiently in a LAN-emulated ATM network and how application specified QoS can be maintained in a LAN-emulated ATM network. QoS for mobiles is different from that of fixed stations in an ATM network. More precisely, mobile QoS (M-QoS) is an augmented version of fixed station QoS. M-QoS is comprised of the traditional QoS found in wired ATM networks augmented with wireless and handoff QoS. For each of the major traffic types such as ABR, VBR, CBR, and UBR, QoS parameters such as end-to-end delay and desired bandwidth are specified by the application.

This specification must then be mapped into wired, wireless and handoff QoS requirements. The definitions of wired, wireless and handoff QoS are as follows.

- Wired QoS refers to the traditional quality of service supported by the wired ATM networks, such as link delay, cell-delay variation, and bandwidth.
- Wireless QoS refers to the QoS parameters associated with the wireless link, such as link delay, error rate, and channel reservation.
- Handoff QoS refers to the performance parameters associated with the handoff, such as handoff-blocking probability and cell loss.

Hać and Hossain (1999) shows that servers of different VLANs are interconnected through permanent or dynamically established virtual connections. The base stations/LECs are also connected through PVC or SVC to their parent LES/BUS pair, thus forming the mobile handling backbone (MHB). Both the base stations and LES/BUS pairs are equipped with buffering and switching capability to reduce handoff processing times. The MHB is connected to the ATM backbone network. Such an architecture can effectively handle connection, location, and membership management.

### 2.4.1 Mobile Quality of Service (M-QoS)

An M-QoS framework is used by the Cambridge Wireless™ ATM network. This framework can also be applied to other wireless ATM network architectures. M-QoS consists of the traditional QoS found in wired ATM networks augmented by wireless and handoff QoS.

Mobile communications in a wireless ATM-LAN environment can take place in two different forms: mobile to fixed and mobile to mobile. For either of these communications, a mobile connection is a concatenation of wireless and wired links. This undoubtedly defines M-QoS to be attributes of the wireless and wired links. The term wired QoS includes traditional QoS parameters used in ATM networks while wireless QoS is related to bandwidth variation owing to mobile host roaming, link delay, etc. While the switching, buffering, and traffic handling policies built into ATM switches have a direct influence on the wired QoS, the channel access, buffering, and traffic-handling policies implemented at the base stations have a similar influence on the wireless QoS.

An essential aspect of M-QoS is handoff QoS. As a mobile host migrates from one wireless cell to another, a connection can be forced to terminate due to the failure of a mobile handoff. Such failures are characterized by the handoff blocking probability and they can result in an inability to provide the desired wireless QoS in the new wireless cell, a failure of a base station, an inability to hand over in time, a failure of radio links, and an inability to fulfill the desired wired QoS in the wired links.

Hence, handoff blocking can be regarded as another M-QoS metric. In addition, the need to support seamless handoffs has led to a new M-QoS metric, which is known as handoff urgency or deadline. The ability to control handoff urgency can be impacted by a number of factors, for example: speed of mobile host migration, size of overlapping wireless cells, direction of migration, and types of ongoing mobile traffic.

Handoff urgency is interpreted in terms of handoff request priority. A mobile handoff request consists of connections identifiers, an intra-/intercell flag and a handoff urgency flag. The handoff QoS parameters related to per-virtual circuits are handoff urgency, probability of handoff blocking, a traffic-disruption period during handoff, cell loss during handoff, ATM cell sequencing during handoffs, and the speed of handoff operations. Multiple virtual circuits related parameters include the percentage of calls degraded, QoS consistency during handoffs, and the virtual circuit degradation priority.

In particular, for a mobile host with multiple active virtual circuits, the priority as to the degradation of specified virtual circuit's QoS can be introduced. For example, data streams can typically be allocated the lowest priority, followed by video and audio streams.

While the ATM cell-loss ratio is not part of the currently defined per-virtual circuit QoS vector and is treated as a network-wide performance parameter, this approach is probably unsuitable for mobile applications since care must be taken to selectively retransmit cells over the wireless link. Hence, cell loss in general and during handoff is identified as a per-virtual circuit QoS parameter.

At the higher level of the network protocol stack, QoS is specified by application through an Application Programming Interface (API). Depending on the requirements of the application, different classes of traffic–transport services, such as ABR, VBR, CBR, and UBR are chosen. The service class is then used to define the channel reservation policy, buffering policy, cell dropping policy, etc., during connection set-up. For each of the major traffic types, other QoS parameters such as end-to-end delay and desired bandwidth are specified by the application. This specification must then be mapped into the wired, wireless, and handoff QoS requirements.

The desire for adaptation in a mobile network environment is outlined by Noble. Satyanarayanaan (1995) argues that the resource poverty nature of mobile elements shows the need for more reliance on static servers. However, the need to cope with unreliability and low performance of wireless networks with mobile elements having power constraints suggests that such elements should have greater self-reliance. Hence, it was proposed that mobile clients must be adaptive. One extreme is individual applications while the other extreme is application-transparent adaptation, with the responsibility for adaptation solely born by the system. Application-aware adaptation, however, lies between these two extremes. This concept is further expanded to include bandwidth and delay adaptation.

Conventional approaches towards specifying desired connection bandwidth have always been a single numeric unit. For bursty traffic such as video, specification of burstiness has been proposed along with mean and peak bandwidth to allow further efficient use of the existing network bandwidth.

The concept of bandwidth window ($BW_{window}$) is applied and introduced by Toh (1996) for mobile applications. This approach allows the application to adapt to variations in the available network bandwidth as a mobile host migrates. The adaptation process is feasible since applications can scale down their QoS requirements without being forced to terminate. Examples can be video applications, where reductions in frame rate and picture resolution size are possible. Bandwidth adaptation can also be applied to internet applications. An example is the video mosaic (VSOAIC) application. With an adaptive video datagram protocol, VSOAIC can adapt to variations in available bandwidth over the internet.

In addition to throughput adaptation, delay adaptation for multimedia applications is possible provided that the delay is reasonably stable. The use of correctly sized buffers can avoid signal distortions as a result of varying delays. Hence, variations of delay due to the longer path after a handoff can be resolved in this manner, provided that this variation is small enough for a practical buffer size.

Consequently, we can tradeoff network bandwidth and delays for greater roaming capability and flexibility without being forced to terminate frequently. However, a handoff can be forced to terminate when the available M-QoS is less than the minimum operable value required (or that can be accommodated) by the application. Hence, handoffs of connections that cannot fulfill the minimum-operable QoS should be avoided. Adaptation, on the other hand, can also result in better utilization of network resources.

There are possibilities that future mobile applications have the mechanisms to adapt to a varying QoS environment. There are other interpretations of application QoS adaptation. In addition to application adaptation, the network must also be able to adapt to changes in network resource allocation.

If the new base station is able to support the wireless and handoff QoS requirements and the newly selected partial path can fulfill the wired QoS and handoff QoS requirements, the handoff process can then be completed successfully without the need for an M-QoS adaptation.

### 2.4.2 Architectures of High-Speed Wireless ATM Networks

Several high-speed wireless network architectures have evolved over the past few years, including several prototype implementations of these networks. According to Toh (1997), designing a high-speed wireless network architecture requires careful consideration of the type of services to be supported, the mobility profiles of users, the communication and computation ability of mobile hosts/peripherals, the need for internetworking, the ability to scale, the availability and limitations of wireless technologies (infrared or radio), and many other factors.

#### 2.4.2.1 Architecture Based on Distributed Control: AT&T™'s Architecture

A distributed control strategy for wireless ATM networks was proposed by AT&T™. Three types of applications requiring different kinds of control have been identified as follows.

| Types of Application | Types of Control |
|---|---|
| cellular-phone based | requires call and connection control |
| laptop data based | primarily connection control |
| multimedia based | requires call, connection and user control |

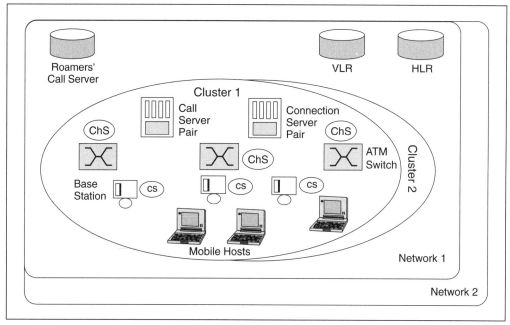

**Figure 2–13**  The AT&T™ Cluster-Based Multinetwork Architecture

Although the architectural work is focused on cellular phone applications, there are several features that are noteworthy. The architecture separates the functions associated with a mobile call and a connection control. Here, call control refers to the invocation and coordination of services during a call while connection control refers to routing and the derivation of a communication path over which user information can be exchanged. The architecture proposes a cluster based wireless LAN architecture and defines distributed servers and algorithms to effectively support mobility management. The resulting connection setup latency is reduced by employing parallelism during connection establishment.

Referring to Figure 2–13, the call server performs call-control functions, which include (a) maintaining the call state, (b) triggering explicit services, and (c) handling multiple connections per call. The connection server, on the other hand, derives routes and establishes end-to-end connections that fulfill the desired QoS, modifies and releases connections, and handles segments of a connection. Every ATM switch and base station has a channel server (ChS), which provides VC management functions and supports handoff operation. Channel servers implemented at the base stations map VPI/VCI to air interface resources, via a media access protocol. The home location server (HLS) is responsible for keeping track of its registered and assigned mobile hosts while the visitor location server (VLS) keeps track of the cluster where the visiting mobile is located in its network. The roamers' call server is responsible for managing calls by mobile users roaming into another network.

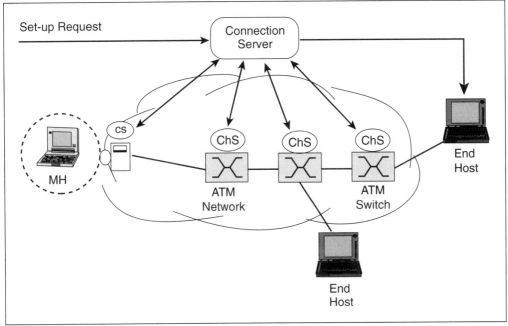

**Figure 2–14** A Parallel Connection Setup Scheme Using Separated Connection and Channel Control

While traditional call setup is based on a hop-by-hop call admission and resource reservation scheme, the AT&T™ distributed control architecture employs parallelism during call setup. Parallelism is achieved by separating channel-control functions from connection-control functions. A mobile that wishes to initiate a connection sends a connection request to the cluster-connection server, where a route for the connection is computed. Messages to reserve channel resources are then multicast, in parallel, by the connection server to the channel servers which constitutes the computed route. These channel servers then verify that the demanded resources can be met before they assign the VCIs, reserve the desired resources and reply back to the connection servers. Figure 2–14 illustrates the parallel-connection scheme.

Basically, handoffs are invoked by the mobile hosts, resulting in a new segment to be established from the new base station to an ATM switch which also has a path to the old base station. The handoff functions are embodied as a part of the channel server functions. During a segment establishment, the channel servers belonging to this new segment are identified and made known to the connection server. However, it is unclear how the common switch is located, how the route associated with the new segment is derived, and how connection rerouting is achieved. Handoffs occurring within a network, but involving different clusters, require interactions between the connection servers in each cluster, so that information regarding the old segment can be relayed across. In internetwork mobile handoffs, for the call server in the new network to exchange information with its peer at the old network, the old base station has to relay its connection server identifier to the new base station.

### 2.4.2.2 Architecture Based on Hierarchical Organization: NEC™'s Wireless ATM Architecture

An ATM-based transport architecture for the next generation of multi-service wireless personal communication network (PCN) has been proposed by NEC™. The architecture advocates the need to retain QoS classes and provides extensions to the ATM protocol stack to support mobility. To allow for the ease of interfacing with wired ATM networks, a PCN packet has a 48-byte payload, with an additional PCN header and trailer. Once the packet is received by the base station, the PCN header and trailer are removed and replaced by the standard ATM 5-byte header. In this manner, uniformity is maintained as packet units traverse from the wireless PCN into the wireline ATM backbone network.

The architecture shown in Figure 2–15 depicts a hierarchical organization of switching, multiplexing, and wireless-support system components. The PCN is meant to support a rich variety of computing and telecommunication services over a wide area. Each PCN node is comprised of large telecommunication-type ATM switches, small LAN-type ATM switches, multiplexers, and base stations. Associated with each ATM switch is a Mobile Service Unit (MSU), which contains the necessary switch-control software to support connection, handoff, and location management functions.

In NEC™'s scheme, end-to-end virtual circuits are divided into fixed and dynamic segments, with dynamic segments reconfigured during handoffs. New partial segments are established from a chosen handoff switch (HOS) to the target base station. This HOS will then perform a VC remapping operation during a handoff.

In the NEC™ wireless PCN architecture, ATM compatibility is achieved by defining a PCN data-link packet as a 53-byte ATM cell encapsulated by an additional PCN header. The wireless header and trailer result in additional protocol overhead. This overhead is reduced by compressing the ATM header. When the wireless PCN packet is received by the PCN network interface unit (NIU) at the base station, the packet header and trailer are removed and the compressed ATM header is exploded, leaving the payload intact. The resulting 53-byte ATM cell is then forwarded by the ATM NIU to the backbone ATM network.

### 2.4.2.3 Architecture Based on Intelligent Network: British Telecom™'s Architecture

In Global System for Mobile Communication (GSM), mobility management and handoff functions have been integrated into the mobile switching center (MSC). As a result, the processing requirements at the switch are substantially increased, which severely reduces the subscriber-handling capability of the MSC, compared to the other switches in the fixed-backbone network.

There is a need for harmonization of system components via integration of mobile and fixed capabilities into a common architecture. In particular, mobile call-handling functions are provided by the fixed network while mobility management functions are supported by the intel-

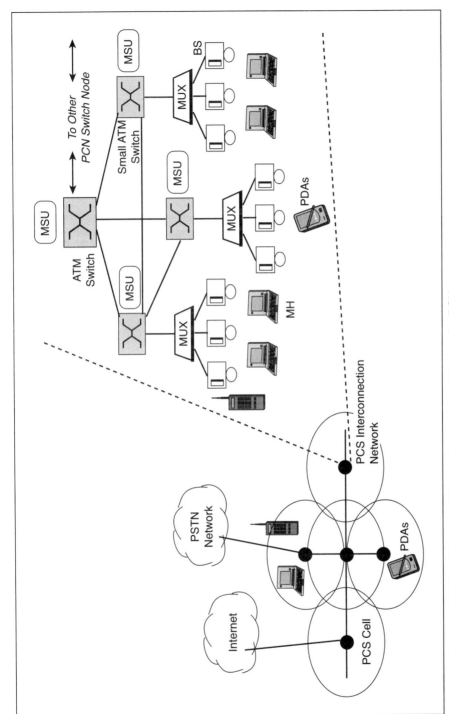

**Figure 2–15** The NEC™ ATM-Based Wireless Transport Architecture for PCN

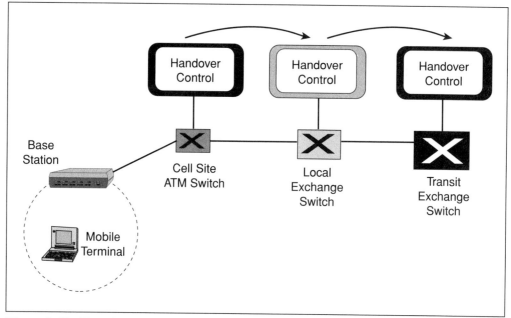

**Figure 2–16** Transferring of Handoff Control in UMTS

ligent network. The application of intelligent networks can be found in both switching and radio access components. The British Telecom™ mobile architecture uses a distributed and separated call and connection control and IN capabilities to support a variety of mobility-related functions.

An interesting feature is the transfer of handoff control, as shown in Figure 2–16. As a Universal Mobile Telecommunication System (UMTS) terminal migrates across coverage areas overlooked by different network entities, this may require the transfer of service states (including call-state machines) from one control point to another, hence achieving distribution of handoff control functionality over different control entities within the network.

In conventional mobile systems, such as the analogue-cellular systems, handoffs are only used to satisfy speech and data quality requirements as a result of switching to a new radio path. Within UMTS, a handoff may be requested by the user (for example, specific subscribers' services) or the network operator (for example, network management purposes), irrespective of the radio-link quality. Handoffs may involve interactions between the base stations, local exchange, or even transit exchanges (as shown in Figure 2–17) and a common point has to be determined to reroute the ongoing calls. The processing burden caused by handoffs may be minimized by dividing the functionality between the fixed network and the control components. One possibility is the use of IN for handoff control, thereby removing the processing burden of handoff control from the switches.

# Guaranteed QoS in Wireless ATM Networks

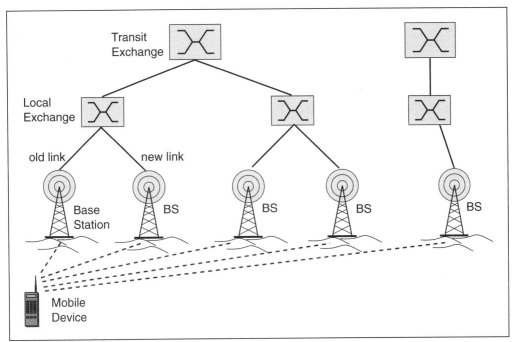

**Figure 2–17** Handoff Possibilities and Treelike Network Structure in UMTS

### 2.4.2.4 Cambridge Wireless ATM LAN Architecture

The architecture shown in Figure 2–18, is based on a combination of hierarchical organization, distributed control, and uniformity (i.e., the ability to fulfill a variety of mobility-related requirements in a consistent manner). To provide compatibility with ATM, a wireless ATM header similar to that of NEC™ is adopted.

The wireless cell architecture assumes that handoffs based on partial reestablishments are employed. Partial reestablishments refer to establishing only a partial path during handoffs. The partial path, also known as the convergence path, converges with the existing path prior to a handoff. The anchor point is a switch known as the crossover switch (CX). Similar to the cluster switch, the CX has the ability to perform connection rerouting. This implies that a handoff management entity is associated with each switch. Since connection rerouting involves real-time VCI re-mapping, cell buffering, and synchronization, this operation is totally handled by the CX entity. Call and connection control can be handled by external entities such as call and connection servers. This separates channel, call, and connection control.

As a result of wireless cell clustering, handoff management is distributed. Switches in the network are required to have the minimum ability to perform VCI re-mapping to support mobile handoffs. In addition, as a mobile host migrates from one wireless cluster to another, the responsibility of handoff management is changed from a cluster switch to a CX, hence attaining trans-

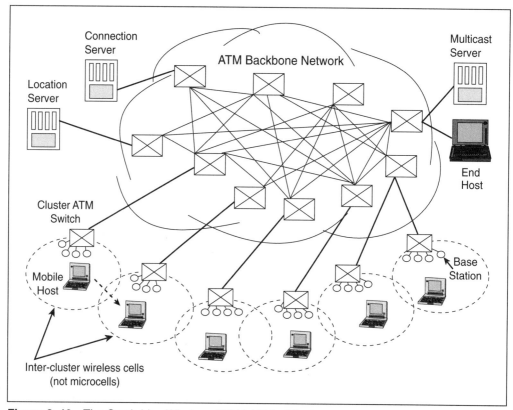

**Figure 2–18** The Cambridge Wireless ATM LAN Architecture

fer of handoff control. Depending on the connectivity of the switches, a CX is or is not a cluster switch.

The crossover switch-discovery mechanism is embodied into and invoked by the handoff protocol, and utilizes existing routing and QoS information to derive a suitable partial path and an appropriate CX. If the selected partial path fails to be established for whatever reasons, a fallback strategy is used to select the next most appropriate CX. Several CX discovery algorithms are used for wireless ATM LAN employing both the centralized or distributed connection scheme. Figure 2–19 illustrates the concept of CX discovery.

Although handoffs result in additional signaling traffic over the wireless and wired networks, idle handoffs (i.e., handoffs of a mobile host that does not have an ongoing connection) still require location information to be updated. Hence, location updating traffic contributes to the total signaling traffic, regardless of whether the handoff is active or idle. The location signaling traffic is a function of the rate of mobility, the population of the mobile users, the size of the wireless cell, and the mobility profiles of mobile users. In a wireless ATM LAN employing pico-sized cells, several hundred base stations can exist within a building.

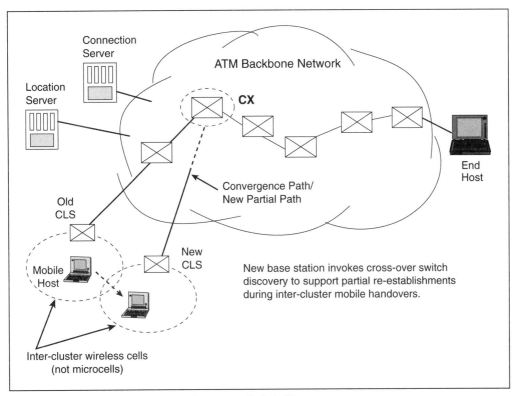

**Figure 2–19**  The Concept of the Crossover Switch Discovery

This location management architecture adopts a hierarchical approach. Fine and coarse granularities of location tracking are also defined. Different entities contain different accuracy of location information, as shown in Figure 2–20. Since the location updating traffic is directly related to the granularity of location tracking, by splitting this granularity, signaling traffic is localized and the location management burden is now shared by different entities. Hence, this makes the overall system efficient and scalable to the increasing number of mobile users.

2.4.2.5   LAN Emulation Model for Wireless ATM

The ATM forum LE on ATM specifications defines the architecture of a single-emulated LAN only. The LE service has the following components: LEC, an LE service consisting of an LE configuration server, an LE BUS, and an LES, as illustrated in Figure 2–21.

The LEC is the entity in the end stations that performs various control functions such as data forwarding, and address resolution. The LES implements the control coordination function for the emulated LAN that includes registering and resolving the MAC address. The BUS provides services to support broadcast and/or multicast traffic and the initial unicast frames that are sent by an LEC before the data target ATM address is resolved. The LE configuration server

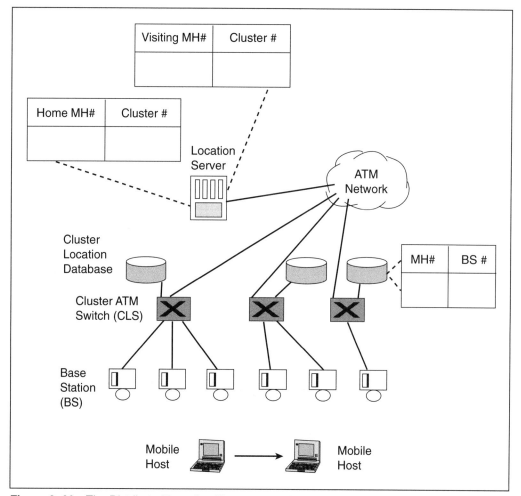

**Figure 2–20** The Distributed Location Management Architecture

implements the assignment of individual LE clients to different emulated LANs. There are six types of ATM connections used in LES (shown in Figure 2–21):

- A configuration-direct VCC is set up by the LEC to perform the configuration protocol exchanges with the LE configuration server.
- A control-direct VCC is set up by the LEC as part of the initialization phase and it is maintained by both the LEC and the LES as long as LEC participates in the emulated LAN.
- A control-distribute VCC is set up by the LES as part of the initialization phase. The LEC is required to accept this VCC. It is maintained by both the LEC and the LES while participating in the emulated LAN.

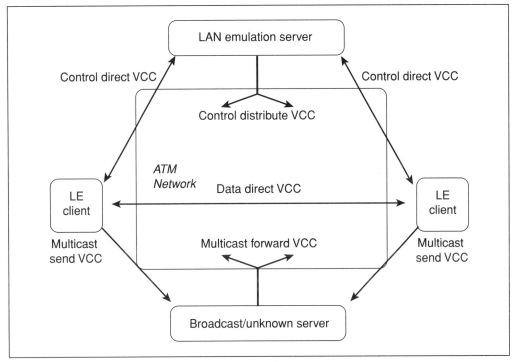

**Figure 2–21**  Components of the LAN Emulation Service

- A data-direct VCC is used for an LEC to send user frames to another LEC after the original LEC learns the address of the destination LEC. If the end station does not have resources to establish a data-direct VCC, it uses a multicast send-data VCC to request the BUS to forward its frames to the destination LECs.
- A multicast-send VCC is established from an LEC to the BUS. It is maintained while the LEC is participating in the emulated LAN.
- A multicast-forward VCC is a point-to-multipoint VCC with the BUS being the root and all of the LECs being the leaves.

An LEC is identified by two addresses: an individual Institute of Electrical and Electronics Engineers (IEEE) 48-bit MAC address and a one or more 20-byte ATM address. An LEC uses its primary ATM address for establishing a control-direct VCC, multicast-send VCC, and as the source address in the control frames. An LES is assigned one or more ATM addresses. This address may be shared with the BUS. Similarly, a BUS may have one or more ATM addresses assigned to it, which may be shared with the LES.

The LEC is a part of each end station participating in the LE service. Each LEC represents one or more users that are identified by their MAC addresses. The main function of an LEC is to provide its users with a MAC-connectionless service by using the services of AAL5 to communicate with other LECs and remote LE service components. These communications can take place over permanent, switched, or mixed (both permanent and switched) ATM connections.

Various functions performed at the LEC include the following:

- Establishing a connection to the LES.
- Establishing a connection to the BUS.
- Registering with the LES.
- Generating and responding to address resolution requests.
- Transmitting and receiving MAC frames.

Before joining an emulated LAN, the LEC is connected to an LE configuration server for its configuration information, which includes the ATM address of the LES, one or more MAC addresses, emulated media type, and maximum frame size for the media being emulated.

Following the configuration, the LEC should have the ATM address of the LES. Using this address, the LEC establishes a control-direct VCC with the LES. This connection is mainly used for address resolution. When the LEC has a frame to send to another client and does not know its ATM address, it sends a message to the LES requesting the ATM address of the destination station. The LES returns the requested address using the same VCC. When the LEC has the ATM address of the destination LEC, it can establish a switched VCC (i.e., a data-direct VCC) and transmit its frames using this connection. Alternatively, the LEC can send its frame to the BUS at the same time it sends the address resolution request to the LES. The LEC determines the address of the BUS using the address resolution procedure. After obtaining the address of the BUS, the LEC establishes a multicast-send VCC and uses this connection to transmit frames to the BUS. Upon receiving a frame from a multicast-send VCC, the BUS multicasts the frame to all LECs in the emulated LAN.

An LEC first goes through an initialization phase to obtain its configuration information and the addresses of both the LES and BUS. Upon receiving a frame from a higher layer, an LEC first tries to send the frame directly to the destination LEC over data-direct VCC. This means that each LEC maintains an address resolution table that provides a mapping between the destination MAC address and data-direct VCC. If the destination MAC address is not in the table, the LEC checks whether or not the ATM address corresponding to the MAC address is known. If it is known, the LEC establishes a data-direct VCC to the destination LEC using UNI signaling. If not, the LEC sends an address resolution request to the LES. At its option, the LEC can, at this time, send the frame to the BUS for multicasting to all LECs in the emulated LAN. After receiving a response to its address-resolution request, the LEC can use signaling to establish a data-direct VCC to the destination client.

An LEC needs to obtain the ATM addresses of both the LES and BUS, and establish connections to each server. Obtaining the ATM address of the BUS is relatively straightforward once a connection to the LES is established. It requires sending an address-resolution request for the broadcast MAC address. The reply from the BUS includes its own address, which is used by the LEC to establish a multicast-send VCC to the BUS.

Determining the ATM address of the LES is not as simple. An ATM network does not provide a broadcast service similar to that of legacy LANs. One alternative to make the ATM address of the LES known to the LECs is to configure it manually at each LEC in an emulated LAN. This is not a preferred solution in ATM networks that support a large number of emulated LANs and/or a large number of LECs. The LE configuration service allows emulated LANs to be built dynamically.

To remove any configuration burden, an LEC locates the LE configuration server (LECS) using one of the following methods, attempted in the order listed:

- LECS address via the Interim Local Management Interface (ILMI): The LE client must issue an ILMI message to obtain the ATM address of the LECS for the UNI to which it is attached.
- Well-known LECS: If the ILMI attempt fails or the LEC fails to establish the connection with the ATM address obtained through the ILMI, then a well-known ATM address can be used to open a configuration VCC to the configuration server.
- Configuration-direct SVC: This can be established using the broadband low-level code point for the ATM LE configuration-direct VCC.
- LECS PVC: If the LEC cannot establish a VCC to the well-known ATM address of the LECS, then the well-known PVC with $VPI = 0$, $VCI = 17$ is used for the configuration-direct VCC.

The LES provides the address-resolution mechanism for resolving MAC addresses. Various functions performed at the LES include responding to the registration frames, forwarding address-resolution requests, and managing LEC address resolution information.

If the ATM address of the destination LEC is not known at the originating LEC, the address-resolution request is sent to the LES over a control-direct VCC. The LES forwards this request to all clients in the emulated LAN over a control-distribute VCC. This address resolution request includes, among other information, the source and destination MAC addresses and the ATM address of the LEC originating the request. All LECs in the emulated LAN are required to receive and process address-resolution requests from the LES. This processing at the LEC involves checking the destination MAC address included in the request and, if possible, responding from the particular MAC address with an address-resolution reply that contains its own ATM address. These responses are sent to the LES over a control-direct VCC. The LES forwards this reply to the LEC that originated the address-resolution request over its control-direct VCC.

The control-distribute VCC is a point-to-multipoint ATM connection with the LES as the root and all LECs registered in the emulated LAN as the leaves. An LES maintains an address-resolution table that maps MAC addresses of LECs with the corresponding ATM addresses and control-direct VCC identifiers. These MAC addresses are primarily the ones that are assigned to ATM end stations participating in the emulated LAN. Some of these end stations may control another set of MAC addresses (i.e., bridges) and not all of these MAC addresses can be registered at the LES. Hence, even if all ATM end stations participating in an emulated LAN are known to the LES, a control-distribute VCC may still be necessary to reach stations that are attached to some of these ATM stations.

The BUS mainly broadcasts LAN frames with unknown MAC addresses (at all origin LECs) to all LECs participating in an emulated LAN. The various functions performed at the BUS include receiving frames with unknown MAC addresses from a multicast-send VCC, assembling the frame, and broadcasting over a multicast-forward VCC.

An LEC obtains the ATM address of the BUS via an address resolution request using its control-direct VCC to the LES and establishes a multicast-send VCC to the BUS. The multicast-forward VCC is separate from the multicast-send VCC. A multicast-forward VCC is a point-to-multipoint ATM connection, with the BUS as the root and the LECs as the leaves. This point-to-multipoint VCC can be established as a set of point-to-point connections or it can be a tree connection, depending on whether or not the network supports multicast.

If an LEC does not know the ATM address that corresponds to the destination MAC address in a LAN frame, it sends an address resolution request to the LES. As an option, an LEC can transmit the frame to the BUS in parallel. A number of frames can be sent over a multicast-send VCC to the BUS until a response to the address-resolution request is received. The main objective in using the BUS to forward these frames is to reduce the total delay in sending LAN frames to an unknown ATM LE client.

The BUS framework is simple to implement. However, it does not scale well as the size of the emulated LAN increases. In particular, it is a centralized scheme and cells of different frames cannot be interleaved on the multicast-forward VCC, since AAL5 does not allow VC multiplexing. This means that frames are transmitted as cell trains, which necessitates that all the cells of a frame arrive at the BUS before the frame can be transmitted. The main consequence of this is the delay it takes to receive the entire frame. As the number of LECs and/or traffic in the emulated LAN increases, the delay at the BUS increases, causing the BUS to become a performance bottleneck.

Similarly, as the number of frames broadcast in the network increases, the end stations start to receive a large number of frames. Each frame is required to be processed at each end station to decide whether to copy (i.e., the destination MAC address is the same as one of the MAC addresses the station supports) or discard the frame (i.e., no match). Since an end station discards a large portion of the frames it receives, a considerable processing capacity is wasted. This also adds to the cost of the end station, since high speed filtering and/or large buffers are required to process a high volume of arriving frames.

Emulated LAN consists of a number of LECs and a single LES. Communication among LECs and between LECs and the LE service is performed over ATM virtual connections (VCCs).

An emulated LAN has two logical multicast servers to emulate shared medium characteristics of legacy LANs. When a station is connected to an emulated LAN, it establishes a circuit (or uses a preestablished circuit) and registers it with the LE service, its MAC, and ATM addresses. The station also registers itself to send and receive broadcast MAC frames and may register for group and functional addresses as well. Once a station establishes a connection to the LES and another one to the BUS, it is ready to process incoming and outgoing frames.

If an LEC knows the ATM address of the destination of its frame, a direct connection can be established between the source and destination nodes. If the ATM address of the destination is not known, the LEC forwards the frame to the BUS. It also sends an address-resolution request to the LES. If the LES knows the ATM address that corresponds to the destination MAC address, it replies to the LEC without any additional processing. Otherwise, it broadcasts the address-resolution request over a control-distribute VCC. When an end station responds to the address-resolution request to the LES, the LES forwards the response to the LEC that requested the address resolution.

If a frame with unknown ATM address is sent to the BUS, the BUS uses its multicast forward VCC and broadcasts the frame to all LECs in the emulated LAN or uses its optional intelligent forwarding capability to forward the frame to its destination over a point-to-point VCC.

#### 2.4.2.6 The Mobile Handling Backbone (MHB)

In an ATM network, certain QoS parameters have to be satisfied when a call is set up, and have to be maintained during the lifetime of the connection. Since in wireless ATM, the end points are mobile, and therefore, may change connectivity to the fixed network during the lifetime of a connection, it is necessary that the QoS parameters also be satisfied no matter where in the network the mobile migrates. As the mobile migrates in a different part of the network, the resources (like bandwidth, etc.) may not be available in that part of the network. In this case, it is necessary to have an application quality of service adaptation (i.e., bandwidth and delay adaptation). However, certain traffic types like CBR or VBR may not withstand any changes in the promised QoS parameters and therefore, can be forced to disconnect.

To solve this problem, the necessary resources must be reserved (i.e., bandwidth) in the parts of the network where the mobile visits during the lifetime of its connection. Still, this reservation is an inefficient utilization of network resources. To ensure the necessary resources for a mobile wherever it visits, it is required to make reservations. The goal is to do this efficiently with little waste of network resources. In Hać and Hossain (1999), a MHB, a logical entity, is proposed which is responsible for providing the necessary bandwidth to a mobile as it roams among different cell areas in a wireless ATM network. The MHB complies with the ATM Forum's emerging LAN emulation standard for mobility management in a wireless ATM network.

The MHB is responsible for implementing the mobility-related issues such as location management, connection management, call admission control, membership management, handoff management, and guaranteed QoS. Huang and coworkers show that the LAN emulated wireless ATM can efficiently handle location, connection, and membership management.

A MHB scheme is proposed by Hać and Hossain (1999) that is responsible for ensuring application QoS, and the mobility issues described by Huang and coworkers (1997). The backbone consists of the LES/BUS pairs and the LECs (base stations). Veeraraghavan et al. (1997) show that base stations with switching and buffering capability result in faster handoffs. Hać and Hossain propose that a similar feature be included in the base stations (LECs) and in the LES/BUS pairs.

A set of control-direct VCCs between LECs (base stations), a corresponding LES, and a set of control-distribute VCCs between the LES and corresponding LECs (base stations) makes up the MHB. The ATM Forum LE on ATM specification defines only the architecture of a single-emulated LAN. Hać and Hossain (1999) propose additional VCCs between different LES/BUS pairs such as those described by Huang et al. (1997) be incorporated into the MHB. These three types of VCCs are used to make available the required application QoS parameters, like bandwidth, to the mobile as it migrates to a neighboring cell. The MHB can efficiently make available the required bandwidth, and reserve the network resources with minimum waste of bandwidth.

The handoff proposed is a mixture of base station assisted handoff (as in a code division multiple access (CDMA) cellular system) and mobile assisted handoff as in an Advanced Mobile Phone Systems (AMPS) cellular system. It is necessary to reserve network resources like bandwidth to all the neighboring cells of a mobile to ensure the QoS in order to have free roaming of the mobile between cells during the lifetime of a connection. The question is how efficiently the reservations can be made so that the network utilization is high.

The mixture of two kinds of handoffs is proposed to reserve the bandwidth for as little time as possible so that the overall network utilization improves. In the handoff scenario, the base station constantly monitors the power level of the mobiles connected to this base station as it is in CDMA. Whenever the base station determines an ensuing handoff (base assisted handoff), it notifies the parent LES/BUS pair through a control-direct VCC and also starts sending frames to the LES/BUS pair. The LES/BUS pair then reserves the necessary bandwidth in the neighboring cells through the control-distribute VCC and also starts sending the frames to the neighboring base stations. This continues until the mobile prompts the exact neighbor base station (mobile assisted handoff) that is in its coverage area.

The new base then sends a message to the LES/BUS pair confirming the arrival of the mobile so that the LES/BUS can discontinue transmitting the frames to the other neighbor bases. The MHB continues to supply the frames until a data-direct VCC is established between the new base station and the other end of the connection.

This results in reservation of bandwidth in only that part of the network where it is necessary and only for the optimum period of time that is necessary. This also results in an almost seamless transmission/reception of frames to the mobile without interruption due to handoff, and fast handoff (a plus for the high-speed network). In the regions of overlapping cell areas, the mobile continues to receive frames from multiple base stations which cause self-interference. We suggest that a CDMA type of multiple-access protocol be used where the frequency reuse factor is 1 (i.e., no frequency/channel change is necessary for the mobile as it migrates) and also where the self-interference and multipath effects are an advantage.

The location-management scheme is distributed and is performed by each LES/BUS pair. We adopt the algorithm developed by Huang and coworkers (1997). Each LES/BUS pair maintains two databases for tracking the location and membership of the mobiles, namely the out-membership tracking database (out-MTD) and the in-membership tracking database (in-MTD). The out-MTD is simply a count of the number of mobiles that have moved out from its LECs. The in-MTD is used to record the information of the mobiles moved into the base stations supported by the particular LES/BUS pair. The in-MTD also stores the ATM address of the corresponding base station and the VCC to the base station. In addition it stores the ID of the LES/BUS pair from which the mobile is currently visiting.

The LES/BUS also works as a connection manager. Whenever a call is targeted to a mobile, a query is made to the nearest LES/BUS of the MHB. Initially, upon power on, every mobile registers to the BS, which in response registers to the parent LES/BUS of the mobile. When a query is made to an LES/BUS, a multicast message is sent throughout the MHBs to all the LES/BUS pairs. The responsible LES/BUS pair then responds by sending a message to the corresponding base station through a control-distribute VCC which contains the ATM address of the ATM end station and also the QoS parameters. The LEC then establishes a connection with the end station through a data-direct VCC.

A mobile call refers to a new originating call or an ongoing call which requires handoff from time to time. Consequently, mobile call admissions refer to both new call and handoff admissions. The following discusses call admissions related to a wired ATM, mobile ATM, and handoff calls.

- Wireline ATM Call Admissions

    In wireline ATM networks, call admissions are performed at each individual switch in order to ensure that the QoS associated with the call can be fulfilled prior to call acceptance. If a switch decides to admit a call, a resource reservation is made. The LES/BUS pair which acts as a connection manager derives a possible route and requests the switches in the route for possible call admission.
- Mobile Call Admissions

    In a mobile call supported by both wireless and wired links, call admissions not only occur at the switches but also at the base stations. Consequently, a mobile host-to-fixed-host call can only be admitted if the base station and the subsequent switches in the route can fulfill the desired QoS.
- Handoff Call Admissions

    Call admissions can also apply to ongoing mobile calls. When there is a hint from the base station that one of its mobiles is going to be handed off, the LES/BUS parent must be able to reserve the necessary resources in the neighboring cell areas. If this fails in any of the neighboring cells, the handoff is blocked.

CHAPTER 3

# Models of Telecommunications Networks

## 3.1 Synchronization in Multimedia Systems

Multimedia refers to the integration of text, image, audio, and video in a variety of application environments. Recent developments in fiber optics and broadband communication technology such as B-ISDN and ATM, provide bandwidth and delay characteristics necessary to support these new media services. A multimedia system can either transmit live video or audio (e.g., teleconferencing), or use stored video, audio, and image information (e.g., a distributed multimedia information system (DMIS)). The DMIS system supports integration and coordination of audio, video, text, and image originating from different servers interconnected by high-speed networks. A server can be dedicated either to the storage of data of a single-medium type or multiple-media types. Information from multiple-media types is retrieved by a client from servers and presented to the user in a meaningful fashion. New networks, protocols, and operating systems are necessary to meet the requirements of multimedia applications.

One of the requirements of a multimedia system is the need to provide synchronization of multiple-media data elements which are sent via different channels and experience random delays during the transmission period. Traditional data communication networks are used to provide error-free transmission, and are not able to meet the real-time delivery requirements or to achieve synchronization. The task of synchronization is to eliminate all the variations and delays incurred during the transmission of the multiple-media streams and to maintain synchronization among the media streams.

End-to-end delay in a distributed multimedia system consists of all the delays created at the source site, network, and receiver site. These delays usually consist of the packetization delay, the network access and transmission delay, protocol processing delay, and presentation delay. Synchronization is much more complicated, and asynchrony is more likely to occur, if the media data involved in the same presentation originated from different sources in the network. Another source of asynchrony is clock drift, the mismatch between the sender's transmission rate and receiver's presentation rate. If the playout rate of the receiver is faster than the transmission rate of the sender, the receiver may suffer starvation. The receiver may suffer overflow if the receiver's transmission rate is slower than the sender's transmission rate.

There are two types of media data: continuous media data and discrete media data. Continuous media data, such as voice and video, have time-dependent values. They require real-time service delivery and fine-grain synchronization. An example is the synchronization of the mouth movements in video sequences with audio presentation, which is also known as "lipsync." This synchronization is more difficult to achieve when the video data and the corresponding audio data are sent from separate locations over a network. Discrete media data, on the other hand, have time-independent values, such as text and graphics. They pose less stringent end-to-end delivery requirements yet need superior reliability in transmission. However, if this data has synthetic timing relationships with other data, as in the case of a slide show with blocks of audio allotted to each slide, it should be delivered on time without error. Based on these two types of media data, synchronization of multimedia includes both intramedia and intermedia synchronization. While intermedia synchronization needs to maintain an intermedia relationship, intramedia synchronization guarantees that continuous media are played back at their fixed rates.

To achieve media synchronization, several synchronization requirements have to be identified and solved. First, it is desirable to have some useful tools or means to express complex temporal relationships of multiple-media data in applications involving the simultaneous presentation of multiple-media types. Second, real-time communication protocols and operating systems are essential. These systems assign the utmost priority to meeting deadlines for a set of scheduled time-dependent tasks. Specific scheduling is required for storage devices, processors, and communication resources in order to overcome the latency of data, from its source to its destination. Third, efficient buffer management is needed to smooth out variations in end-to-end delay for media units. End-to-end delay varies according to the transmission path lengths, network traffic load, and the host's central processing unit (CPU) latency, which is also load dependent.

Research has been done in the media synchronization area. Steinmetz presents a general concept of programming constructs for expressing time relationships among media. Little and Ghafoor (1990 and 1991) use petri nets to define a synchronization model called the object composition petri net (OCPN) for the description of presentation sequences and relationships of multimedia objects. Nicolaou (1990) presents a scheme for implementing synchronization among related streams by inserting synchronization points in each individual stream. Leung and coworkers (1990) presents a scheme using a multimedia virtual circuit. The transport layer of the sender multiplexes all the interrelated medium data into a single virtual circuit. Ferrari (1992) presents a delay jitter-control scheme that insures media synchronization as long as a bound on delay can be guaranteed, and the sender and receiver clocks are kept in synchrony. Every node on the transmission path in the network is involved in synchronization control by regulating the random delay, which has been introduced to an arriving packet by the previous hop. This method may be hard to implement in an ATM network due to too much workload on the intermediate nodes that carry thousands of connections. The adaptive feedback method presented by Rangan and coworkers (1993 and 1995) solves the asynchrony in multimedia on-demand services. It requires an additional connection for each sender and receiver pair to transmit feedback units. The centralized multimedia server uses feedback units transmitted by the receiver to detect the asynchronies among them. Correction of the asynchrony is made at the server by speeding up or slowing down the traffic using skip or pause media data on the various media streams. Adaptive feedback is not suitable for WANs because its synchronization guarantees are limited by the maximum network delay. Little et al. (1991, 1992) propose a scheduling protocol for stored media applications. They use an OCPN to capture the temporal relationship among media units. According to the OCPN and end-to-end delays, the playout schedule and transmission schedule for receiver and sender can be calculated respectively. Senders and receivers will then just follow the computed schedules to transmit and playback the media data. However, due to the nondeterministic random delay of the network, scheduling and predicting the traffic is not sufficient to maintain a simultaneous data delivery. Also, the scheme does not consider the clock drift problem. Lamont and Georganas (1994) present a stream synchronization protocol that tries to provide multimedia news services. The proposed protocol adds "intentional delay" to those media streams which experience less end-to-end delay without violating the tolerant asynchrony bound specified by the application. The local clock drift problem is not considered nor is a way of correcting transmission rates of different servers. This method is hard to apply to WAN area where the network delay jitter is relatively large.

Hać and Xue (1997) present a DMIS in which data applications are stored. DMIS uses some preorchestrated, stored multimedia data, and requires management of heterogeneous data with vastly different storage, communication, and presentation requirements. In DMIS, synchronization is a complex problem because several streams from independent sources can require synchronization to each other, in spite of the asynchronous nature of the network. Hać and Xue present a complete synchronization scheme for multiple data streams generated from distributed media database servers without using a global clock. Besides using the OCPN model to describe the complex temporal relationship among coupled medium streams in applications, media scheduling at the server side and buffer reservation at the client side are needed during the connection set-up phase. A buffer size configuration method is presented by Lue and coworkers (1994), but it only considers intramedia synchronization. To reach intermedia as well as intramedia synchronization, the determination of buffer size for one medium needs to consider the effect of other related media. The medium retrieval time plays an important role here. If the coupled media are retrieved at the same time, the buffer size of one medium has to be configured based on the maximum delay experienced by related media streams instead of its own medium delay. If they are retrieved at each corresponding control time, then there is no need to consider the delay effect of other media. Synchronization recovery is performed at the receiver end before the playback. To manage the buffer level behavior efficiently, control methods are introduced. A scheme to control the average level of the receiver frame buffer to a nominal value is proposed by Kohler and Muller (1994) to reach the intramedia synchronization goal. Since the buffer is supposed to compensate for delay jitter, the dynamic level change of the buffer is normal. Instead of fixing the nominal value at one point in the buffer, we propose three different nominal values according to three different buffer statuses: underflow, normal, and overflow. Skew tolerance parameters are used here as control constraints. In order to realize the intermedia synchronization goal, a master media-based control method is also provided.

### 3.1.1 The Architecture of DMIS

An ideal network for multimedia communications is a constant-delay network, which could be characterized also as a zero-jitter network, where jitter (more precisely, delay jitter) is a variation of stream packet delays. With this ideal network and real-time operating system available both at the server and client sides, the synchronization of multimedia data becomes quite simple, because both the network transmission delay and operating system latency are bounded. However, this is usually not the case in practice, and it does not consider the clock drift problem.

The computers and the high-speed packet-switched networks that connect different servers and clients involved in multimedia communications may introduce large amounts of jitter because of the fluctuation of their loads. We assume that the broadband network interconnecting the multimedia servers with clients is a high-speed network, and on a wide area range. Therefore, the DMIS imposes nondeterministic delays in data retrieval due to (a) query evaluation, seek, and access delay at the server side; (b) transmission delays due to network transmission, packetization, buffering, and depacketization; and (c) presentation delays due to decompression of data at the client side.

Implementing a synchronization algorithm for a specific multimedia application requires specifying the QoS for multimedia communications. The QoS is a set of parameters that characterize communication services, including bandwidth, the bounds on tolerable end-to-end delays, jitter, and the level of reliability in transmission (e.g., bit error rate and packet error rate).

QoS may be classified into three different parts: application QoS parameters, which describe the requirements for the application services; system QoS parameters, which describe the communication and operating system requirements resulting from the application QoS; and finally, network QoS parameters, which describe the network performance requirement such as bandwidth and jitter. We will focus on the network QoS. With the network QoS specification available, the transport requirements for various types of media traffic are defined, and the network can allocate and reserve resources for multimedia applications.

In a high-speed network like B-ISDN, multiple-medium streams, such as text, image, voice, and video are first chopped up into fixed-length packets, then multiplexed into a single-bit stream which is transmitted across a physical medium. How to support multiple QoS classes for differing application requirements on delay and loss performance is a critical issue. ATM supports multiple QoS classes for differing application requirements on delay and loss performance.

B-ISDN, based on ATM, supports a wide range of all services such as voice, packet data (IP, etc.), video, imaging and circuit emulation.

Continuous media, such as video and voice, are very time sensitive. The information cannot be delayed for more than a blink of the eye, and the delay cannot have significant variations. Discrete media, such as text and image, are not nearly as delay sensitive as continuous media. However, they are very sensitive to loss. Thus, the B-ISDN network must have the ability to discriminate between continuous media and discrete media, giving continuous media traffic priority and bounded delay, while simultaneously assuring the high reliability and low loss for discrete media traffic.

In the distributed multimedia information systems, media servers and clients are connected by the high-speed network. This network is characterized by hosts (servers, clients) and switches interconnected by physical links in an arbitrary topology. The switches operate according to a corresponding design principle, introducing queuing delays while the links introduce only propagation delays.

For DMIS, in which data applications are stored, the system has flexibility in scheduling the retrieval of data, since each multimedia application service has an associated temporal presentation scenario describing the temporal relationships among the data objects involved. This scenario is stored in the multimedia database. Each medium may reside in its own medium server or in the database itself. The medium servers are of different types, such as an image server, voice server, text and graphic server, and video server. The server has the ability to adjust the data-retrieval rate of the media. Data in different media are transferred over different channels according to their traffic characteristics and transmission requirements as specified by the QoS. Through this kind of distributed multimedia information system, activities such as multimedia data retrieval and playback are performed continuously in realtime.

### 3.1.2 Multimedia Synchronization Scheme

The main cause of asynchrony in multimedia communication is jitter. Jitter consists of three parts. First is retrieval delay, which is the time needed for the server to collect media units and segment them into packets for transmission. The second part is the network delay from the network boundary at the server end to the boundary at the client end. The third part is the delivery delay which is the time the client needs to process the media units and prepare them for playback. Note that none of these delays are practically constant in the real world. In high-speed networks, like B-ISDN, the network delay varies because of the variable transmission rates and different queuing delays in network switches, which are caused by the unpredictable traffic burstiness in the network. The retrieval and delivery delay also vary from time to time due to different processing times of media units. The processing time is dependent on the type of CPU, architecture, input/output (I/O) and CPU workload of the corresponding host. Therefore, to support the transmission of time-dependent multimedia data, specific scheduling is required for storage devices, processors, and communication resources to overcome the latency of data, from its source to its destination.

### 3.1.2.1   Concept Model Description of Media Synchronization

The problem of multimedia synchronization is to satisfy temporal precedence relationships under real-time constraints. There is intramedia synchronization and intermedia synchronization.

**SIU (synchronization information unit) defined**

Multimedia consists of several medium types, like voice, video, image, and text. For presentation of continuous media data, we specify the transmission and playout of the objects at a finer-grain media rather than at the level of an object, to facilitate constant rate presentation and maintain temporal relationships among various media. The finer-grain synchronization is achieved by dividing each object into a sequence of subjects, each with its own synchronization interval, which we call a synchronization information unit. Each SIU holds a part of the original object. The transmission of an object then consists of a stream of SIU over a channel. For the purpose of interstream synchronization, related objects (e.g., videoclip and audio samples that require lip synchronization) are divided into SIUs of equal duration. Each SIU is marked with a synchronization interval number as a part of its packet header information. These sequence numbers are used at the destination to determine the skew among multiple streams so that appropriate steps can be taken for realigning them. This way, SIUs allow fine-grained synchronization. The duration of an SIU is dictated by the type of object. For instance, a video object can be divided into multiple SIUs, with each SIU holding information for a complete video frame. If an audio stream is related to a video object, both streams can be divided into SIUs having the same duration. Although the duration of an SIU for each of the two streams is the same, the byte size of an SIU can vary.

**OCPN model**

Intra- and interstream synchronization schemes involved in multimedia systems are usually described by placing multimedia streams along a common time line, as is done by a digital production studio. Figure 3–1 shows a temporal relationship in an application consisting of a sequence of video and audio. This synchronization scheme specifies intrasynchronization constraints such as the duration and relative rates of presentation of each video and audio SIU (marked as VIU and AIU, respectively). The intermedia synchronization constraints such as "the audio must start with the video stream" and "every two video SIUs have to be synchronized with their four related audio SIUs" can be deduced from such a description. Note that such a method does not have well-defined interstream synchronization semantics. Indeed, interstream synchronization constraints are defined indirectly from the time line and thus are not explicitly and clearly specified.

To specify synchronization and the relationship between medium clearly, Petri nets extended with time have been chosen for the formal modeling of multimedia-process synchronization. The timed Petri nets model (TPN) is a Petri net with a time duration assigned to places. This model allows an easy and structured specification of intra- and intersynchronization constraints of multimedia synchronization. In fact, a medium stream is modeled as a sequence of timed places. Therefore, when using TPN for stream modeling, SIUs are modeled as timed

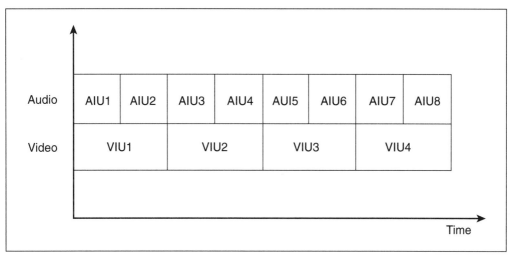

**Figure 3–1**  Temporal Relationship of Multimedia

places. The time value associated with each place specifies its nominal duration. A more general method to specify the time relationship of multimedia objects called OCPN is given by Little and Ghafoor (1990 and 1991). The OCPN has shown the ability to capture all possible temporal relationships between objects required for multimedia presentation. The OCPN augments the conventional Petri net model with values of time as durations, and resource utilization on the places in the net.

*Definition:*  The OCPN is a directed graph, defined by the sextuple $\{T, P, A, M, D, R\}$, where:

$T = \{t_1, t_2, ..., t_n\}$,

$P = \{p_1, p_2, p_n\}$,

$A: \{T \times P\} \cup \{P \times T\} \rightarrow I, I = \{1, 2, ...\}$,

$M: P \rightarrow I, I = \{0, 1, 2, ...\}$,

$D: P \rightarrow R$,

$R: P \rightarrow \{r_1, r_2, ..., r_n\}$,

$T$, $P$, and $A$ represent a set of transitions (bars), a set of places (circles), and a set of directed arcs, respectively. $M$ assigns tokens (dots) to each place in the net. $D$ and $R$ are mapping from the set of places to the real numbers (durations), and from the set of places to a set of resources, respectively.

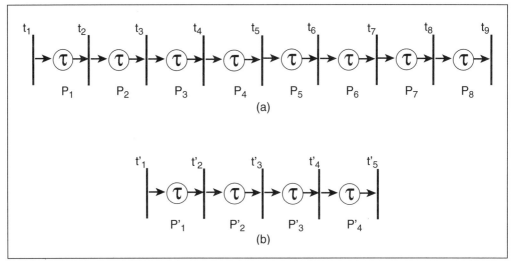

**Figure 3–2** Modeling Intramedia Synchronization with OCPN

The firing rules governing the semantics of the OCPN model are summarized as follows:

1. A transition $t_i$ fires immediately when each of its input places contains an unlocked token.
2. Upon firing, the transition $t_i$ removes a token from each of its input places and adds a token to each of its output places.
3. After receiving a token, a place $P_j$ remains in the active state for the interval specified by the duration $t$. During this interval, the token is locked. When the place becomes inactive, or upon expiration of the duration $t$, the token becomes unlocked.

Figures 3–2 and 3–3 are the examples of OCPN representing sequential and concurrent presentations of objects for synchronous live audio and video as described in Figure 3–1. The OCPN template allows for characterizing live sources with representative or worst-case playout durations and object sizes.

Here, resource $r_k$ indicates a certain medium obtained from a specific media server and communication channel. Hence, the description of this corresponding data traffic can be based on this source and type, and can be associated later with specific channel delay and capacity requirements in the network data transmission. With this kind of OCPN model available, it is relatively easy to choose the granularity of synchronization by assigning the time durations to the places.

### 3.1.2.2 Playout Schedule and Synchronization

In distributed multimedia information systems, different media can start at different times and last for different durations, and they may be stored at different servers. The playout time of each medium data unit is specified so that the playout rate in terms of data units per second, deter-

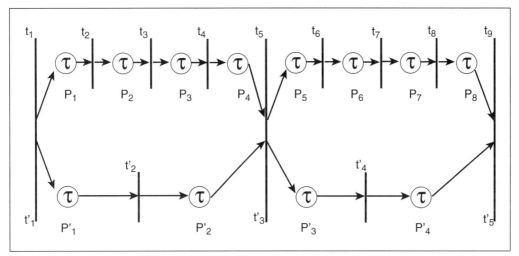

**Figure 3–3** Modeling Intermedia Synchronization with OCPN

mined by the nature of the medium, can be maintained. We call this a temporal playout schedule. For stored data applications, every multimedia application service has the associated temporal presentation scenario that describes the temporal relationships among the media objects involved in the application. This scenario is specified exactly by the OCPN and stored in a central database.

The purpose of a synchronization mechanism is then reduced to meeting the media playout schedule specified by this presentation scenario. Both inter- and intramedia synchronization are achieved by ensuring playout of each medium and each SIU at its scheduled playout time.

**Meeting playout schedule of each medium**
Suppose the playout schedule for each medium of a multimedia application has been worked out. The media synchronization will then be met if each medium unit can be presented according to the scheduled playout time.

First, let us consider a simple case where a medium stream has only one SIU with a playout time instant $T$. Sufficient time must be allowed to overcome the delay $L$ caused by various processes, such as data generation, packet assembly, network access, transmission, etc. Otherwise, the deadline $T$ will be missed. Delay $L$ is, therefore, the time interval from the moment the server sends an SIU until the client receives the SIU. If we choose time, called the control time $C$, such that $C \geq L$, then the scheduled retrieval at $C$ time units prior to $T$ will guarantee the SIU is presented to the client at time $T$, and successful synchronization is guaranteed. The packet retrieval time is defined as $R = T - C$. The case of synchronization becomes a problem of meeting a single real-time deadline, which is a real-time scheduling problem. We can always find a control time $C$ large enough to meet the deadline. However, if an SIU is retrieved too early, it will be buffered at the clients for an unnecessarily long time before it is played out. Figure 3–4 shows the relationship between $L$, $C$, $R$, and $T$.

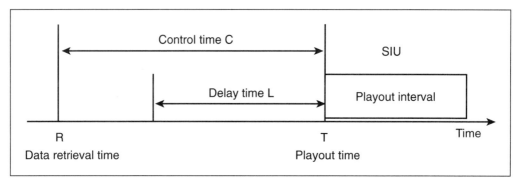

**Figure 3–4**  Definition of Control Time

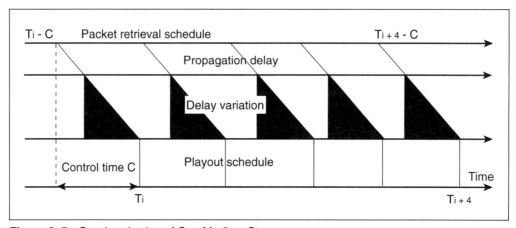

**Figure 3–5**  Synchronization of One Medium Stream

For a medium stream consisting of more than one SIU, since the control time takes into account the maximum delay an SIU can experience, all the data units will meet their respective playout deadlines if the medium stream is sent $C$ time units before its scheduled time. A general scenario for synchronization of a media stream in a network is shown in Figure 3–5. This single medium synchronization case can be extended to multiple media streams including continuous media as well as discrete media. If we send each medium at $C$ time units prior to its scheduled playout time, it will arrive on time for presentation, and synchronization between different media streams will be maintained.

**Scheduling with resource reservation**

With both the playout and control time available, it is now relatively easy to do real-time scheduling for multimedia applications at the server side.

First, we decompose the playout schedule as specified by the OCPN to each subschedule for each resource or medium $r_k$. The resource decomposition algorithm is as follows:

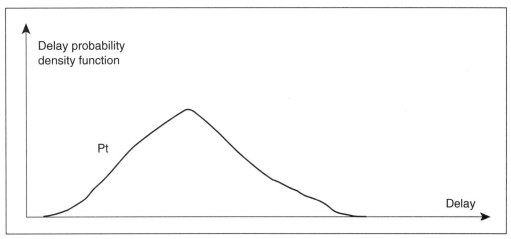

**Figure 3–6**  Delay Distribution

for all places $P_i$ in $P$

if $R(P_i) = r_k$ then

$$\prod(r_k) = \prod(r_k) + T_i$$

$i(r_k) = i(r_k) + 1.$

With this subschedule available for each different resource, we then reserve enough channel bandwidth for each different medium. For continuous media, the packet retrieval time simply tracks the playout times with the introduced control time, that is $R_i = T_i - C_i$, for all $i$. The playout sequence is merely shifted in time by $C$ from the retrieval time. Similarly, the retrieval time for discrete media, such as image and text can be defined according to its corresponding control time, that is $R = T(\text{deadline}) - C$.

### 3.1.2.3    Buffer Configuration at the Client Side

The client receives packets of data units, created by the server and transmitted over the network, with some random delay that could be described by a delay distribution $p(t)$ (Figure 3–6). To reduce the delay variance of the arriving packets, a buffer has to be provided at the received site. It is important to note that different medium streams can have different control times and buffer requirements because streams can be transferred by different channels in the network, and different media may have different data size. Moreover, data may originate from different servers at different sites.

To prevent data starvation at the receiver, the estimated delays used to derive the control time are the maximum delays in the worst case scenario. Because of the existence of network delay jitters, the actual delay experienced by each SIU may be smaller than the estimated value. Those earlier arrivals are buffered for a time duration so that each SIU will experience the same

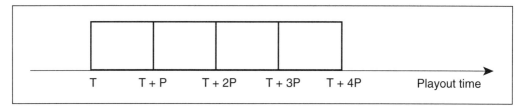

**Figure 3–7**  Playout Schedule of a Medium Object

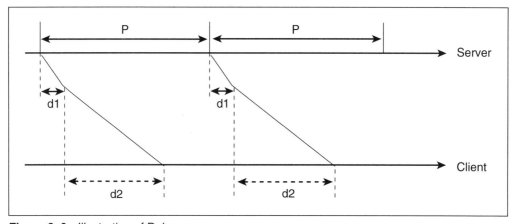

**Figure 3–8**  Illustration of Delays

maximum delay as far as the client presentation layer is concerned. Therefore, a suitable buffer size for each medium is required at the receiver to prevent buffer overflow.

**Size determination**

Suppose we have a playout schedule for a certain medium as shown in Figure 3–7. Each SIU has the same duration of $P$ units of time. The delay time of sending a SIU from a server to its arrival at a client consists of the following elements (Figure 3–8):

1. Time taken from the server's upper layer to fetch, assemble and send the SIU to its transport layer, which is called retrieval delay $d_1$.
2. Network transmission delay from the server's transport layer to the client's transport layer, which is called transmission delay $d_2$.

Therefore, the first data unit is sent and arrives at the client at time $d_1 + d_2$. The second data unit will be fetched and transmitted $p$ time units after the first SIU was fetched and transmitted, and will arrive at the client at $(d_1 + d_2 + p)$ (Figure 3–8). We assume $d_1$ and $d_2$ of the first SIU are statistically the same as $d_1$ and $d_2$ of the subsequent SIUs. It can be shown that the time difference between the time a data unit is presented and the time it arrives is $C - (d_1 + d_2)$ (i.e., $(d_{1\max} + d_{2\max}) - (d_1 + d_2)$). This time difference is the buffering time required at the receiver to make each data unit experience the same total delay before being presented. The buffering time reaches the maximum when $d_1 = d_{1\min}$, $d_2 = d_{2\min}$. Thus, the maximum buffering time is obtained by

$$T_b = C - (d_{1\min} + d_{2\min}).$$

The maximum delay jitter between two SIUs is obtained by

$$J = (d_{1\max} + d_{2\min}).$$

The latest time a packet of a stream can leave the buffer is equal to the sum of its departure time from the server, the maximum delay from the server to the client, and the maximum buffering time $T_b$. In practice, if a packet has experienced the maximum delay from the server to the client, it will not be buffered for the maximum buffering time $T_b$. Only the packet sent at the earliest departure time and experiencing minimum delay from the server to the client will be buffered for the maximum buffering time $T_b$. The earliest time a packet can arrive at the client is equal to the sum of its earliest departure time and the minimum delay from the server to the client. Assume the SIUs are numbered in an ascending order 1, 2, 3, etc., and the buffer size is allocated in $N$ data units in size. To prevent a buffer overflow, the SIU numbered 1 must have left the buffer before the arrival of the SIU numbered $N + 1$, even in the worst case. This will be the case in which SIU 1 leaves the buffer at the latest possible time and data unit $N + 1$ arrives at the earliest possible time. Therefore,

$$d_{1\min} + d_{2\min} + T_b \leq d_{1\min} + d_{2\min} + PN$$

$$N \geq \frac{T_b}{P}. \qquad (3\text{--}1)$$

The preceding expression shows the minimum buffer-size requirement to prevent buffer overflow is $T_b/P$. It is obvious that when the delays are constant (that is, there are no delay jitters), then there is no need for buffering (i.e., $T_b = 0$). This is true because when there is no delay jitter, the SIU will arrive at the client with fixed delay and can be played out directly without buffering.

If the maximum end-to-end delay between server and client, and the minimum end-to-end delay between the server and client can be known in advance, then the perfect synchronization can be reached if the buffer size at the client side is configured as follows:

$$B = \left\lceil \frac{D_{\max} - D_{\min}}{P} \right\rceil. \qquad (3\text{--}2)$$

Here, $\lceil x \rceil$ stands for the smallest integer greater than $x$. $P$ stands for the playout period of a medium at the client side. With this perfect buffer configuration at the client side, given no packet loss and no clock drift, one can always reproduce identical medium streams at the client side, and the goal of synchronization of one medium is reached.

**Intermedia synchronization**
The perfect buffer-size method only considers one medium, in other words, only considers intramedia synchronization. The solution is much more complicated when considering intermedia synchronization. Reaching the goal of intermedia synchronization does not purely depend on the buffer size for different media, but also depends upon the different net delay of different media and the retrieval time at the server side. The reason for this is that the size of the perfect buffer for one particular medium is related to the maximum buffering time of frame units of this medium. The basic idea behind the perfect buffer size is to let each frame unit experience the same total amount of delay, which is equal to the maximum delay, from the retrieval time at server side to the presentation time at client side. Those experiencing a longer time delay will be put in the buffer for a shorter time, while those experiencing a shorter time delay will be put in the buffer for a longer time. Frame units that have the maximum delay will be sent directly from the buffer without staying in the buffer for any length of time. Frame units that have the minimum delay will stay in the buffer for the longest time, which is equal to $D_{max} - D_{min}$. The playout time for one medium is shifted from its retrieval time by the maximum delay experienced by this medium under these perfect buffer-size constraints. Assuming two related media have different maximum delay, but are retrieved at the same time, there will be contradictions on the start playout time as each medium selects its own playout time by its own maximum delay, according to this perfect buffer-size configuration method. The medium retrieval schedule is a critical issue.

However, it is not easy to estimate the maximum and minimum end-to-end delay for both the operating system and network. Due to the lack of real-time operating system and real-time transport protocols, none of these two values are bounded in the real world. Experimental results showed the delay statistic varies with data-unit sizes, the current traffic load of the network, and the number of media streams requested simultaneously at the server. Therefore, to have a good estimation value of control time $C$, we have to analyze the characteristic of these two delay parts.

3.1.2.4   Determination of Control Time in Stored Data Retrieval

One of the key solutions to the media synchronization problem is to find a reasonable control time for independent classes of media.

Control time $C_i$ is the skew between the time an object or packet is retrieved in the server and the time it is played out at the client side, that is, $C_i = T_i - R_i$. $C_i$ depends on how accurately $d_1$ and $d_2$ can be estimated, where $d_1$ is the retrieval delay of one SIU and $d_2$ is the transmission delay of one SIU. Thus, $C = d_1 + d_2$.

**Estimation of control time**
The retrieval control time $d_1$ consists of the host's processing delay, which depends heavily on the workload of the host. $d_1$ can be divided into two parts: $d_{1min}$ and $d_{1var}$ (i.e., $d_1 = d_{1min} + d_{1var}$), where $d_1$ is the minimum delay when there is no other workload at the server, and $d_{1var}$ is the extra delay incurred when there are other workloads. $d_{1var}$ reaches the maximum value and $d_1$ becomes $d_{1max}$ when there are maximum acceptable loads at the server.

$d_2$ is the end-to-end packet transmission delay incurred in the network. We can break this end-to-end transmission delay $d_2$ into three components: (1) a propagation delay $D_p$, which is a constant and related with the distance between the two end hosts, (2) a transmission delay $D_t$, which is proportional to the packet size, and is equal to $P_m/C$, where $C$ is the channel capacity and $P_m$ is the packet size for the medium $m$, and (3) a delay variation $D_v$, which depends on the end-to-end network traffic. Therefore, the end-to-end transmission delay for a single packet is given by

$$d_2(packet) = D_p + D_t + D_v \qquad (3-3)$$

A typical multimedia object may consist of many packets. Let $[x]$ be the size of object $x$ in bits. Then, $r = [x]/P_m$, gives the number of packets required to constitute the object $x$. The end-to-end transmission delay for an object $x$ is then given by the following:

$$d_2(object) = D_p + rD_t + \sum_{j=1}^{r} D_{vj} \qquad (3-4)$$

$D_v$ is a variable which is determined by the channel traffic and the network congestion control method. In the absence of real-time transport protocols, $d_2$ has to be estimated during the connection set-up phase based on the reserved resources and current network. Several approaches have been proposed to find the control time $C$. The network delays can be estimated by either measurement techniques or by guaranteed QoS offered at the access to the network. For example, statistical data characterizing throughput, delay, arrival rate, interarrival rate, and buffer occupancy can be allocated using some observation mechanism. With these collected data available, the network admission control mechanism can make resource allocation decisions when it sets up a connection. Connections can be accepted or refused based on the available bandwidth or delay characteristics. Once accepted, the admission control guarantees that the network is able to provide the requested QoS for a connection since system congestion is prevented by the resource allocation mechanism. Therefore, Little and Ghafoor (1991 and 1992) and Stone and Jaffay (1994) assume that the underlying network transport and multiplexing mechanism provides a characterization of interarrival distributions of consecutive packets through service interface on a per connection basis. At the time of connection establishment, given a class of traffic QoS, a server and client pair, the admission control returns the channel delays, the mean, and variance of the interarrival time of consecutive packets. This admission control mechanism can also be extended to operating system resources including database and storage subsystems. Similarly, Lu (1994) suggests that with a suitable admission control policy and resource management policy, we can assume that the load on the network, when the first packet is transmitted, is statistically the same as the loads when subsequent packets are retrieved and transmitted. Then $d_2$ of the first packet will be statistically the same as $d_2$ of the later packets. Therefore, the meaningful control time $d_2$ can be established based on the maximum delay of the first packet.

#### 3.1.2.5 Synchronization in DMIS

With the previously described mechanisms available, synchronization in DMIS is possible. When a client requests a multimedia application, this request is first transmitted to the media database server. The database server then sends the presentation scenario related to this application back to the client. According to this scenario, the client makes connection requests to all the media servers involved in the service. This is a connection set-up phase, which includes a request, indication, response, and confirmation. During this phase, QoS parameters specified by the application are treated as guidelines for the connection. System resources are managed to provide real-time data delivery. Resource reservation protocols are employed to give sufficient bandwidth and buffer space to meet the statistical characteristics of time-dependent data streams. If the required QoS cannot be met, the request for the connection is refused. The delay variations on each channel and end-to-end delays are estimated during this connection setup, and are used to determine the buffering requirements at the client side. This buffering is used for reshaping the channel delay distribution and server load to reduce variance and to provide the requested level of service, in terms of delay and late-arriving data units. If all the connections to the servers are guaranteed, the client then calculates the retrieval time of each medium according to its corresponding channel delay. We assume that the load on the servers and network during the connection-established phase is statistically the same as the loads during the real-data transfer phase. The client sends each retrieval schedule back to its corresponding server, and the media server involved in this application starts the retrieval and transmits the media packet to the client.

In a real-time system, both short-term and long-term guarantee violations in delay and bandwidth exist due to changes in the channel traffic characteristic or unanticipated temporary overloads in the host operating system. With the above-mentioned session establishment mechanism for scheduling the retrieval time, the reservation of channel resources and a suitable buffer-size configuration, the media synchronization goal may still be disrupted due to buffer overflow or underflow. Media packets may be lost in network transmission (i.e., via the ATM congestion control method), causing discontinuity during the playout period. Also, the perfect buffer-size method gives the size of the buffer in terms of the frame numbers rather than the number of bytes. In reality, it is the number of bytes, not the number of frames, that is reserved in advance during the resource-reservation period. This is fine for constant frame size. With the frame numbers available, it is easy to calculate the number of bytes to be allocated first. It is equal to the frame numbers multiplied by frame size. However, with the development of compression technology, almost all media frames transmitted over the net have dynamically changing sizes that depend heavily on the compression method. Yen indicates the range of frame size can be from 0.3 to 1.5 Mb. With dynamic frame size, the number of bytes to be reserved first is usually only an estimate based on the average frame size, to save some resources. Frame units may be lost due to buffer overflow for large frame sizes. There is a requirement to do some buffer adjustment, for media synchronization at the client side, in the presence of packet losses, temporary guarantee violations, and dynamic frame sizes.

In addition, because there is no global clock in B-ISDN, the clock drift between the servers and the client side must be considered. Assuming the sending rate of the server is faster than the rate of display at the client, theoretically, there will be a buffer overflow at the client side. If the sending rate of the server is slower than the display rate, then the buffer will underflow at the client side. It is obvious that certain buffer management techniques must be applied to control the buffer to overcome the clock drift problem.

Based on the preceding discussion, a good synchronization mechanism consists of the following three steps: first, we need to analyze the end-to-end delay, calculate the control time, and decide on a suitable buffer size for different media. Second, we must schedule the retrieval time for different media based on this control time to eliminate the random latency. Third, we have to provide a suitable buffer management at the receiver side to reach a fine-grained synchronization.

### 3.1.3 Buffer Control System

To solve both the fine interstream and intrastream synchronization problem, one of the possible approaches is based on controlling the client buffer. Uncontrolled buffer management will result in buffer overflow or underflow due to the network and operating system behavior, dynamic frame size, and clock drift. The delay jitter, which strongly affects the buffer behavior, is a factor which dominates the characteristic of media synchronization. Research has been done to overcome the uncertainties of the jitter property. Hać and Xue propose a receiver-based buffer control model that runs independently of the server. The goal is to realize the playout synchronization, which is called both intrastream synchronization and interstream synchronization. The basic idea is to apply an automatic control methodology onto the buffer adjustment of a playout system.

Typically, the way a playback system for continuous media works after initialization is that the media frames are consumed at a fixed rate out of the buffer. The buffer is supposed to compensate for both the jitter effects that are introduced between frame production at the sender and frame delivery at the receiver, and clock drift error. When there is no frame in the buffer due to a continuous large network jitter, the playback system has no frame for playing out, and a glitch happens to the application users. When the buffer overflows, late-coming frames will have to be discarded. We count the number of frames, not the number of bytes, in the buffer. This is reasonable since the frame, which has a variable size, is retrieved and played back. Assuming a constant delay, server and client are synchronized when the number of frames in the buffer is held constant. However, due to network jitter effects, the arrival spacing of the frames varies temporarily. The average spacing is constant and is determined by the server's clock. The buffer is emptied at a constant rate which is driven by the receiver's playout clock.

Hać and Xue focus on buffer level control on the client side, and a master media idea is introduced to reach intermedia synchronization. The concept of nominal value is treated as a reference value for frame numbers in the buffer (i.e., the frame numbers of any asynchronous media should be kept to its corresponding nominal value). The aim of the control scheme is to keep the amount of media frames in the buffer at a nominal value, if possible. The skew-time tol-

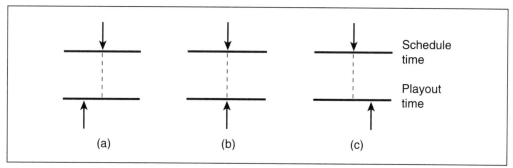

**Figure 3–9** Skew: (A) Leading (B) None (C) Lagging

erance (Figure 3–9) of a human on different media is used as the control constraint of buffer operations.

**Skew tolerance**

Timing parameters can characterize intermedia and real-time synchronization for the delivery of continuous media as well as discrete media. For continuous media data such as audio and video, a data unit can be lost, resulting in a gap in the playout period. Such a loss causes the corresponding media stream to advance in time or to cause a stream lead. Similarly, if a data unit is duplicated, it causes the stream to retard in time, or lag (Figure 3–9). Because audio and video have substantial short-term temporal data redundancy, it is possible to control the rate of playback by dropping or duplicating a data unit without affecting the quality of synchronization. The synchronization errors that can be tolerated by human perception vary in different application scenarios. Steinmetz and Engler (1993) conducted some experiments in this area. For example, 120 msec mismatching in lip synchronization between an audio and video stream can be perceived as disturbance by the application users. On the other hand, with 200 msec delay for voice in a text annotation playback period, the presentation is still well accepted. These application defined parameters are called "skew tolerance parameters."

Skew tolerance can be a measure of intermedia synchronization. Synchronization, which is usually defined as an absolute instant time, can be extended to a time interval (e.g., a tolerance to timing variations). We specify a skew parameter $sk_{ij}$, which is the maximum temporal skew allowed between $i$th and $j$th channels, given in terms of the number of durations of playout period (1, 2, etc.). In other words, the range of data unit values in $[X_i, Y_{j+sk_{ij}}]$ or $[X_i, Y_{j-sk_{ij}}]$, for $X$ media on $i$th channel and $Y$ media on $j$th channel is acceptable for users. For example, in digital TV application, the $sk_{ij}$ can be {2}, meaning that the user can tolerate up to two consecutive frame index skews between the video channel and the audio channel. With the availability of this kind of skew parameter describing the time relationship between different media, an object 1 with the actual playout times $P1 = \{P1_m\}$ on the $i$th channel is synchronized with another object 2 with the playout times $P2 = \{P2_m\}$ on $j$th channel, if, for all $m$, $|P1_m - P2_m| \leq sk_{ij}$. This skew parameter can be used as a constraint for the buffer control management to reach intermedia synchronization as well as intramedia synchronization.

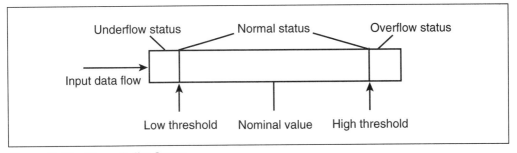

**Figure 3–10**   Three Buffer Status

**Thresholds**

A buffer can compensate for the normal jitter disturbance during playout if the buffer size is large enough. However, if there is a significant amount of jitter or the reserved buffer size does not match the real network delay, two extreme cases of overflow and underflow will result in a loss of information beyond the skew tolerance of users. Overflow happens when the number of arriving frames is greater than the remaining buffer length. Some frames, that are to be discarded, can be larger than the skew tolerant amount since there is no room to store the frames. The number of discarded frames is equal to the difference between the total number of current frames and the buffer length. If this number is higher than the skew tolerance, inconsistent and thus, unacceptable information will be provided to the users. Underflow occurs when no new frames arrive in time due to the jitter. The media in the buffer cannot be updated, thus a blank frame has to be played out. This is an abnormal case which seriously degrades the QoS.

To prevent the occurrence of the preceding case, we define three different buffer states by setting two thresholds close to both ends of the buffer (Figure 3–10). One near the tail of the buffer is called low threshold (LT), while the one near the head of the buffer is called high threshold (HT). When the frame number of the media is greater than the HT (i.e., the amount of data is close to the overflow status), a controller is started to reduce the difference of current frame numbers and nominal value (adjust the frame number below the HT) by throwing out frames within the skew tolerance. When the frame number is less than the LT (i.e., the amount of data in the buffer is close to the underflow status), the controller is started to reduce the difference of current frame numbers and nominal value (adjust the frame number above the LT) by duplicating the current playout frame within the skew tolerance. Within a normal status area, the controller does not do anything. Based on these two thresholds, the number of frames which must be duplicated or discarded can be increased or decreased smoothly by duplicating or discarding them gradually before an underflow or overflow occurs. The blank frame, due to buffer underflow, is replaced by duplication of the last frame in the buffer.

The threshold plays an important role in the buffer control scheme. If the thresholds are close to the ends of the buffer, the number of duplications or discards will be reduced by increasing the risk of information loss through buffer overflow and underflow. On the other hand, if the threshold is set around the middle of the buffer, there will be a higher number of frame duplication and discards since there are more opportunities to stimulate the control operational procedure. The number of buffer overflows or underflows is decreased dramatically by this control procedure. The loss of information by frame duplication and discard is the price paid for the proposed buffer control scheme. The choice is between "intentionally" lost frame information by duplication or discard of frames via the buffer control scheme under skew constraints, or "carelessly" allowing the buffer to stay in an uncontrolled mode so that the loss of frame information depends purely on the delay jitter with the occurrence of buffer overflow or underflow.

Theoretically, the thresholds have to be set according to the characteristic of the buffer-control model and the delay jitter. The "distance" between the thresholds and the buffer ends will reflect the adjusting time before the system reaches steady state.

**Nominal value**

Suppose there is a nominal value (Figure 3–10) to represent the number of frames which should occupy the buffer if there were no jitter, and a real value of frame numbers that have actually occupied the buffer. The goal of the synchronization scheme is to keep the average buffer level (frame numbers in the buffer) to the nominal value. The control model will take the operations of adjustment by duplicating and discarding frames according to the nominal value. Since a buffer is used to compensate delay jitter, the dynamic change of buffer level is within the range of buffer compensation. The nominal value should not be fitted at one point in the buffer. Considering three different buffer statuses set by low and high thresholds as indicated before, we propose a definition of nominal values to reach the intermedia synchronization as well as intramedia synchronization. Various values, that will be selected as nominal values, according to different buffer status for each media, are as follows:

| Threshold | Nominal Value (NV) | Status |
|---|---|---|
| $N_m \geq HT_m$ | $NV = HT_m - 1$ | Overflow |
| $N_m \leq LT_m$ | $NV = LT_m + 1$ | Underflow |
| $LT_m < N_m < HT_m$ | $NV = N_m$ | Normal |

Here, $N_m$ stands for the current frame count of one medium in its corresponding buffer. $HT_m$ and $LT_m$ are the thresholds in two ends of the buffer for this medium. Generally, if the frame count in the buffer is between two thresholds (HT and LT), which is the normal status, the nominal value for this medium is the current frame count number, whereas if the frame count of the medium is close to overflow/underflow, the nominal value for this medium is the frame count before it closes to abnormal status.

Because each medium has its own medium buffer for playing out, buffer overflow or underflow, frame discard, or duplication happens in each separate media buffer without having any effect on other media. This results in an inconsistency of frame index between coupled medium streams. While one medium stream is playing the $N$th frame, the other corresponding medium stream may play the $M$th frame. The difference $(N - M)$ of the current frame index of two media should be less than the skew tolerance. With the accumulation of media discard or duplicate operations, the difference may soon be beyond the tolerance of the skew value.

A master medium-based control scheme is proposed for this intermedia synchronization problem, because a reasonable assumption for the multimedia system is that one of the media takes the role as a master medium, while the others are treated as slaves relative to this master. That means, if any frame index value of slave media is inconsistent with the frame index of its corresponding master medium, the control procedure should be used to reduce this inconsistency by adding the difference value to the current slave nominal value until both medium frame indexes are the same for playing out. Index inconsistency due to frame loss or corruption during transmission can also be fixed in this way.

A single variable controller associated with the nominal value, which is treated as a reference dominated by the buffer status and index inconsistency, is proposed. The control scheme is based on the master media operational stage (i.e., any asynchronous medium should be synchronized by its corresponding nominal value or master media).

**Buffer adjustment**

The adjustment method of the master media-based buffer-control mechanism is to process (count, duplicate, or discard) only the amount of frames. The buffer stuffing—duplication and discard of frames—is the main operation in the buffer control model. When the number of frames in the buffer is increased fast and close to the overflow, some of the frames will be discarded and the ratio of playout is speeded up. The frames that should be fetched in the next time period will be played currently, and the frames that should be played currently if there is no occurrence of overflow will be thrown out. While the number of frames in the buffer is decreased quickly and close to underflow, the last frames will be duplicated, and the frame that is fetched in the last time period will be played again. The ratio of playout is slowed down. The adjustment method is also applied to synchronize the playout index of different media based on the master media index. The slave media, which is lagged behind the master media, has to be discarded in order to "keep pace" with the master media, while the slave media leads before the master media has to be duplicated to "wait" for the master media. All these buffer adjustment operations (duplication or discard) must be done within the skew tolerance of different media.

With the concept of skew tolerance of users, an effective buffer control management scheme can be applied to adjust the buffer level to provide graceful service degradation when the buffer overflow or underflow is pending or the media index is inconsistent.

## 3.2 User Mobility Management in Wireless Networks

A personal communication service (PCS) network allows the users to communicate with each other instantly, no matter where they are or what time it is. This is the result of the development and combination of technologies in computer, cellular and conventional telephones, and the digital communication network. The construction of a PCS network involves the establishment of both wired and wireless communication systems. BSs connected by communication lines constitute a fixed network that covers the service area of the PCS network. Users access the fixed net-

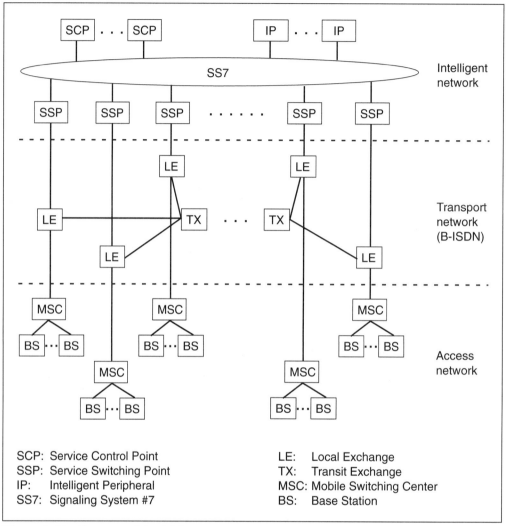

**Figure 3–11** Structure of the PCS Network

work through the BSs using radio links. Unlike conventional communication networks, users are able to move within the service area without losing access to the PCS network. Thus, the network needs a mechanism to keep track of the users. Signaling network and location database are introduced for this purpose. Because of the large number of users in a PCS network, tracking users' movement is difficult.

### 3.2.1 The Architecture of the PCS Network

The PCS network can be divided into three layers (Figure 3–11): the access network layer, the transport network layer, and the intelligent network layer. The transport network and intelligent network are fixed networks, and the access network is a radio network.

#### 3.2.1.1 Access Network

The access network is the user interface of the PCS network, which allows users with mobile terminals (MT) to access the transport network. The network is composed of MSCs and BSs. The PCS network coverage area is divided into cells as in conventional cellular telephone systems. There is a BS in each cell to communicate with MTs in this cell. Several BSs form a group and are connected to an MSC, which is connected to a B-ISDN switch. A B-ISDN network connects all the MSCs to form the infrastructure of the PCS network.

To provide access to users, the access network must be available in all environments. Different environments are characterized by their user density, user mobility level, set of supplied services, etc. There are different requirements for access networks. This leads to a flexible access network architecture, which includes different configurations of cells, such as microcell, macrocell, picocell, umbrella cell, and highway cell.

*Macrocell.* The macrocell, which usually covers areas with diameters of tens of miles, is used in conventional cellular telephone systems to maximize the coverage area and reduce the system's infrastructure cost. It is used in the PCS network to provide services to rural areas with low-terminal density.

*Microcell.* Because of the shortage of radio-frequency spectrums, using current radio technology, the PCS network capacity is limited. The upper bound of radio communication is around 2Mb/sec. By adopting a microcell with a radius of less than 1000 meters, frequency can be reused at a higher level, which increases the number of serviced users and improves the system capacity. The side effect is more handoffs. Microcells are widely used in metropolitan areas where the user density is high. Another advantage of using microcells is that the BS can be made very compact and be easily mounted on utility posts or buildings.

*Picocell.* For areas like shopping malls or airport terminals, a picocell with a radius of tens of meters can be used to handle the extremely large number of users in relatively small areas.

*Umbrella Cell.* An umbrella cell covers several to tens of microcells. It is used to provide efficient handoff for mobile terminals traveling through microcells at high speed, and maintain continuous coverage.

*Highway Cell.* A highway cell is a special kind of microcell whose coverage area is tailored to match a section of the road. This can be achieved by using directional antennas.

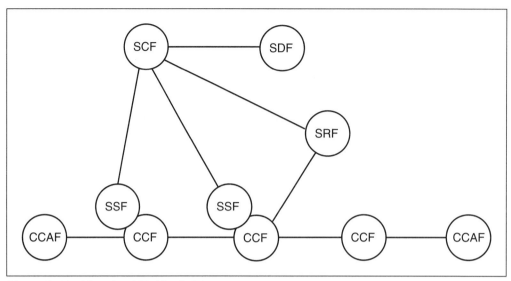

**Figure 3–12**  Functional Entities in IN

### 3.2.1.2    Transport Network

B-ISDN is selected to transport user data in a PCS network. Because the current form of B-ISDN is not designed to support mobile communication systems like the PCS network, it needs some enhancements to handle mobility management, handoff, and other mobile-system specific problems. These can be put into a mobile extension of B-ISDN.

One of the problems in early mobile systems is that so many functions are implemented in mobile switches, that their call processing capacity is low. If the size of the mobile extension is not controlled, B-ISDN switches can become too complex and the same problem can happen in the PCS network. Therefore, the mobile system extension of B-ISDN must be minimized. Only necessary functions are included. Other functions should be moved to either the intelligent layer or the access layer. Some functions that already exist in B-ISDN for fixed services are also useful for mobile services. They must be identified and utilized. This extension involves only the B-ISDN signaling protocol. The ATM layer and ATM AAL are not affected. Thus, the modification is limited.

### 3.2.1.3    Intelligent Network (IN)

IN architecture from its functionality's viewpoint is shown in Figure 3–12. This architecture illustrates the IN functional entities and their relationships.

- A call control agent function (CCAF) provides end users access to call and service processing.

- The service switching function/call control function (SSF/CCF) provide basic switching functions and call control (e.g., routing and charging), and can detect events that require the invocation of IN service logic.
- The service control function (SCF) executes the IN service logic, which are the programs providing various services. It is notified by the SSF when certain events occur, such as an MT being turned on. It can also request the SSF/CCF to perform functions needed to provide services.
- The service data function (SDF) performs the data processing functions.
- The specialized resources function (SRF) manages specialized resources which are added to existing resources in the IN to perform functions such as interactions with users via connections to play announcements and to collect digits.

Figure 3–13 shows how the functional entities are implemented in a physical IN. Various physical entities such as SSP, SCP, Service Data Point (SDP), and an Intelligent Peripheral are connected through the signaling system number 7 (SS7). Comparing this with the IN functional architecture, there are some corresponding relationships between the functional entities and physical entities: CCAF, CCF, and SSF reside in SSP, SRF in the Intelligent Peripheral, SCF in SCP, and SDF in either SDP or SCP.

The IN was first proposed for fast service creation and deployment. But in the PCS network environment, IN is used to create a mobile network out of B-ISDN by providing mobile functions. The characteristics of the new environment are as follows:

1. The mobile functions are well defined and do not change rapidly.
2. The volume of IN traffic and amount of IN processing are very high because of terminal mobility.

In the PCS network, more emphasis is put on the IN's performance than on flexibility.

In order to integrate the IN into the PCS network, new functions like user authentication, access check, location management, and support for handoff need to be added to IN. They usually are put in the SCP.

Some mobile specific information must be stored in the SDP or SCP. The home location register (HLR) contains relevant permanent and temporary data about each user (e.g., location information and a personal profile including the services to which the user subscribes). The visiting location register (VLR) corresponds to one or several MSCs. It contains detailed data for those MTs entering its location area. The VLR is also responsible for allocating temporary mobile terminal identity (TMTI) to visiting MTs.

The IN interacts with the transport network most of the time. Sometimes mobile terminals need to talk with the IN directly for some noncall functions, without going through transport network switches. For example, when an idle terminal crosses a location area border, it needs to send registration and deregistration messages to the new and old MSC, respectively. A transport network is not involved. Thus, the IN must have interfaces with the access network as well as the

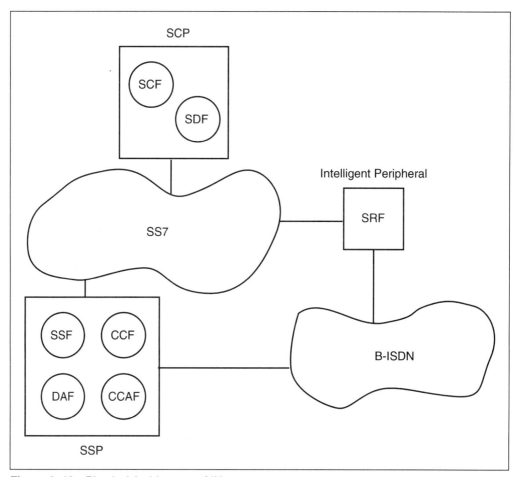

**Figure 3–13** Physical Architecture of IN

transport network. A new functional entity is added into the IN to let mobile terminals interact directly with the SCF and SDF. We call this the direct access function (DAF). This function can be integrated into the SSP. Figure 3–13 shows the IN architecture used in the PCS network.

The basic idea of an IN is centralized service control. The service information is stored in a central database. In conventional telecommunication systems IN traffic is not heavy, and the number of users is not large. A centralized structure may be adequate. But the PCS network is a global telecommunication system providing access to a large number of users. Furthermore, the placement of the users and terminals is not fixed. Higher IN traffic is needed to manage such mobility. If we use a centralized database, its size may have to be too large. Also, it is difficult and expensive to find a computer that can efficiently handle a large number of database operation requests.

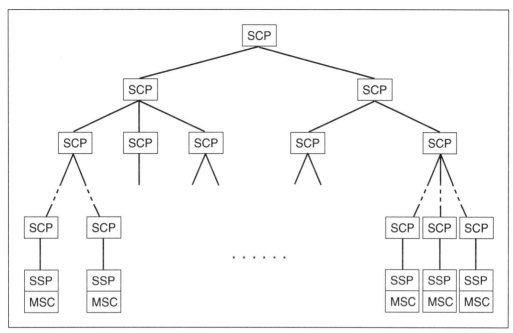

**Figure 3–14** Hierarchical Database in the PCS Network

A distributed IN database can be implemented in several ways. Two popular schemes are horizontal distribution and hierarchical distribution.

The horizontal scheme assumes that a central database is partitioned into several parts which are distributed to local SCPs as a local database. Each local database contains information of those users whose home registration areas are under the control of the corresponding local SCP. All or part of a user's information may be moved or copied to other local databases depending to which registration area this user moves. Because of the high level of user mobility in the PCS network, the cost of moving user information is high. There is also a remote database retrieval and update.

The hierarchical scheme shown in Figure 3–14 is better suited for a PCS network environment because it is capable of localizing the database update.

There are many SCPs and associated databases (or SCP-SDP pairs). They are connected through SS7 with all the other IN physical components like SSPs and intelligent peripherals. Physically all the SCPs are equal and independent. However, logically they are organized in a hierarchical structure. Each database keeps information regarding those MTs under the control of its children databases. The concept of information localization is applied. Database retrieval and update are kept in the subtrees of the hierarchy.

### 3.2.2 Location Database Management in the PCS Network

One of the primary goals of the PCS network is the availability of services to both fixed users and mobile users, independent of time and location. This goal is achieved by setting up a fixed transport network with switching nodes interconnected by communication links, and allowing a mobile user to access the nearest node through radio waves.

In this environment, to setup a call or maintain the established connection, the system must find users' locations and track their movements. This problem is not unique to the PCS network, it exists in telecommunication networks involving mobile users. Generally it is handled through a paging and location database.

*Paging.* When a mobile user is called, a network-wide search is carried out. A message is sent to each BS. On receiving the message, every BS broadcasts it in the cell it manages. If the callee is on and not busy it sends a reply to the BS, which contacts the caller's BS. Then a connection is established. This method is simple but uses too many network and radio resources.

*Location Database.* In this method, the system maintains a database of user locations. Each base station keeps broadcasting its own identification number in its cell through a control radio channel. A mobile terminal periodically monitors the control channel in a short interval and saves the received identification number. If the broadcasted identification number is the same as the number saved the last time, the mobile terminal is in the same cell as before. Otherwise, the mobile terminal has moved across the cell boundary, and a message is sent to the BS, which in turn sends the database update request to the system to change the mobile terminal's location to the new BS. When a user makes a call set-up request to a mobile terminal, the mobile terminal's number is used to look up its current location in the database, then a connection is established between two corresponding base stations.

#### 3.2.2.1 Location Management Strategies Before the PCS Network

Wireless telecommunication network development spans three generations. The first generation is the analog cellular telephone network, which is an extension of the public switched telephone network (PSTN). Location management is handled by the paging method and no location database is used.

The second generation is digital cellular network. There are two standards for this network, Interim Standard 41 (IS-41) in North America, and Global System for Mobile Communications (GSM) in Europe. Both of them adopt a two-level hierarchical location management strategy, which divides the location database into two parts, HLR and VLR.

First generation wireless telecommunication networks usually covered one metropolitan area, whereas the second generation wireless network can cover several countries and serve many more users. The pure paging method will consume a lot of resources, because one call set-up request will result in one message to each cell, the number of which is large in the second generation wireless network. A pure location database method does not work in this environment either. It is not practical to perform location update in terms of cell. The network coverage area is divided into location areas. Each location area is composed of a number of cells. There is

MSC in each location area connected to all the BSs managing these cells. Moving between cells of the same location area does not cause a database update. Location database is updated only when mobile terminals cross a location area boundary. Thus, database inquiries give us only the location areas and not the cells. To find the exact location of the mobile user, the corresponding MSC needs to page all the cells in this location area. This location management method is a combination of paging and location database.

One HLR and one VLR are associated with each location area. The HLR keeps the information of those mobile users permanently registered in this location area. The VLR keeps information about visiting mobile users whose homes are not in this location area. The following steps describe the location update and the function of HLR and VLR in IS-41.

1. Once a terminal enters a new location area, the terminal sends a registration request to the MSC in that area.
2. The new MSC sends an authentication request to its VLR to authenticate the terminal, which in turn sends the request to the terminal's home location area HLR. The HLR sends its response to the new MSC.
3. If the terminal is authenticated, the new MSC sends a registration notification message to its VLR.
4. The VLR relays the registration notification message to the terminal's home location area HLR. The HLR updates the terminal's location entry to point to the new MSC, sending a response, which contains parts of the terminal's profile to the new VLR. The VLR stores the profile in its database and responds to the new MSC.
5. If the terminal was previously registered in a different location area, its HLR sends a registration cancellation message to the old VLR. On receiving this message, the old VLR erases database entries for the terminal and notifies the old MSC.

#### 3.2.2.2 Distributed Location Database for the PCS Network

The third generation of wireless telecommunication networks provides similar services to the users as described above. But the coverage is worldwide. This requires a new location management method because the method of the second generation network is not efficient in this environment.

The two-level hierarchical location management strategy using HLR/VLR is essentially the centralized strategy. Although the location database has multiple partitions residing in different nodes, a user's information is always stored in the database of his/her home location area. No matter where the user is, database inquiry and update requests about this user are always directed to the home location area. This is adequate for networks with a relatively small coverage area and a limited number of users. But if it is used in the PCS network, sometimes the resulting network behaviors might be unacceptable. For example, if a user visits a city thousands of miles away from his or her hometown, a database update is requested as a result of travel in that city having to go thousands of miles in the network back to the home location area. Similarly, if another user in that city dials the number, a database inquiry is sent to the home location area to get the current location, even if the user is in the same cell as the caller. Situations like these put unnecessary signaling traffic on the network and use network bandwidth.

We describe the hierarchical distributed location database as a location management strategy for the PCS network. In this strategy, database partitions are organized in a tree structure. The root corresponds to the entire coverage area, the world. Under the root, there is a node corresponding to each country. Moving one more level down, there are nodes corresponding to regions in the countries. This way, the tree expands downward until the nodes correspond to the location areas. The structure of the distributed database does not have to be a strict mathematical tree. Cross connections between nodes are allowed. A node at/or close to the root may be composed of several fully-connected subnodes to provide better reliability and efficiency.

Each leaf node is associated with a location area, and keeps information about mobile terminals currently in that location area. The terminals may be permanently registered in this location area or from other location areas. Each nonleaf node keeps information about mobile terminals currently in location areas under its control. Contents of databases in leaf and nonleaf nodes are different. A leaf node database contains detailed information of users like a user profile. A nonleaf node database contains pointers to lower level nodes where other pointers or user location information are stored.

The purposes of adopting a hierarchical distributed location database strategy is to increase performance, reduce signaling, improve reliability and availability, and allow incremental growth of the database. This is achieved by storing information about mobile terminals as close as possible to their current location. Most database inquiries and updates can be performed locally, only a small part of them involve a remote database. Even the operations requiring remote database accesses can be handled within a finite number of steps, which depend on the depth of the tree.

Initially, for each mobile terminal, pointers are created at the node corresponding to that terminal's home location area, its parent, and its parent's parent, all the way up to the root of the signaling network. Each pointer refers to its child node in the path to the mobile terminal. When the mobile terminal moves around in the PCS network coverage area, these pointers need to be updated to track the movement. No matter where the mobile terminal is, there is always a chain of pointers from the signaling network root to the node covering the mobile terminal's current location area.

When a mobile terminal moves to a new location area, a location update request is sent to the network. The request is processed using the following steps:

1. Start from the node covering the new location area, which goes up one level each time. At each node, a pointer referring to the mobile terminal is created which points to the previous node in the path of its upward movement.
2. Stop at the first node with a pointer referring to this mobile terminal. This node is the lowest common ancestor of nodes managing the old and new location areas.
3. From the lowest common ancestor, go down toward the node covering the old location area. At each node, the pointer referring to the mobile terminal is deleted.

When a user dials a mobile terminal's number, a call setup request is sent to the network. The request is processed in the following steps:

1. Go upward from the node managing the caller's current location area until a node with a pointer referring to the callee is found.
2. Follow the pointer down to the node managing the callee's current location area.
3. Page the callee and return the result.

### 3.2.3 Placement of a Hierarchical Database

Usually the design of a signaling network structure in the PCS network is based on location areas, which are defined primarily according to geographical features of the coverage area. Databases are put in every node of the network. Important factors like user density, user movement patterns, and calling rate are given little consideration.

Studies on user mobility management using the hierarchical distributed network are presented by Wang and coworkers (1992, 1993) and Awerbuch and Peleg (1991). A user-tracking strategy using a distributed database is proposed by Wang. Every node has a database. Wang and coworkers present a location registration method which combines hierarchical and HLR/VLR schemes. Its performance is analyzed using a queueing system model. A fixed database configuration is assumed. A graph partitioning concept setting up a distributed hierarchy for mobility tracking in which the user movement and call rate are not used to improve the system performance is presented by Awerbuch and Peleg (1991).

The hierarchical distributed database has the advantage of localizing database inquiries and updates, thus reducing the cost of communication and distributing the load of database operations. An associated disadvantage is that a single location update or call setup request may cause updates or inquiries at several databases. By properly selecting a database placement scheme, a good distributed database system should be able to balance these two sides.

The influence of database placement in a hierarchical distributed database can be viewed from two perspectives: database operation and communication in the signaling network. Let us find which databases are involved in user location update and inquiry. If a user moves from node $i$ to node $j$, supposing the lowest (closest to leaves) common ancestor of $i$ and $j$ in the tree-structured signaling network is $k$, the location update message first goes up from $i$ until reaching $k$. Each database on this "upward" path needs to be updated to delete the entry about the mobile user. The location update message then goes down from $k$ until reaching $j$. Each database on this "downward" path also needs to be updated. But this time, an entry about the mobile user is added that points to the lower-level database in the path. For call setup requests from users in node $j$ to users in node $i$, the resulting location inquiry messages follow the same route as location update messages generated by user movements from $i$ to $j$. Because user movements and call setup requests cause database operations in every database on paths taken by corresponding location update and inquiry messages, the number of database operations in the signaling net-

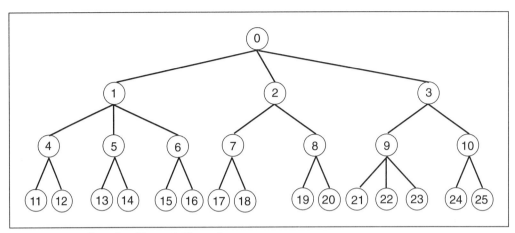

**Figure 3–15** Signaling Network Example

work is determined by the placement of databases, as well as the pattern of user movements and call setup requests.

Considering the total number of database operations, if too many databases are put in a region with little user movement and a low call rate, collectively there will be more than the necessary database operations, and individually we will have underloaded databases. Figure 3–15 shows an example of a signaling network. If there are not many movements and call setup requests for mobile users under control of nodes 11, 12, 13, 14, 15, and 16, one database in node 1 will be able to handle the required database operations. If we put another database in node 4, user movements or call setup requests between users in nodes 11 and 12 will cause database updates or inquiries in node 4 instead of node 1. Because of the low user movement and call setup request rates among nodes 11, 12, 13, 14, 15, and 16, databases in nodes 4 and 1 will be underloaded. Furthermore, if a database is added in node 4, for user movements or call setup requests between node 11 (or 12) and any other node except node 12 (or 11), there will be one more database on the path taken by location update or location inquiry messages. Thus, the total number of database operations is increased.

On the other hand, if too few databases are put in a region with many user movements and a high call rate, we will have overloaded databases. Suppose an area covered by nodes 21, 22, 23, 24, and 25 has a very high mobile user density, and there are a lot of user movements and call setup requests among these nodes. If only one database is put in node 3, it may not be able to handle the huge number of database operation requests per unit time. Node 3 becomes a bottleneck in the signaling network. By putting databases in node 9 and 10, database loads are distributed among three databases. The database in node 3 will not see the database operations caused by user movements and call setup requests among nodes 21, 22, and 23, or 24 and 25. This example illustrates how the location update is localized in the hierarchical database.

The overall communication cost of location inquiry and update is also affected by database placement. In Figure 3–15, with a database in node 1, adding a database in node 4 will reduce communication traffic on the link between nodes 4 and 1. This is because for user movements and call setup requests between nodes 11 and 12, location update and location inquiry messages will not go up to node 1, they are handled in node 4. The communication load on the link between nodes 4 and 1 is reduced. But for user movements and call setup requests between node 11 (or 12) and any other nodes except node 12 (or 11), adding a database in node 4 means one more database on the paths taken by location update and location inquiry messages. On receiving a message, a node without a database simply forwards it, while a node with a database must process it before letting it go. Thus more databases on the path of the message lead to more delays of messages in the signaling network.

The method to find the optimal database placement scheme is presented by Hać and Sheng (1998). It uses a dynamic programming concept to calculate the number of databases needed and where to place them in the tree-structured signaling network. For a given estimated user movement and call setup request rate, the method minimizes the delay of messages caused by mobile user activities. Both database operation delay and communication delay are considered.

## 3.3 Tracking Strategy for Personal Communications Networks

PCS introduces three capabilities: terminal mobility provided by wireless access, personal mobility based on personal numbers, and service mobility through management of user service profiles. INs are greatly successful in storing and accessing the information required to route calls according to subscribers' desires so that the implementation of an intelligent network concept in mobile communications supports wireless PCS.

Mobility management is an important issue in PCS. One of the main tasks of mobility management is location tracking. Location tracking functions provide PCS with the methods to keep the location of mobile users and terminals current. In PCS, subscriber mobility is controlled through location registration and update. Strategies have been proposed to locate mobile users to reduce signaling traffic, and to reduce resources used in fixed networks at a stationary level in relation to database access.

### 3.3.1 Structure of PCNs

Successful implementation of mobility management in cellular communications, availability of low-power lightweight mobile terminals, and emerging intelligent network technologies, has resulted in significant activities in wireless personal communications. Although within both radio and networking aspects there are commonalities between traditional cellular mobile communications and wireless personal communications, the differences are due to radio propagation and fading effects, interference environment, smaller cell sizes, type and pattern of mobility, and call delivery.

In the structure of PCNs there are two levels, stationary and mobile level. At stationary level, the MSC is a central office, which provides the call processing and channel setup functionality. The MSC interconnects the MS with other MSs through BSs or interoperates with other MSCs and public fixed networks. Networks are interconnected by way of a signaling system

(SS7), which is used as the transport mechanism for call control and database transactions. For subscriber tracking and locating, network elements are used, signal transfer points (STPs) and SCPs, such as HLR and VLR.

HLR contains permanent subscriber parameters and features for a group of subscribers within a network. For each subscriber, there is a pointer to the VLR to assist in the routing of incoming calls. VLR is a dynamic database, whose entries change as subscribers move. VLR contains detailed information on location and service data regarding those MSs entering its coverage area for routing of incoming and outgoing calls. VLR also contains subscriber parameters from HLR. When an MS arrives in a new location area (group of cells), it updates HLR regarding the status of supplementary services. VLR performs authentication and allocates temporary location-specific telephone numbers. Functions of VLR, in addition to location update, include allocation of the temporary mobile station identity (TMSI) for location confidentiality, and storage of the mobile station roaming number (MSRN). Moreover, frequency coordination, channel management, and handoff must be considered in different cell environments (picocell, macrocell, and microcell), and this is done in the base station controller (BSC).

The mobile level consists of mobile units, which communicate with BSs via wireless links. MSs are either vehicle-mounted or hand-held terminals used by subscribers to send and receive information. The location area (LA) is divided into groups of cells with each BS covering an area with a radius of a fraction of to tens of kilometers. Each cell is served by a BS. An MS has radio communications capability to the base station.

Subscriber location tracking enables unrestricted mobility of the MS within the service area. For example, calls can be automatically routed to their destinations, regardless of calling and called parties' location. Location tracking is supported by the location updating (LU) procedure, which determines the LA in which the MS operates and updates the HLR and VLR, accordingly. When an MS destined call arrives, the MS can be paged in the appropriate group of cells.

First-generation mobile systems provided analogue radio voice communication systems. Second-generation systems provided digital radio voice communication with limited data communication capabilities. Third-generation systems, such as the UMTS, offer a wide variety of communication services called PCS.

PCS introduced three capabilities: terminal, personal, and service mobility. Terminal mobility refers to the ability of a mobile terminal to access telecommunication services from any location while in motion, and the capability of the network to locate and identify a mobile terminal as it moves. Personal mobility is the ability of end users to originate and receive calls, to access subscribed telecommunication services at any terminal in any location, and the ability of the network to identify end users as they move. Personal mobility is based on the use of a unique personal identity (i.e., a personal number). Service mobility refers to the network capability to provide subscribed services at the terminal and/or location designated by the user.

The IN is a telecommunications network service control architecture. The goal of an IN is to provide a framework so that the network operator can control and manage services effectively, economically, and rapidly. INs have become the primary provider of a wide variety of new services. An IN can separate specification, creation, and call control of telephony services from physical switching networks. INs store and access information required to route calls according to the subscribers' desires. The IN concept is extended to provide a framework to support personal and terminal mobility, and enhanced services in wireless mobile and PCS.

MSs store and access the users location data in the home and visitor databases, which are the HLR and VLR. The home database (HDB) contains subscriber information with pointers to the last known LA of the user. An incoming call will query the HDB to determine the LA of the user. The visitor database (VDB) contains the most up-to-date information about the user location within an LA designated by the HDB.

The HDB and VDB are two basic elements of the IN's data architecture. Other databases are also required in the IN. The following data are stored in IN to support three mobilities in personal communication services. In terminal mobility, the HDB is used to store subscriber information and routing information; the VDB is used to store the mobile roaming number, temporary subscriber ID, location, and supplementary service parameters. IN personal mobility databases are required to store a personal number, the user's location, and authorization information. IN service mobility databases are needed to store service profile, charging and billing information, and service access information. For convenience, users automatically register at their terminals as they move. Each user carries a small, wireless personal device, which transmits the user's personal ID by radio signals. The network can determine a user's location by polling the devices in each area and automatically registering the user at the terminals in the area.

### 3.3.2 Locating Strategies in PCN

One of the main tasks in mobility management is location tracking. Subscriber mobility is controlled through location registration and update. Specifically, information about the MS location is updated in the VLR whenever the MS crosses LA boundaries. The LA indicates a paging domain and consists of a number of cells related to an MSC. Location registration, based on LA, provides a compromise between updating and paging traffic.

Awerbuch and Peleg provide a formal model for tracking mobile users. The strategy is based on a hierarchy of regional directories, where each directory is based on a decomposition of networks into regions. On a move, only nearby directories are updated and point directly to the new address. To provide access to users in remote directories, a forwarding pointer is left at the previous location to do a direct search to the new location. The proposed framework is elegant and provides answers as to how to balance search cost against update cost in a network with mobile users.

The question of how to locate users in a PCN is also addressed by Ng et al. A hierarchical architecture based on the metropolitan area network (MAN) with three levels of databases is considered. The three levels are: access MAN, backbone MAN, and metropolitan MAN. The root is the metropolitan area (MA) database which manages subscribers within the area. Several MAs are connected by a central office. Each subscriber has a home access MAN and when the user is outside the home MA, the home access MAN has a pointer that leads to the general location of the user. To locate a callee, the ripple search scheme is proposed. Search always proceeds in a sequence of searches involving hierarchy of databases. The search is confined to the caller's local MA, and then to the caller's home MA. If this search sequence fails to locate the user, then the directory database is searched. The directory database contains information about the home access MAN for every user. Hence, this database is very large and access is expensive. Once the home access MAN is found, the user is located. This scheme is based on the principle of call locality—the callee is located close to the caller. However, the update and search scheme is fixed independently of calls to the user's mobility patterns.

Imielinski and Badrinath discuss an issue of database queries for tracking mobile users by partitioning the users' location knowledge across the network. This approach has two drawbacks. The partition of user mobility patterns to maintain a certain degree of knowledge about a user's whereabouts is the optimal partitioning problem that minimizes the expected cost of querying and updating. The optimal partitioning problem makes this approach difficult to implement. Query processing is another disadvantage in the presence of imprecise knowledge about the users' locations.

Upamanyu and Madhow et al. consider a distance-based strategy. The users are paged at a high rate, and therefore, at the time of paging the system is not in a steady state. Whenever a user is paged and found and the user's location is updated, the strategy considers the evolution of the system between two successive pagings. Using a dynamic programming approach, an iterative algorithm for computing the optimal value of distance is proposed.

Tabbane introduces an intelligent method for locating users called an alternative strategy (AS). This proposal is based on the observation that mobility behavior of a majority of users can be predicted. The strategy can reduce signaling messages involved in mobility management procedures, leading to savings in system resources. This approach reinforces information capabilities of the system, and provides tools for prediction by integrating more information (in the system) about the external world. The main advantage of an AS scheme is to save location updates by storing the user's mobility information in profiles. Drawbacks are a possibility of increased paging delays (i.e, when the system must page a user in different locations in a sequential manner), increased memory requirements, and profile updating costs for users who do not have long-term periodic mobility patterns.

### 3.3.3 Tracking Strategy

In a cellular architecture, the fixed network is informed about the MS's movements, and the current location of an MS is stored in a database. The information flow to the database is an LU and it can be performed based on the time, movement, and distance, or by the zone method. The zone method requires splitting the system area into zones called LAs. The MS updates its location in the database only at each LA boundary. In the case of an incoming call, the MS is searched by paging its current location area.

There are two types of resources required for actions: updating and searching. Updating is the cost of database access in the fixed network at a stationary level. This cost depends mainly on the LU. Searching is the use of wireless channels for the LU and cell access. The occupancy of paging channels is a function of LA size (i.e., the number of cells within the LA).

The zone method adapts two extreme approaches: always update and always search. In an always update approach, the MS updates its location each time it moves into a new cell, and a search causes a simple database lookup. The update cost is very high, but the searching cost is negligible. In an always search approach, the MS updates its location while it moves. When a call arrives, all cells are searched to locate the user. This results in a very high searching cost and zero update cost.

Location registration based on LA provides a compromise between the updating and paging traffic. The LU procedure results in the reduction of signaling traffic caused by reduced paging traffic, but introduces new signaling LU traffic. The size of the LA should be optimized to minimize the total signaling traffic.

Bar-Noy and Kessler selected a subset of nodes in unweighted graphs (ring and grid) and weighted special graphs (line and tree), which can be used to represent a signaling network to optimize the use of wireless resources for tracking mobile users. However, the topology of wireless network is neither unweighted nor as simple as a tree, and is usually represented by cells with six edges to connect with neighboring cells. Also, from a networkwide consideration, the method does not define the searching cost in detail, so that the total signaling cost to be minimized is not properly considered. Thus, the final solution cannot be applied to solve a real problem.

Hać and Zhou's approach is based on cellular architecture, when a partition scheme is designed for the system and customer characteristic to reduce total signaling costs in terms of location update and searching cost. The basic idea is to allow the MS to transmit update messages only at specific cells (the number of which is small compared to the total number of cells in the network), while restricting every search for a mobile user to a small subset of cells. The scheme minimizes the total cost for the entire network rather than for each individual user. In this approach, the cell size is isolated from the problem. We find an algorithm to solve the problem, that is, to get a good performance of the result. The problem is only related to searching and updating traffic.

In a simple cellular system, there is one BS per cell, and a large number of BSs in the system. A subset of BSs called reporting centers is selected among all BSs. The cells associated with these BSs are referred to as reporting cells while other cells are called nonreporting cells. The reporting centers periodically transmit short messages on the wireless channel to identify their roles. Thus, an MS can learn whether it is in a reporting cell or not by listening to the message.

The vicinity of reporting cell $i$ is the collection of all nonreporting cells that are reachable from cell $i$ without crossing another reporting cell. Also, cell $i$ is in the vicinity of itself. Thus, each reporting cell has a number of cells as its vicinity. Therefore, we can specify two operations under these circumstances: update and search. Update is used only when an MS enters a new reporting cell. It transmits an update message to the BS associated with the cell to report its new

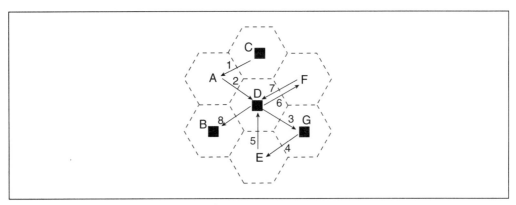

**Figure 3–16** An Example of Tracking Strategy

location. The search operation is used when a call arrives in the MS. The search is restricted within the vicinity of the reporting center to which the MS last reported. An example in Figure 3–16 shows this tracking strategy. Suppose we choose reporting centers as cells: $B, C, D, G$, which are shown as solid squares in Figure 3–16. Cells $A, E$, and $F$ are nonreporting cells. Cell $B$ is directly connected to nonreporting cells $A$ and $E$, and there are no reporting cells between them. Thus, the vicinity of cell $B$ comprises cells $A, E$, and $B$ itself. For the same reason, the vicinity of cell $C$ is composed of cells $A, F$, and $C$; the vicinity of cell $G$ embraces cells $E, F$, and $G$; and the vicinity of cell $D$ comprises cells $A, E, F$, and $D$.

If a user starts in cell $C$, then he or she travels along the path

$$C \xrightarrow{step\ 1} A \xrightarrow{step\ 2} D \xrightarrow{step\ 3} G \xrightarrow{step\ 4} E \xrightarrow{step\ 5} D \xrightarrow{step\ 6} F \xrightarrow{step\ 7} D \xrightarrow{step\ 8} B$$

in each step. The user reports his or her location when he or she arrives in cells $C, D, G, D$, and $B$ since these are the reporting centers the user reaches during the journey. Note that in Step 2 the user reaches cell $D$, and the user needs to transmit the location update message to the BS in cell $D$. The reason is that the user has just come from reporting center $C$ via nonreporting cell $A$. Reporting center $C$ is the same one the user reported last time, and reporting center $D$ is the new center the user reaches. For the same reason, in Step 5, the user needs to report this location at reporting center $D$ when he or she comes from reporting center $G$ via nonreporting cell $E$. In Step 7, when the user reaches cell $D$, the user does not transmit the update message to the BS in cell $D$. The reason is that the user has just come from reporting center $D$ via nonreporting cell $F$. Reporting center $D$ is not the new center for the user at this time.

To search for the user, when he or she is in cell $A$, for instance, the system searches cells $A$, $F$, and $C$ since these are in the vicinity of reporting center $C$ that the user has just reported. At the moment the user is in cell $E$, the system searches cells $E, F$, and $G$ since these are in the vicinity of reporting center $G$, which the user has recently reported.

The update cost increases with the volume of update messages that are transmitted by mobile users. This volume is related to reporting cells only since there are no update messages sent to BSs that are not reporting centers. The volume for each reporting cell $i$ increases with the frequency that the mobile users enter cell $i$ (excluding reentries into the same cell at consecutive times) that we call weight $w_{mi}$ for cell $i$. Here $m$ represents the term mobility (i.e., movement). Thus, the total updating cost is the sum of weights over all reporting cells. The searching cost for cell $j$ increases with the frequency of query into the cell multiplied by the size of the vicinity in which the search is performed. If $j$ is the reporting cell, there is only one value of vicinity size related to $j$. If $j$ is a nonreporting cell, there can be several different values of vicinity size in each of the vicinities to which cell $j$ belongs. For example, in Figure 3–16, cell $A$ is a nonreporting cell, and it is also one of the vicinity cells in either reporting cells $C$, $B$, and $D$. The size of the vicinity of reporting cell $C$ is 3 which includes cells $A$, $F$, and $C$. The size of the vicinity of reporting cell $B$ is also 3 which includes cells $A$, $E$, and $B$. The size in the vicinity of reporting cell $D$ is 4 which includes cells $A$, $F$, $E$, and $D$. Thus, the upper bound of the searching cost for cell $j$ is taken to be the frequency of query into cell $j$ which we call the weight $w_{qj}$ multiplied by the largest size of the vicinity among all vicinities to which cell $j$ belongs. Here $q$ represents the term query (i.e., call into a cell). The total searching cost is bounded by the sum of the cell's upper bound of the searching cost.

The problem is how to choose the reporting centers to minimize the total signaling cost (i.e., both updating and searching costs). If we select a large number of cells as reporting centers, we get a low searching cost since each reporting cell has a small number of cells as its vicinity. However, the update cost is high since the users reach the reporting cells more often as they move. If we select a small number of cells as reporting centers, we get a high searching cost and a low updating cost. To minimize both costs at the same time is not an easy task. The strategy we use is to bound one cost and to minimize another cost. Hać and Zhou bounds the searching cost and minimizes the updating cost in our algorithm.

The total searching cost is bounded by the cell's upper bound searching cost. Also, the upper bound of the searching cost for a cell is taken to be the frequency of query into the cell multiplied by the largest size of vicinity among all vicinities to which the cell belongs. The frequency of query is based on the customer characteristic in the network. The largest size of vicinity for each cell is decided by the result of the reporting center's selection. Thus, in the first step, we bound the largest vicinity in the network (i.e., the upper bound of vicinity size for all cells), and we minimize the update cost. We have therefore, bounded the searching cost for the network to a certain degree. We define the largest size of vicinity in the network as an integer $Z$. For the network with $N$ cells, the value of $Z$ ranges from 1 to $N$. In the second step, we minimize the total cost based on the solutions in the first step for different values of $Z$.

We use a graph representation to illustrate the problem. We define the wireless network in a mobile level as graph $G$. Each node in $G$ represents a cell. The links between the node with the six neighboring nodes represent a cell connecting with the six neighboring cells. A graph is shown in Figure 3–17, where the nodes are drawn as solid circles. There are two kinds of weight associated with each node: the mobility weight $w_m$, which represents the mobility rate into the node and the query weight $w_q$, which represents the query rate into the node. Links are drawn as

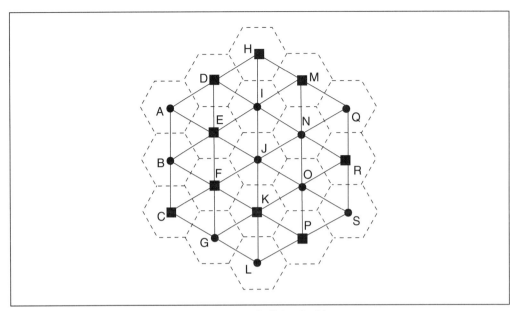

**Figure 3–17** A Typical Graph Representing a Cellular Architecture

thick solid lines. Dashed lines consist of cells in the mobile cellular architecture. There is a BS in each cell.

We choose a subset of nodes among $G$, called centers. Each center has a set of nodes as its vicinity. The vicinity of the center $i$ is the set of all nodes, which are not in the centers, and are reachable from the center $i$ by a path containing no centers. Also, the center $i$, itself, is included in its vicinity. For example, in Figure 3–17, we choose the following nodes as centers: $C$, $D$, $E$, $F$, $H$, $K$, $M$, $P$, and $R$, which are shown as solid squares. The vicinity nodes for each center in Figure 3–17 are as follows:

| Center Node | Vicinity Nodes |
|---|---|
| C | A, B, G, L, C |
| D | A, B, I, J, N, O, Q, S, D |
| E | A, B, I, J, N, O, Q, S, E |
| F | A, B, I, J, N, O, Q, S, G, L, F |
| H | I, J, N, O, Q, S, H |
| K | G, L, I, J, N, O, Q, S, K |
| M | I, J, N, O, Q, S, M |
| P | G, L, I, J, N, O, Q, S, P |
| R | I, J, N, O, Q, S, R |

Using this representation, we can formalize the reporting centers problem. We use the following notation:

- $G$ — the graph representing cellular architecture
- $N$ — the number of nodes in graph $G$
- $w_{mj}$ — mobility weight related to node $j$ (also called mobility rate for cell $j$); it represents the frequency with which the mobile users enter the cell (i.e., the average number of moves (into a cell) per unit of time)
- $w_{qj}$ — the query weight related to node $j$ (also called query rate for cell $j$); it represents the frequency with which users need to be located in the cell (i.e., the average number of queries (into a cell) per unit of time)
- $Z$ — the integer which represents the upper bound size of the vicinity
- $Z_j$ — the size of the vicinity to which node $j$ belongs. If $j$ is a center node, there is only one value of vicinity size related to $j$; if $j$ is a noncenter node, $Z_j$ is the largest size of vicinity among all vicinities to which node $j$ belongs
- $S_i$ — the set of center nodes under current selection in $i$th solution
- $z(S_i)$ — the largest size of the vicinity among all vicinities under the current selection of center nodes $S_i$
- $w_m(S_i)$ — the sum of mobility weights over all center nodes $S_i$

The reporting centers problem is defined as follows. Given a set of integers $\{Z\}$, $1 \leq Z \leq N$, and a weighted graph $G$ with $N$ nodes, for each of these nodes, there are two weights associated with it, $w_m$ and $w_q$.

1. For each of the integers $Z = 1, 2, \ldots, i, \ldots, N$, select a set of centers $S_i$ such that $z(S_i) \leq Z$ and $w_m(S_i) \leq w_m(S_i')$ for all $S_i'$ that meet $z(S_i') \leq Z$. Here $z(S_i) = \max(Z_j), j \in S_i$, and

$$w_m(S_i) = \sum_{j \in S_i} w_{mj}.$$ Similar definitions hold for $z(S_i')$ and $w_m(S_i')$.

2. Find solution $i$ from among all solutions in step (1), which makes $\left(\sum_{j=1}^{N} w_{qj} Z_j + \sum_{k \in S_i} w_{mk}\right)$ minimal. $Z_j$ is the largest size of vicinity among all vicinities to which node $j$ belongs, and $S_i$ is the set of centers in solution $i$.

In step (2), the term $w_{qj} Z_j$ is calculated as the upper bound of the searching cost for cell $j$ (i.e., it is equal to the frequency of query into cell $j$ multiplied by the largest size of vicinity among all vicinities to which cell $j$ belongs). Term $\sum_{j=1}^{N} w_{qj} Z_j$ uses the cell's upper bound of the searching cost to get the total searching cost in the entire network. Term $\sum_{k \in S_i} w_{mk}$ summarizes the mobility weights over all reporting cells to calculate the total updating cost. Thus, the formula $\left(\sum_{j=1}^{N} w_{qj} Z_j + \sum_{k \in S_i} w_{mk}\right)$ in step (2) represents the total signaling cost in the network, which is the sum of both the total searching and updating costs.

For $Z \geq 2$ the optimal solution in step (1) is an NP-complete (Nondeterministic Polynomial time complete) problem. The best solution can be found by a simple enumeration of all possible solutions. There are only two states associated with each node (i.e., either a center node or non-center node). Thus, for a graph with $N$ nodes, there are $2^N$ possibilities.

A heuristic method to find a solution uses these three steps as follows.

1. Use an approximate algorithm to find each solution in the reporting centers problem (1) for different values of $Z$.
2. Select the best solution which has the lowest total signaling cost.
3. Refine the solution by running a refinement procedure to find an even better solution.

For a mobile network with $N$ cells, we minimize the total signaling cost in terms of update and search. We control one cost and minimize another cost. Integer $Z$ represents the upper bound size for each vicinity (i.e., group of cells) to control the searching cost within the network. Each vicinity represents the domain of query when a call arrives. For $Z$ between 1 and $N$, the value is the constraint for our approximate algorithm. We assume that the frequency in which the mobile users enter the cell is known in advance. The information about this rate is easily recorded by the user's movement pattern.

All cells are selected as reporting cells so that the size of the vicinity for each cell is 1. This means that if a call arrives, the system immediately knows in which cell the user is located. The query goes into that cell. The current value is the smallest value of $Z$. Mobile users transmit an updating message whenever they move into a new cell. Therefore, the updating cost is the largest. To minimize the updating cost, we reselect some cells that are nonreporting cells so that mobile users do not need to update their location every time they enter a new cell. To decide which cells should be reselected as nonreporting cells, the algorithm uses the following reselection scheme.

The frequency in which the mobile users enter a reporting cell corresponds to the updating cost for the cell. The higher the frequency, the higher the updating cost. If we reselect a cell with the highest frequency as a nonreporting cell, the users do not need to update their location whenever they enter this cell. This reduces the total updating cost by the largest amount for the entire network. This action is under the constraint of value $Z$ (i.e., the upper bound size for each vicinity). We continue this step and reselect the cell with the highest frequency among all remaining reporting cells. If the resulting situation after cell reselection still meets the $Z$ constraint, the step is successful. Otherwise, the step is not successful (i.e., the cell must be a reporting cell). Thus, we have to try another cell. This procedure continues until all cells are tried.

We use the following analysis in our algorithm. If for any integer $1 \leq Z \leq N$, we put all nodes in centers $S$, we certainly meet the condition $z(S) \leq Z$, but $w_m(S)$ is the largest at this moment. Then, we take node $x$ away from $S$ (i.e., we make node $x$ a noncenter node) which has the largest mobility weight among all nodes in $S$ only if the condition $z(S - x) \leq Z$ holds. This action is heuristic, but it can help since $w_m(S - x)$ is reduced mostly when taking another node away. We continue this step if taking away another node $x$ still makes the condition $z(S - x) \leq Z$ hold, which means that the action is successful. We update $S$ to become $S - x$, mark $x$ as a noncenter node, and put it into the noncenters list $V$. If after taking away node $x$ from centers $S$, the remaining situation conflicts with condition $z(S - x) \leq Z$, then we should return this node to centers $S$. However, we can continuously try another node with the next largest mobility weight. We repeat this procedure until all nodes are tried. Finally, the results are two lists with nodes in centers $S$ and noncenters $V$.

The approximate algorithm is as follows.

1. Build two lists: first for centers list $S$; second for noncenters list $V$.
2. Initial $S$ with all nodes in it.
   Sort $S$ in the descending order of the node's mobility weight.
   Mark all nodes that are not tried yet.
   Initial $V$ empty.
3. Pick node $x$ with the largest mobility weight from $S$ (i.e., the first node in $S$).
4. Make $x$ a noncenter node.
   If $z(S - x) \leq Z$ then do:
      1) $S = S - x$.
      2) Put this node into $V$.
   Otherwise
      1) Put this node back to the tail of $S$ (i.e., $x$ is still the center node).
      2) Mark the node as tried.
5. If there is still a node not tried yet, then go to Step 3.
6. Otherwise, return list $S$ as centers and list $V$ as noncenters.

A better solution, in which the total signaling cost is reduced, is based on the solution selected from the reporting centers problem (2). Using the approximate algorithm to choose noncenter nodes in order of descending mobility weight, forcing the noncenter nodes in the solution from the reporting centers problem (2) to become center nodes, and reconsidering the latter nodes status, the total signaling cost of the solution can be reduced when the latter nodes become noncenter nodes.

The refinement procedure is as follows:

1. For a given graph with $N$ nodes $x_1, x_2, ..., x_N$, find each solution in reporting centers problem (1) for different values of $Z$ by using the approximate algorithm.
2. Select the best solution in which the total signaling cost is the lowest.
3. Copy the best solution into the best result so far.
4. Arrange the nodes of the result in descending order of mobility weight from left to right into a set $\{x_k\}$, $1 \leq k \leq N$. Noncenter nodes are $x_{l_1}, x_{l_2}, ..., x_{l_i}, ..., x_{l_m}$, $1 \leq l_i \leq N$, $m < N$
5. for $i = m$ to 1 Step $(-1)$
6.     for $j = i$ to $m$ Step 1
7.         Keep the status of the nodes at the left side of node $x_{l_i}$ in set $\{x_k\}$ unchanged
8.         Change the status of nodes from $x_{l_i}$ to $x_{l_j}$ to become center nodes
9.         Reconsider the status of nodes at the right side of node $x_{l_j}$ in set $\{x_j\}$ under the constraint of the same value $Z$
10.         If the new result is better than the best result so far in the sense of reducing total signaling cost, then update the best result to be the new result
11.     end
12. end
13. Return the best result.

Hać and Zhou proposed a scheme of partition and tracking mobile users, and its implementation based on cellular architecture. According to the tracking strategy, MS transmits only update messages at specific cells called reporting cells, while the search for a mobile user is done at the vicinity of a reporting cell to which the user has just reported. In this algorithm, the cost analysis model is used to demonstrate feasibility of updating the cost and searching the cost to reduce the resources in using the wireless channels.

## 3.4 Location Management Schemes for Mobile Communications

The GSM allows the users universal and worldwide access to information and the ability to communicate with each other independently of their location and mobility.

Tracking mobile users and routing calls are basic functions of the network system. Upon request, the system needs to update and provide information about the location of mobile users. Queries can be formulated by both the users and the network resource management facility (e.g., aggregate information for adaptive channel allocation).

The criterion for efficient update is a low signaling cost incurred by the relocation of users between cells. The cost should be kept small enough not to affect network performance. PCN cellular technology is expected to progress to much smaller cells for greater bandwidth sharing and reuse. The signaling load for location updates is higher due to more frequent relocation for small cells.

Location Management Schemes for Mobile Communications 125

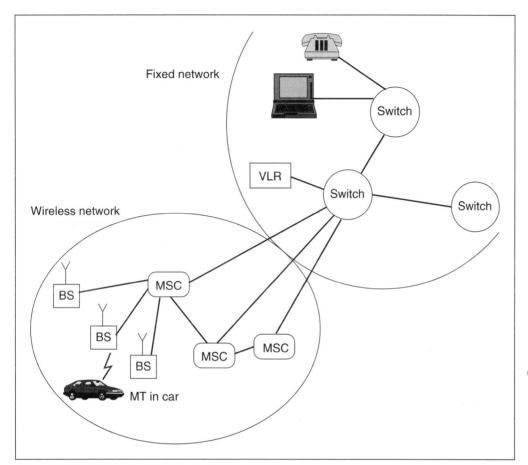

**Figure 3–18** Model of a Cellular System

Providing information about user location is coupled with updating. Because of the frequent relocation of mobile users, especially for small cells, it is impractical to keep track of their exact locations at all times. The more general approach is to keep the location information updated upon query, which requires extra search work. Keeping more complete information updated requires a higher signaling cost, while less search work is needed upon query. Finding efficient methods or schemes can balance the effects of these factors and achieve optimal overall performance.

An architectural model of the cellular system is shown in Figure 3–18. The part of the system directly serving the mobile users can be viewed as a network extension providing wireless access to the fixed information network. The network extension essentially consists of three elements: wireless MTs, BSs, and MSCs. Each BS serves a group of MTs currently residing in a cell. An MSC typically connects from up to 100 BSs to trunks of the PSTN or ISDN. Neighboring MSCs can also connect to each other through direct links. The databases related to location control and management are usually integral parts of their MSCs, but databases may also be stand-alone entities. Typical databases are the HLR, VLR, and equipment identity register (EIR). These facilities combined with their counterparts in the fixed information network, normally at routers or switches, provide routing and location management functionality for the system. The switches and databases that perform the functions of tracking users and delivering enhanced services to the users communicate by using the SS7 in current implementations. However, because SS7 was designed before there were large-scale mobile networks, the ability of the existing systems to meet the emerging needs should be examined. In a simple model, it is often convenient to use a single abstract notion to denote these functional entities, which we call a location server. In some cases, a location server can include an MSC and the HLR, with VLR directly attached to it. It is also possible for one VLR to control the service areas of several neighboring MSCs. In such cases, a location server can include a major switch node of the PSTN, the HLS, the VLS, and those MSCs under the control of the VLR.

The system elements are as follows. The MTs terminate the radio link on the user side of the user-to-network interface. They are either vehicle-mounted or hand-held terminals used by subscribers to send and receive information. BSs terminate the radio link on the network side of the user-to-network interface. The basic role of a BS is to manage radio channels and process information directly related to the establishment, maintenance, and release of radio links. The MSC performs the bulk of the wireless and wired call processing in existing cellular systems. For call connections, the MSC processes MT registrations, queries location registers, manages paging processes, selects a BS to handle a call, and establishes connections between the PSTN and selected BS. When the MT crosses cell boundaries during a call, the MSC also coordinates handoff activities. The HLR contains master records of the information for a group of MTs within a given (regional) network. The record includes a mobile identification number (MIN) to which the MT responds; the electric series number (ESN) burned into the MT during manufacturing; the user's profile indicating the service features available; the MT's state (active or not); and a pointer to the last VLR that has reported the user's whereabouts. The VLR is used when a mobile user roams out of his or her home area. It contains the information for all subscribers currently residing in its service area. The information includes copies of users' profiles obtained from their HLRs when these users enter and register in the VLR's service area. The VLR assists in keeping track of a visiting MT and delivering calls destined for the MT.

The total signaling load generated by mobile communications in GSM results from the following four system activities:

1. Initiating calls from mobile users. The procedures are similar to those for an ISDN call originating in the fixed network. The MSC currently serving the MT first checks the mobile terminal identifier and the authentication information of the MT through the VLR. If the information is not available, the VLR gets it from the MT's HLR. A TMTI is assigned to the MT. The MSC sends out the initial address message (IAM) to transit switching nodes to set up the call.

2. Delivering calls destined for mobile users. Besides the basic procedures for an ISDN call originating in the fixed network, there are additional procedures. First, the system determines the registration area where the mobile user currently resides. The network then pages the mobile user to determine the current cell of the mobile user. Following that, the system authenticates the MT to ensure that its access to the network is valid. Then the network assigns a cipher key to the radio link and may assign a new temporary local number (TMTI) to the MT. The MSC obtains the user profile from the VLR, which contains the information about features and special services to which the user has subscribed. The signaling traffic incurred is much higher than in the fixed network.
3. Updating the location information upon users' moves. A location update is initiated when a level boundary is crossed by an MT. There are several different situations. If the only cell boundary crossed is inside the service area of an MSC, a local update may or may not be performed, depending on the location management scheme used. If an MT enters the area of a new MSC but under the control of the same VLR, an update throughout the registration area may or may not be performed, depending on the management strategy. If a boundary of areas controlled by different VLRs is crossed by an MT, the current VLR must obtain the authentication parameters from the previous VLR. The VLR authenticates the MT and allocates a new TMTI. Then the user's HLR is updated in the new VLR, and the HLR informs the previous VLR that the MT is no longer in its control area.
4. Handovers (handoffs). When an MT crosses the cell boundary during a connection, the new BS takes over the task of maintaining the radio link between the MT and the system.

The total contributions of the above activities to the signaling volume in terms of the number of bytes per call can be expressed as follows:

$$Svolume_{percall} = Svolume_{m.o.} \times f_{m.o.} + Svolume_{m.t.} \times f_{m.t.} + Svolume_{l.u.} \times r_{u.c.} + Svolume_{h} \times r_{h.c.} \quad (3-5)$$

where $Svolume_{m.o.}$ is the signaling volume (in bytes) for a mobile user originating a call, while $Svolume_{m.t.}$, $Svolume_{l.u.}$, and $Svolume_{h}$ are signaling volumes (in bytes) for a call destined for an MT, with an average of one location update, and one handover, respectively. The other factor in each term is the corresponding frequency of occurrence. In particular, $f_{m.o.}$ and $f_{m.t.}$ are the percentages in total calls for calls originated by an MT and calls destined for an MT, respectively, while $r_{u.c.}$ and $r_{h.c.}$ are the average number of updates per call and handoffs per call, respectively.

The frequency of occurrence for each activity is determined by the users' behavior, either by their mobility patterns (for $r_{u.c.}$ and $r_{h.c.}$) or by communication patterns (calling patterns of mobile users for $f_{m.o.}$, and calling patterns of their communication partners for $f_{m.t.}$). The signaling volume per call for each activity ($Svolume_{m.o.}$, $Svolume_{m.t.}$, $Svolume_{l.u.}$, and $Svolume_{h}$) is the summation of the signaling volumes for all the basic procedures taken in the activity.

### 3.4.1 The Two-Scope Concept

A location server can include an MSC and the HLR with the VLR directly attached to it. It is also possible for one VLR to control the service areas of several neighboring MSCs. In such cases, a location server can include a major switch node of the PSTN, the HLS and VLS, and the MSCs under the control of the VLR.

When a user moves within the area under the control of the same location server, we say the relocation or movement is a "bound roaming." Otherwise, it is an "unbound roaming." Location control and management can be viewed as the combination of two different scopes: global scope and local scope. For example, to locate an unbound roamer, first the locating mechanism of the global scope needs to find the location server for the area in which the roamer currently resides. Then the location server finishes the job in the local scope. In the local scope, the signaling traffic occurs within the service area of a location server. In the global scope, the signaling traffic also occurs between location servers.

#### 3.4.1.1 Global Scope

The global scope focuses on providing efficient and timely call delivery to a user who roams out of the area of his or her HLS (with HLR).

When an unbound roamer moves from the serving area of one location server to another, a message is sent to the roamer's HLS to update its location information in the HLR. The HLR keeps an updated record of the location server that controls the area in which the roamer resides. The HLR is always involved in delivering calls to roamers outside their home area in the most straightforward call delivery scheme, which is called "dogleg routing" and "triangle routing." (These two terms will be used interchangeably.) Basically, to set up a call destined for a roamer outside its home area, the call request is first routed to its HLS, and then to the location server currently serving the area where it resides. Figure 3–19 shows a scenario of establishing a connection for a call destined for an MT roaming out of its HLS using dogleg routing.

Although in some circumstances, this method, which is used in current mobile systems, can be acceptable, there are many situations where the overhead of passing through the HLS is too high. For example, this happens when the caller is located geographically very close to the roamer location and far from the roamer's home area. Another situation is when communicating partners frequently exchange a high volume of data with a user when the user roams away from his or her home area. For these situations, dogleg routing introduces an unreasonable penalty to the network cost.

The general approach to improve the efficiency of call delivery over dogleg routing is to distribute the location information (replication) in the network, and to eliminate certain overheads of passing through the HLS (with HLR). Under certain conditions, caching the location information obtained from previous calls at selected network switches can be an efficient way to deliver subsequent calls to the same roaming user. The location information distributed network-wide needs to be selectively updated as users roam over the boundaries of areas served by different location servers, to achieve optimal efficiency. The cost of updating should be evaluated together with the gain on call delivery for a net improvement. To limit the updating cost, parti-

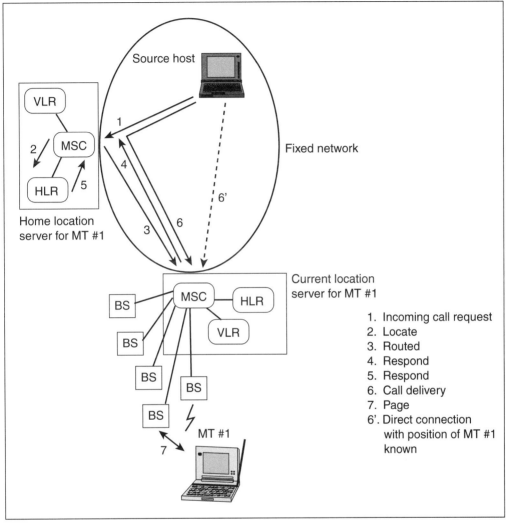

**Figure 3–19** Scenario of "Dogleg (triangle)" Routing

tioning location information in a hierarchical database structure spread geographically over the network is a good solution.

### 3.4.1.2 Local Scope

In a local search, the location server has the database (typically a VLR) containing the whereabouts (ideally the exact cell location) of all mobile users currently residing in the area. When an MT makes a move within the area from cell to cell, a message can be sent from an MT to the location server to update the location records. The cost of the user informing the BS and/or the location server of the move is higher than the cost of a BS or location server contacting the user, due to the contention among users and the required authentication procedure. That is, the cost to update the location is several times higher than the cost of a simple call delivery with either the authentication procedure or searching. This makes keeping constant track of users' exact locations uneconomical. Limiting the signaling volume for the location update decreases cost and keeps location information incomplete but reasonably accurate. However, the searching (e.g., broadcasting) cost incurred due to the lack of complete location information must be balanced with the updating cost to make the total cost optimal with consideration of the response time for queries.

The three basic schemes are as follows.

1. Broadcasting. When the exact location of an MT is not known, the location server broadcasts the searching message to a group of BSs, which includes the BS currently providing a radio link to the MT.
2. Forwarding pointers. When an MT moves into another cell, it leaves a pointer to the new cell at the previous cell. Searching is done by following a chain of pointers from the last known position to the current position.
3. List. When an MT moves following a certain pattern, a list of next possible locations with known likelihood can be stored in order. Searching is done from the last known position by trying the new locations successively on the list.

### 3.4.2 Location Management in Local Scope

Several protocols are proposed for mobile internetworking. The mobile internetworking control protocol (MICP) is used by the mobile support stations (MSS) to exchange control information (i.e., the information about mobile user location), and is used by mobile hosts (MHs) to signal their MSSs when entering a new cell. A method of combining caching and forwarding pointers is evaluated for efficient location management. Several locating strategies are compared by Imielinski and Badrinath.

Two general characteristics of users' mobility, spatial locality and temporal locality, can be exploited for the basic principle of limiting the location updating volume and achieving efficiency. The concept of spatial locality means that a limited number of consecutive moves of a user, which correspond to a set of contiguous cells, can be covered by a local zone centered at the starting point of the first move. The updating actions can be limited to only those times when a user crosses the current zone boundary. With the radius of a zone suitably determined, the searching cost (e.g., broadcasting cost) is also limited. The concept of temporal locality is actually another interpretation of spatial locality. That is, within a certain period of time, a user can only roam a limited distance away from the starting point, due to the finite speed. We can expect an upper bound for an update frequency.

Making use of the location information obtained from the previous call helps to lower the updating cost. When a roamer is connected to the network during the call, the location information is updated with a much lower cost than purely the user-initiated location updating.

**Location Management Schemes for Mobile Communications** 131

Because a particular location management scheme usually performs well under certain conditions and not as well under other conditions, designing a combination of schemes promises a better overall performance than is possible when using any one scheme.

### 3.4.2.1 Method of Static Partitioning and Method of Dynamic Neighbor Zone

The service area of an MSC can be divided into partitions composed of a number of cells in the static partitioning method. The service area of a location server is partitioned and location update events are limited to cases of crossing partition boundaries. On call delivery, the location server broadcasts only to the BSs within the partition, which covers the latest known position of the user.

In a static partitioning method, the boundaries never change. Savings on both the updating volume and searching cost are possible, and a resource cost is incurred in this method due to the additional computation. The partition information is stored in each MT (in addition to the copy at the server), because an MT has to decide if it should initiate a necessary location update, depending on whether a boundary crossing has occurred. The extra memory and computation cost of each MT is considered reasonably worthwhile. Because technology drives CPU processing costs and memory costs down, and more importantly, because the network is often the performance bottleneck, especially in the GSM environment, using computation and memory costs for savings in network signaling and database loads is cost performance efficient.

Hać and Liu proposes a new scheme to further reduce update costs. Each MT keeps a local zone comprised of a set of BSs. The center of the zone is located at the cell, which is the last position of the user reported to the location server. Whenever a call is delivered to the user or the current zone boundary is crossed by the user, the current position is updated and this new position becomes the new center of the zone. The zone boundary is also updated. To determine the zone boundary (the set of cells) for the new center position in a consistent way, the identification codes of BSs reflect their geographic relations. An MT can determine if the current zone boundary was crossed after the latest move by receiving an identification code from the new BS.

The dynamic neighbor zone algorithm is as follows:

1. For each move:
2.     Compare identification code of new cell with current zone boundary.
3.     If crossing occurs,
4.         Initiate location update for location server;
5.         Update zone center as current cell;
6.         Calculate new zone boundary.
7. For each call received:
8.     Update zone center as current cell;
9.     Calculate new zone boundary.

### 3.4.2.2  The Model

The basic activities of mobile users with MTs, and of the system are as follows:

1. user requests for calls (more generally, connections) to other hosts,
2. system updating the location information upon users' moves,
3. system making connections to mobile users that possibly involve searching,
4. call handovers handled by the system when a cell crossing occurs during a call connection.

The first activity on the mobile users' side does not add complexity to the fixed network, except for the connection made through exchanging radio signals between an MT and a BS. We focus on two types of events, "location updating" and "call delivery." "Updating" events are systematic activities initiated purely for tracking mobile users. In addition to these events, location information is obtained through the first activity, mobile user initiated calls, and is also updated through the fourth activity, handovers. The first activity is on the users' side and independent of the "location updating." To predict the occurrence pattern of mobile user initiated calls and make use of the pattern in a systematic way involves additional information about the users' behavior, and we consider this effect separately. The duration pattern of connections affects the fourth activity. Because we do not want to make the model dependent on the regularity of users' behavior and the availability of this information, we analyze the additional impact that the effects of the first and fourth activities have on the results from the model.

As discussed by Imielinski, Badrinath, and Yuan the occurrence pattern of a user's move and the calls destined for the user are the main factors determining both the updating and searching costs for a particular management scheme. Typically, the call-mobility ratio is critical in quantitative performance analysis. We define "calls per move" as the number of calls received by a mobile user in the period between two consecutive moves.

The performance criterion is minimizing the combination of the updating and searching costs. We define "cost per event" (an event can be a move or a call) as the total cost (equal to the sum of updating and searching costs) divided by the total number of moves and calls delivered. Another criterion, "cost per call," is defined as the total cost divided by the total number of calls delivered, and describes the impact of mobility on call costs.

In the local scope, the system model consists of the location server under consideration connected to the PSTN and other location servers. A mobile user moves according to "bound roaming" within this area. The characteristic of the user's call-mobility behavior is described by the parameter "calls per move." A move is defined as a relocation of the user from one cell to a neighboring cell. With the same number of calls per move, there can be different details in the user's moving pattern, that is, moving around in a small group of contiguous cells, or crossing a number of cells in one direction. This can also have an impact on the management cost.

We select the following cost metrics:

1. The cost of call delivery when the location of the MT (i.e., current cell) is known and the authentication procedure has been done. This is defined as one basic unit, "one message" cost.
2. The cost of location updating. Because the contention among users and the authentication procedure involved require bidirectional messages sent between a user and the BS, and correspondingly, between the BS and the location server, the total cost is defined as the factor $\alpha$ multiplied by the basic unit "message." When a forwarding pointer is left at the previous BS by the user upon moving, for location update there is the cost incurred, which is defined as factor $\beta$ multiplied by the basic unit "message."
3. The cost of searching. We consider two basic searching mechanisms, broadcasting and pointer forwarding. The location server broadcasting to a set of $\gamma$ BSs yields a maximum cost of $\gamma$ "messages." In pointer forwarding, one forwarding operation yields a cost of "one message." A forwarding chain of length $k$ costs $k$ "messages."

### 3.4.2.3 Performance Evaluation of the Dynamic Neighbor Zone Method

In the dynamic partitioning method, a zone is a partition of a fixed number of cells, centered at the last known position (cell) of the mobile user and moving with the user after a location update. In the dynamic neighbor zone method, we use a pair of integers (similar to two-dimensional x–y coordinates) as the identification code for each BS. The boundary cells are those with a distance to the center cell equal to a predefined radius.

The dynamic neighbor zone method yields a substantial improvement over the method of static partitioning in the entire range of the "calls per move." The cost saving is obtained by further reducing the updating events. The limited broadcast scheme performs better than the forwarding pointers scheme only at the low "calls per move" end. The broadcast cost is usually several times higher than the updating cost. If the "calls per move" increase, the broadcast cost accounts for the larger portion of the total cost due to the higher number of call delivering events. Unless the "calls per move" is small, the forwarding chain is short because the dynamic neighbor zone algorithm does more updates.

In the GSM environment, users' mobility impacts location management costs. At the low "calls per move" (i.e., high mobility) end, the cost of location updating (or setting the forwarding pointers) accounts for the main portion of the total cost for location management. Because one of the key design goals of the network system is to accommodate increasing demands on call delivery capacities, the portion of the updating costs should be kept small. Also, the number of users making network connections in the same period of time is limited. The users in a connectionless state (but still active) account for a certain portion of the total subscribers. Thus, even a small improvement at the low "calls per move" end can lead to significant performance improvements.

### 3.4.2.4 Combining the Limited Broadcast and Forwarding Pointer Schemes

The forwarding pointers scheme performs well in a wide range. However, there is performance degradation at the low "calls per move" end. The pointer chain is longer when no boundary crossing occurs, and there is no call delivery.

Hać and Liu propose an algorithm combining the limited broadcast and forwarding pointers schemes. At the low "calls per move" end, the algorithm switches to the limited broadcast scheme from the forwarding pointers scheme to improve performance. A value $k_m$ for the maximum chain length can be determined from the following equation.

$$\gamma = \beta \times k_m, \qquad (3\text{--}6)$$

where $\gamma$ is the broadcast cost and $\beta \times k_m$ is the accumulated cost of forwarding pointers. To determine from Eq. (3–6) when to switch from forwarding pointers to limited broadcast, an MT keeps a count of the pointers left at the previous BSs for the current chain. The chain is reset whenever a boundary crossing or call delivery occurs. The algorithm is as follows (assume the forwarding pointers scheme is in use):

```
FOR EACH MOVE, DO
  IF CURRENT MOVE CAUSES ZONE BOUNDARY CROSSING
    BEGIN
      INITIATE LOCATION UPDATE;
      SET NEW ZONE CENTER AS NEW POSITION;
      CALCULATE CORRESPONDING BOUNDARY CELLS;
      RESET POINTER COUNT k = 0;
    END
  ELSE
    BEGIN
      DETERMINE k_m FROM EQUATION (3-6);
      IF k > k_m
        INFORM SERVER OF SWITCHING TO BROADCAST;
      ELSE
        INCREMENT k;
    END

FOR EACH CALL RECEIVED, DO
  BEGIN
    SET NEW ZONE CENTER AS NEW POSITION;
    CALCULATE CORRESPONDING BOUNDARY CELLS;
    RESET POINTER COUNT k = 0;
  END
```

When the current scheme is a limited broadcast, we test the instantaneous value of "calls per move" to determine whether we should switch to a forwarding pointers scheme.

$$M < \frac{\gamma}{\beta + 1} \quad (3\text{--}7)$$

where $M$ is the number of moves in the period from last call to the current call, and is equal to the reciprocal of the instantaneous value of "calls per move." Note that pointer count $k$ is used in the forwarding pointers scheme.

If Eq. (3–7) holds, we switch the schemes. Equation (3-7) has the following reasoning: Without taking into account the effect of the zone boundary crossing, we have the following approximations from our model.

$$\text{COST}_{l.b.} \approx \gamma \quad (3\text{--}8)$$

$$\text{COST}_{f.p.} \approx M \times (\beta + 1) \quad (3\text{--}9)$$

Note that Eq. (3–9) is an approximation of the cost from Eqs. (3–7) and (3–8) when switching the schemes. Dependence Eq. (3–7) provides some margin to make it safer for the switching to be taken in the range where the forwarding pointers scheme performs better. This helps to reduce unnecessary switching.

When a limited broadcast is used, an MT keeps count $M$ for moves, and resets $M$ to 0 when a call is delivered. This part of the algorithm is as follows (assume the limited broadcast scheme is in use):

```
FOR EACH MOVE, DO
  BEGIN
    INCREMENT M;
    IF CURRENT MOVE CAUSES ZONE BOUNDARY CROSSING
      BEGIN
        INITIATE LOCATION UPDATE;
        SET  NEW ZONE CENTER AS NEW POSITION;
        CALCULATE CORRESPONDING BOUNDARY CELLS;
        RESET POINTER COUNT k = 0;
      END
  END

FOR EACH CALL RECEIVED, DO
  BEGIN
    IF (3–7) HOLDS
      INFORM SERVER OF SWITCHING TO POINTERS SCHEME AND ITS
      NEW POSITION;
    RESET M = 0;
  END
```

Hać and Liu show a performance comparison of the limited broadcast, forwarding pointers, and the combination scheme. The performance improvement in the combination scheme is significant. The peak cost occurs at the point where the probability of switching from forwarding pointers to the limited broadcast is the highest. This is because the MT informs the server of the switch from the forwarding pointers to limited broadcast.

Transmitting any message costs the MT heavily in terms of energy. However, after the MT informs the server that it is switching from the forwarding pointer to broadcast, it does not need to send a message to the server on every move as long as the broadcast mode is in use. Of course, the location can be updated when switching from the forwarding pointer to broadcast with a negligible extra cost. The algorithm is not symmetric for the two directions of switching. Informing the server of switching from broadcast to the forwarding pointer is done when a call is received and practically no extra cost is incurred (only one more bit during the message exchange).

#### 3.4.2.5 Impact of Zone Size on Performance

There are more boundary crossings for the same set of moves in a small zone than in a large zone. The overall cost of location updating in a small zone is also higher. The cost of searching is expected to be higher in a large zone than in a small zone. In the limited broadcast scheme there are more BSs, to which to broadcast in large zones. In the forwarding pointers scheme, the large zones mean longer pointer chains. The impact of increasing zone size on the searching cost should be much smaller for the forwarding pointers scheme than for the limited broadcast scheme. The number of cells in the zone is proportional to the square of the zone radius while the length of the pointer chain increases more slowly than the zone radius. The net impact is the combination of all these factors.

### 3.4.3 Location Management in the Global Scope

There are two approaches to location management in the global scope. One approach keeps the addressing scheme of the IP suite currently used in the fixed network, which means that each mobile user (for his or her MT) has a private (static) address assigned in a way consistent with the user's local (home) network. Naturally, the HLR keeps constantly updated location information about the MT, at least the location server that serves the area in which the MT currently resides. This information can be retrieved by the HLS, as in the case of dogleg routing. To handle the situations where pure dogleg routing is too expensive, replicas of the location information are distributed network wide.

In particular, Rajagopalan and Badrinath and Yuan distinguish from the rest of the network, the portion that has frequent communications with a certain user. In addition to keeping accurate location information in the HLR for the user, the information is also distributed in this portion of the network (called the "friend network" by Yuan) with a possibly variable degree of accuracy and update frequency. The performance of three updating policies was compared. These policies are "complete update," in which each relocation is recorded in all location servers; "selective update," in which each relocation is recorded only in a "friend network;" and "lazy update," in which other location servers query the HLS when the information is needed.

Rajagopalan and Badrinath studied an on-line adaptive updating algorithm using simulations. For each communicating partner of a mobile user, the algorithm dynamically predicts the cost of updating and the cost of searching without an update, based on the on-line statistics of the user's communication and recorded mobility pattern.

The second approach to location management in global scope does not use a static address for each MT compatible with the current IP addressing scheme. For a particular user, no location server is assigned to be the user's HLS. The distribution of the location information for a mobile user is determined by the user's mobile history and the current location. The model of the hierarchical location server architecture or similar structures is discussed by Awerbuch, Peleg, Masanobu and coworkers, and Badrinath and coworkers, and no dogleg routing is used. The savings on updating location information and delivering calls to roamers is achieved by properly partitioning and selectively updating the information in the hierarchy.

It is beneficial to keep the home address for each MT, relative to the roaming address, which is dynamically assigned by the location server when an MT roams out of its home area and resides in this server's area. This provides a compatibility with the current addressing scheme of the IP suite. Also, with a home address and HLS assigned to mobile users, the updated location information can be stored in the HLSs (with the HLR), which lowers the cost for the hosts communicating with a mobile user within the home area. A number of partners in a user's home area may frequently communicate with the user. In this case, the first approach to location management in the global scope is preferred.

Due to different users' behavior (calling behavior and mobile behavior), and various geographic situations of an MT with respect to its HLS and its communication partners, a simple location management scheme without adaptive features for varying cases cannot provide a good overall performance. Thus, a compound strategy, which incorporates different schemes efficient in different situations yields a higher overall efficiency.

Updating the location information of a roamer locally, within a certain distance from the user's current location, is used in the hierarchical location server architecture. This can be useful for situations where a number of hosts in an area frequently communicate with a roamer residing in the same area but far from his or her home area. The method of locally updating makes the updating cost relatively small and provides high efficiency for these hosts to communicate with the roamer.

The portion of the network that communicates frequently with a mobile user should be distinguished from the rest of the network for the purpose of location information updating. Rajagopalan uses an adaptive updating strategy to distribute the location information of an MT at the source hosts, which frequently initiate calls to the MT. Another possible way is to update the information at the location servers of these hosts, or alternatively, only at some common nodes on the route to the MT's previous location, to further reduce the updating cost. Confining the distribution of the users' location information to a small number of points not only helps to reduce the updating cost, but also helps to preserve the privacy of the mobile users' whereabouts.

#### 3.4.3.1 The Model

The service areas of all location servers provide geographically complete coverage. A location server does not necessarily have a direct link to all location servers with the service areas with which it has common boundaries. On the average, a location server is directly connected to five other servers. Each move of the MT is defined as a boundary crossing between service areas. Different schemes are compared by using the same set of movements.

There are two possible methods to measure the network cost for packets to be sent (or call delivered) from one host to another. We use Cost ($H1$, $H2$) to denote the cost from host $H1$ to host $H2$. One method uses the distance as the cost metric. This is often a good approximation because establishing a connection over a long distance often involves more transmission and processing costs. When using a distance as the cost metric, each MT can store the distances (presumably the shortest paths) between each location server pair. An MT needs the cost information for an adaptive location management scheme. Additional network traffic incurred from obtaining this information should be minimized.

Another method to measure the network cost uses the number of hops traversed by the packets. This metric is more practical. Due to fluctuation of the network traffic, the packets are not necessarily sent along the shortest path because of possible local congestions. Thus, the metric of a hop count is a more accurate measure of the current cost. This information can be obtained by an MT from a packet field similar to the IP's "time to live" (TTL) field. However, the cost information for a particular source host stored in an MT can only be updated when a call is delivered from that source.

To avoid additional network traffic for obtaining the cost information, we use an approximate estimate from the prestored distance parameter in an MT and the previous hop count for the source host during the period when no active connection is maintained. We assume that each MT has reasonably accurate information about the current network cost for the purpose of using adaptive location management schemes. We use distance as the cost metric. For each location server, the shortest-path spanning-tree is calculated with Dijkstra's algorithm to obtain its distance to other servers. We also assume that the distance (cost) from any MT or source host to its current location server is one basic unit.

As in the local-scope example, we use "calls per move" as the parameter for performance evaluation. The performance criterion is "cost per call," where the cost includes both the cost of delivering the call and the possible updating cost.

#### 3.4.3.2 Comparison of Basic Schemes

We consider three basic schemes as follows: dogleg routing; no updating upon the MT's move; basic updating; direct routing with updating at each source host upon the MT's move; and pointers; similar to the forwarding pointer scheme. An MT leaves a pointer to its current location server at the previous server. To deliver a call to the MT, a pointer chain is followed. The start of the chain is reset after a call is delivered.

Hać and Liu discuss the following four situations.

**Case I: The Source Host Is Not Close to the MT's Home Server**
The source host (SH) communicating with an MT is not close to the MT's home server (with respect to the mean distance from the outer servers to the center of the coverage of cellular network).

When the "calls per move" is not small, the pointers scheme performs well. Because Cost (source, home) is not small and Cost (MT, home) is not small either on the average, the dogleg scheme is not cheap in a wide range of "calls per move." Only at the low end of "calls per move," does the cost of pointers scheme and basic updating scheme soar, well above the dogleg line. For basic updating, any cost of updating is wasted if there is no connection over many moves. For the pointers scheme, the performance deteriorates rapidly with the increase of the chain length.

**Case II: The Source Host Is Close to the MT's Home Server**
The SH communicating with an MT is close to the MT's home server (with respect to the mean distance from the outer servers to the center of the coverage of cellular network).

When the "calls per move" is large enough, the performance differences are much smaller than for the last case. In a wide range of "calls per move," the dogleg scheme performs significantly better. The adaptive compound scheme performs better in these situations.

**Case III: An MT Is Moving in Areas Close to the Source but Far from Its Home Server**
We consider a special case where an MT is moving in areas close to the source host but far from its home server. This is an important subset of Case I.

Except for the extremely small "calls per move" end, the basic updating scheme has the best performance. This is consistent with our reasoning for the benefit of "updating locally" at the source hosts close to the MT. We also expect that combining basic updating with the pointers scheme can improve the performance of the pointers scheme because the chain length can be reset to avoid fast deterioration.

**Case IV: An MT Is Moving in Areas Close to Its Home but Far from the Source**
An MT is moving in the areas close to its home server but far from the source host. This is an important subset of Case I.

The dogleg scheme has the best performance. From the MT-home-source triangle, the geometric feature for this case guarantees that the overhead incurred by the dogleg scheme is very small or negligible.

### 3.4.3.3   The Compound Scheme
The pointers scheme should be the base scheme. To avoid fast performance deterioration with the increase of chain length, updating is incorporated into this scheme. The current server has all necessary data of users' profiles. In particular, whenever the following condition is met, the current server is updated at the source host SH of the MT:

$$\text{Cost}(SH, LS_{last}) + \text{Cost}_{chain}(LS_{curr}, LS_{last}) >$$
$$\min(\text{Cost}(SH, LS_{curr}), \text{Cost}(SH, HM)) + \text{Cost}(SH, LS_{curr}) \quad (3\text{–}10)$$

where $LS_{curr}$ and $LS_{last}$ are the current and last known location servers for MT, respectively. The left side of Eq. (3–10) is the cost to deliver a call if the pointers scheme is kept active. The right side of Eq. (3–10) is the cost of delivering the next call plus the cost of updating the host SH. Note that there is only one SH. The update is sent from the home server HM or $LS_{curr}$, whichever costs less based on the calculation of the minimum cost. Because Eq. (3–10) is checked on each MT's move (See the algorithm.), the location server $LS_{curr}$ needs to communicate with HM anyway to update the records in HM. One more bit is needed to indicate who should send the update message to SH. To avoid additional network traffic for obtaining the cost information, we use an approximate estimate from the prestored distance parameter in an MT and the previous hop count for the source host in the period during which no active connection is maintained. We assume that each MT has reasonably accurate information about the current network cost for the purpose of using adaptive location management schemes. The real update cost is much smaller and is discussed in the next section, when a group of source hosts that are close, communicates with the same MT. These source hosts use the same updated source server, one at a time. Nevertheless, Eq. (3–10) provides a criterion for sending updates.

The compound scheme is composed of two modes, the dogleg mode (mode 1) and the pointers-updating mode (mode 2). Because the latter is the base mode, the current mode always starts with mode 2. There are two conditions under which the dogleg scheme is activated: "looking forward" and "looking back." The "looking forward" condition refers to dogleg mode and geographical prediction facility. The "looking back" condition refers to the pointers-updating mode and accumulated data.

Let us consider the "looking forward" condition. When an MT is in Case II or IV as described in the previous section, the dogleg scheme is the choice during the coming several moves. In particular, an MT checks the following inequality after each of its moves and switches to mode 1 if the inequality holds.

$$\text{Cost}(SH, HM) + \text{Cost}(HM, LS_{curr}) < \min(\text{Cost}_{pointer}, \text{Cost}_{updating}) \quad (3\text{–}11)$$

where the left side is the cost for delivering the next call by using the dogleg scheme, while the right side is the cost if the current mode is two.

The "looking back" condition is even simpler. In Case I, the "cost per call" goes up for the pointers scheme and the basic updating scheme at the low "calls per move" end. An MT can record the updating cost since the last call was delivered to the MT from the source host. If the accumulated cost has reached a certain point before the next call delivery, we should activate the dogleg scheme. We adopt the following condition for the mode switching.

$$\text{Cost}_{a.u.} + \text{Cost}_{updating} > \text{Cost}_{dogleg} \quad (3\text{–}12)$$

where $\text{Cost}_{a.u.}$ is the accumulated updating cost.

# Location Management Schemes for Mobile Communications

We also consider the condition for switching from mode 1 back to mode 2. We distinguish the situations where the previous switching is due to the condition shown in Eq. (3–11) or due to the condition in Eq. (3–12). The MT can record the type of previous switching (from mode 2 to mode 1) in "switching-type" as either "looking-forward" or "looking-back." When a call is delivered to the MT, it switches back to the pointers mode if the "switching-type" is "looking-back." Otherwise, the MT checks for inequality as in Eq. (3–13), and if it holds, the MT makes the switch.

$$\text{Cost}(MT, HM) + \text{Cost}(HM, SH) > \delta \times \text{Cost}(MT, SH) \qquad (3\text{–}13)$$

where $\delta$ is a constant. Equation (3–13) gives the upper bound for the overheads allowed when using the dogleg scheme. In Hać and Liu, a reasonable value for $\delta$ is between 1.3 and 1.6. With a smaller $\delta$, there are more unnecessary switchings which add cost. With a larger $\delta$, there is more chance for the dogleg to be unnecessarily active.

The compound adaptive algorithm is as follows:

```
(CURRENT MODE IS SET TO THE POINTERS MODE 2 AT THE START)
FOR EACH MOVE OF THE MT, DO
 BEGIN
  IF MODE = MODE 2
   BEGIN
    IF INEQUALITY (EQ. 3–11) HOLDS
     BEGIN
      MODE := MODE 1;
      SWITCHING_TYPE := "LOOKING FORWARD;"
      INFORM  SOURCE HOST OF MODE SWITCHING AND
      UPDATE RECORD OF CURRENT SERVER OF MT;
     END
    IF INEQUALITY (EQ. 3–12) HOLDS
     BEGIN
      MODE := MODE 1;
      SWITCHING_TYPE := "LOOKING BACK;"
      INFORM  SOURCE HOST OF MODE SWITCHING AND
      UPDATE RECORD OF CURRENT SERVER;
     END
   END /* MODE WAS MODE2 { UPDATING POINTERS } */
  IF MODE = MODE2
    IF INEQUALITY (EQ. 3–10) HOLDS
     BEGIN
      UPDATE RECORD OF CURRENT SERVER AT THE SOURCE HOST;
      INCREASE ACCUMULATED UPDATING COST BY THE NEW COST;
      RESET POINTER CHAIN;
```

```
    END
  ELSE
    LEAVE THE NEW POINTER AT
    THE LAST SERVER; /* DO NOTHING */
END

FOR EACH CALL DELIVERED TO THE MT FROM THE SOURCE HOST, DO
  IF MODE = MODE 1
    IF SWITCHING_TYPE := "LOOKING BACK;"
      MODE := MODE 2;
    ELSE
     BEGIN
       IF (EQ. 3–13) HOLDS
         MODE := MODE 2;
     END
  ELSE
   BEGIN
    RESET POINTER CHAIN;
    RESET ACCUMULATED UPDATING COST TO ZERO;
   END
```

According to Hać and Liu, except for the very low "calls per move" end, the compound adaptive scheme has the optimal performance in various situations. At this end, the cost is reduced to less than one-third over the dogleg line, instead of going up. This improvement is significant; the extra mode switching in this range (and the additional cost incurred) cannot be avoided since the next call cannot be predicted precisely.

### 3.4.3.4 The Environment of the "Friend Network"

We use the concept of the "friend network," mainly composed of the set of source hosts for the MT; it can also include the servers frequently involved in related location management.

When a set of source hosts communicates with the MT, a subset of them receives updates upon each move. Ideally if the minimum spanning-tree starting at the current server of the MT can be computed, the updating cost for the source host SH is the cost from its parent in the tree to the SH. More generally, the updating cost is defined as the cost from the server (between the MT and the SH) that relays the updating information. The larger number of source hosts often results in lower cost per source host, while increasing the total network load. The problem is how to further reduce the total network load.

A group of source hosts resides in the areas of several location servers close together. The entire set of source hosts is geographically distributed in partitions, instead of being scattered sparsely. We introduce the concept of "image home" for such a group of source hosts when they are far from the home server of the MT. If we choose the server serving the SH in the group that communicates with the MT most frequently as an "image home" server, then the other SHs can view it as the real home server and their costs can be reduced dramatically. The original "calls per move" through the "image home" should be high enough to avoid extra cost for constant updating. However, this condition can be relaxed. As long as the cost for additional updating at the "image home" is relatively small with respect to the savings in the source hosts, the method of "image home" performs well.

# CHAPTER 4

# Routing Algorithms

## 4.1 Finding Minimum Cost Multicast Trees with Bounded Path Delay

In multimedia communication, all video, audio, and data messages are concurrently sent to multiple users, all members of the same multicasting group. Such communication is called multicast communication, which simultaneously transmits the same messages to multiple destinations in the network. Multicasting is a very important facility in networks supporting multimedia applications, especially because video and audio applications can use such a multicasting feature effectively.

To support multicast communication efficiently, one of the key issues that needs to be addressed is the issue of routing, which primarily refers to the determination of a set of paths to be used for carrying the messages from a source node to all destination nodes. It is important that the routes used for such communications consume a minimal amount of resources. To use network resources as little as possible while meeting the network service requirements, the most popular solution involves the generation of a multicast tree spanning the source and destination nodes. There are two reasons for basing efficient multicast routes on trees:

1. The information can be transmitted in parallel to the various destinations along the branches of the tree.
2. A minimum number of copies of the information is transmitted, with the duplication of information being necessary only at the forks in the tree.

Multicast routing algorithms for constructing multicast trees have been developed with two optimization goals in mind. These two different optimization goals can be used in the multicast routing algorithm to determine what constitutes a good tree. Two measures of the tree quality are tree delay and tree cost, which are defined as follows:

1. The first measure of efficiency is in terms of the cost of the multicast tree, which is the sum of the costs on the edges in the multicast tree.
2. The second measure is the minimum average path delay, which is the average of minimum path delays from the source to each of the destinations in the multicast group.

Optimization objectives are to minimize the cost and delay. However, the two measures are individually insufficient to characterize a good multicast routing tree for interactive multimedia communication. For example, when the optimization objective is only to minimize the total cost of the tree, we have a minimum cost tree. Although the total cost as a measure of bandwidth efficiency is certainly an important parameter, it is not sufficient to characterize the quality of the tree as perceived by interactive multimedia and real-time applications, because networks supporting real-time traffic need to provide certain QoS guarantees in terms of the end-to-end delay along the individual paths from source to each destination node. Therefore, both the cost and delay optimization goal are important for the multicast routing tree construction. The performance of such a multicast route is determined by two factors:

1. Bounded delay along the individual paths from the source to each destination.
2. Minimum cost of multicast tree (i.e., network bandwidth utilization).

The goal of the multicast routing algorithm is to construct a delay constrained minimum-cost tree. In order to provide a certain quality of service to guarantee end-to-end delay along the individual paths from the source to each destination node, the algorithm sets the delay constraint on the individual path, rather than trying to minimize the average path delay to all destinations. The two measures of the tree quality, the tree edge delay and tree edge cost, can be described by different functions. For example, edge cost can be a measure of the amount of buffer space or channel bandwidth, and edge delay can be a combination of propagation, transmission, and queuing delay.

In real communication networks, we consider the problem of generating such a multicast tree to provide a minimum delay along the tree, which is important for multimedia applications such as real-time conferencing. During a teleconference, it is important that the current speaker be heard by all participants at the same time, or else the communication may lack the feeling of an interactive face to face discussion. Another application is the use of multicast messages to update multiple copies of a replicated data item in a distributed database system. Minimizing the delay variation in this case will minimize the length of time during which the database is in an inconsistent flow state.

Cost optimization means making use of network resources as efficiently as possible. Applications benefiting from cost optimization include news distribution, stock quotes distribution, video-on-demand service, distribution of high bandwidth imagery, etc.

The problem of computing multicast trees has been well studied, and several algorithms have been proposed based on a number of optimization goals. Dijkstra's shortest path algorithm can be used to generate the shortest paths from the source to the destination nodes; this provides the optimal solution for delay optimization. Multicast routing algorithms that perform cost optimization have been based on computing the minimum Steiner tree, which is known to be an NP-complete problem. Some heuristics for the Steiner tree problem have been developed that produce near optimum results. In the KMB (Kou, Markowsky, Berman) algorithm, a network is abstracted to a complete graph consisting of edges that represent the shortest paths among the source node and destination nodes. The KMB algorithm constructs a minimum-spanning tree in the complete graph, and the Steiner tree of the original network is obtained by achieving the shortest paths represented by edges in the minimum-spanning tree. The KMB provides a near optimal solution for cost optimization. However, the KMB algorithm does not produce the optimal result for delay optimization. Other heuristic algorithms have been developed to discuss optimization for both cost and delay. One of these heuristic algorithms such as the KPP (Kompella, Pasquale, Polyzos) algorithm presented a source-based algorithm which can construct a delay-bounded minimum Steiner tree (DMST). Regarding the dynamic multicast tree problem, Waxman examined the dynamic update of the tree if destination nodes join or leave the multicast tree occasionally.

Hać and Zhou present a heuristic algorithm called dynamic delay bounded multicasting algorithm (DDBMA) for constructing minimum-cost multicast trees with delay constraints. The algorithm sets variable delay bounds on destinations and can be used to handle the network cost optimization goal; minimizing the total cost (total bandwidth utilization) of the tree. The algorithm can also be used to handle a dynamic DMST, which is accomplished by updating the existing multicast tree when destinations need to be added or deleted.

During the network connection establishment, DDBMA can be used to construct a feasible tree for a given destination set. For certain applications, however, nodes in the network may join or leave the initial multicast group during the lifetime of the multicast connection. Examples of these applications such as teleconferencing, mobile communication, etc., allow each user in the network to join or leave the connection at any time without disrupting network services to other users.

DDBMA is based on a feasible search optimization method which starts with the minimum-delay tree and monotonically decreases the cost by iterative improvement of the delay-bounded tree. Then, the algorithm starts to update the existing tree when nodes in the network request to join or leave. The algorithm will stay steady when there is no leaving or joining requests from the nodes in the network.

In the multicast routing problem, the network can be viewed as a weighted graph $G(V, E)$, with node set $V$ and edge set $E$. Edges in set $E$ correspond to the communication links connecting various nodes. Nodes in set $V$ can be of the following three types:

1. *Source node*: The node connected to a source that sends out data streams, denoted by $s$.
2. *Destination node*: The node connected to a destination that receives data streams, denoted by $D \subseteq V - s$.

3. *Relay node*: The node comprising an intermediate hop in the path from the source to destination.

Positive real-valued functions are defined on edge set $E$ as follows:

1. *Link Cost Function*: The cost of a link is associated with the utilization of the link. Link cost indicates bandwidth utilization of channels or the amount of buffer space. A higher utilization is represented by a higher link cost.
2. *Link Delay Function*: The delay of a link is the sum of the received queuing delay, transmission delay, and propagation delay over the link. For each path from source $s$ to destination $D \in V$, the delay of the path is defined as the sum of link delays along the path.
3. *Delay Bound Function (DBF)*: The DBF assigns an upper bound to the delay along the path from source to destination in $D$. The delay bound for each different destination node can be different since each communication link in the network can have different delay constraints as long as the service requirement is satisfied.

The optimization problem for constructing the multicast tree which is a DMST can be described as follows:

> Given a graph $G = (V, E)$ with a source node $s$, a set of destinations $D$, a link cost function, a link delay function, and DBF, construct a DMST spanning $D \cup s$, such that the cost function of the tree is minimized while the DBF is satisfied. The DMST problem based on minimizing the sum of link costs in tree T can be reduced to a minimum Steiner tree problem with the delay bound set to $\infty$, which is known to be an NP-complete problem, and only heuristics are of practical interest for its solution. DDBMA is heuristic and designed to solve the DMST problem.

The KMB Algorithm is a fast heuristic algorithm that is used to solve the Steiner tree problem. Since the algorithm produces a near optimum result, it has been frequently used for comparison with other proposed heuristic algorithms.

The KPP algorithm is a heuristics algorithm for constructing a minimum cost multicast tree with delay constraints. The KPP algorithm extends the KMB algorithm by taking into account delay bound in the construction of the shortest paths.

The delay-bound shortest path during the construction of a delay-bounded multicast tree can be found by using the $M$th shortest path algorithm. The algorithm determines the $M$th best route from node $s$ (the source) to node $d$ (the destination) when $(M - 1)$th shortest path found violates the delay constraint.

### 4.1.1 DDBMA

The DDBMA assumes that adequate global information is available to the source (i.e., the source node of the network has complete information regarding all network links to construct a multicast tree). DDBMA is based on the feasible search optimization method. It starts with an initial tree $T_0$, a minimum-delay Steiner tree constructed by Dijkstra's shortest-path algorithm, and iteratively refines the tree for low cost while meeting the delay-bound requirement.

DDBMA consists of four major steps:

Step 1: Construct a minimum delay Steiner tree ($T_0$) using Dijkstra's shortest-path algorithm with respect to the multicast source.
Step 2: Iteratively refine the initial tree ($T_0$) for lower cost.
Step 3: Check for a destination node join or leave requests.
Step 4: Update the existing multicast tree if there are any join or leave requests.

The algorithm terminates when there are no join or leave requests.

In some cases, delay bounds given by the DBF may be too tight (i.e., the bounds cannot be met even in the initial tree $T_0$). In that case, some adjustment is required to loosen the delay bound before any feasible tree can be constructed. To guarantee that the output tree is delay bounded, Step 1 has to reassign delay bounds after negotiation has been accomplished. We assume that DBF gives the delay bounds that have been met by the DMST, $T_0$.

Step 2 iteratively refines $T_0$ for lower cost. A tree obtained during the $i$th refinement is the $i$th tree configuration, as denoted by $T_i$. The algorithm goes through Step 3 and Step 4 checking for the dynamic tree status and updating the existing minimum-delay Steiner tree if new destination nodes request to join, or destination nodes in a multicast tree group request to leave. Figure 4–1 shows a network example and its corresponding minimum delay Steiner tree for a multicast group in the network.

An important part of DDBMA is the refinement from tree $T_i$ to tree $T_{i+1}$ (initially, $i = 0$), which is done by an operation called path replacing. Path replacing in DDBMA means that a path in tree $T_i$ is replaced by another new path which is not in tree $T_i$, resulting in a new tree configuration $T_{i+1}$. The delay-bounded path replacing operation guarantees that tree $T_{i+1}$ is always a delay-bounded tree.

An example is shown in Figure 4–2 where the goal is to obtain a DMST satisfying the delay bound (DB) (DB = 5); accordingly, a path AB (Figure 4–2(A)) of the tree is replaced by a new path BC. A new tree configuration resulting from the above path replacing is shown in Figure 4–2(B). Comparing the tree in Figure 4–2(A) with new tree in Figure 4–2(B) after the path replacing, the cost of tree in Figure 4–2(B) is 23 while the cost of tree in Figure 4–2(A) is 31. This means that we have reduced the cost of the tree from 31 to 23 by path replacing (path AB replaced by path BC), while the maximal destination path delay becomes 4 (Figure 4–2(B)), which is still bounded by the delay bound (DB = 5).

To ensure an effective delay-bounded path replacing procedure during the $i$th tree refinement, using DDBMA involves: choosing the path to be taken out of tree $T_i$, and selecting the new path in G but not in tree $T_i$ which replaces the path to be deleted from tree $T_i$.

To present the candidate paths chosen in the path replacing, we use the superpath. The superpath ($\phi$) in tree $T_i$ is defined as the longest path in tree $T_i$ in which all internal nodes, except

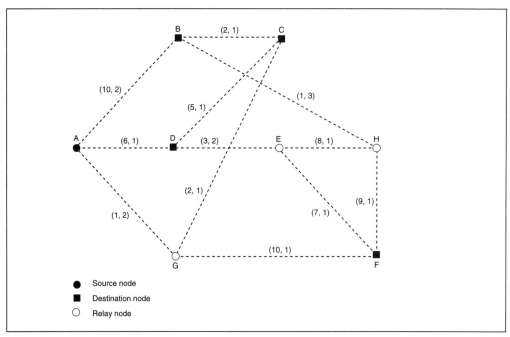

**Figure 4–1 (A)** Example Network and its Link Parameter (cost, delay)

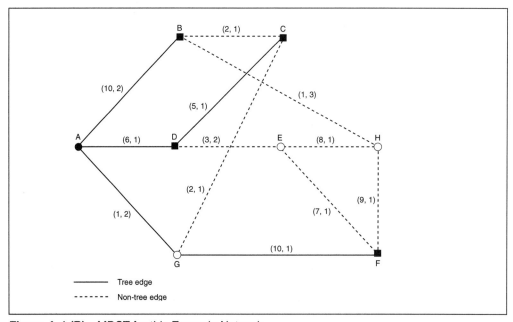

**Figure 4–1 (B)** MDST for this Example Network

# Finding Minimum Cost Multicast Trees with Bounded Path Delay

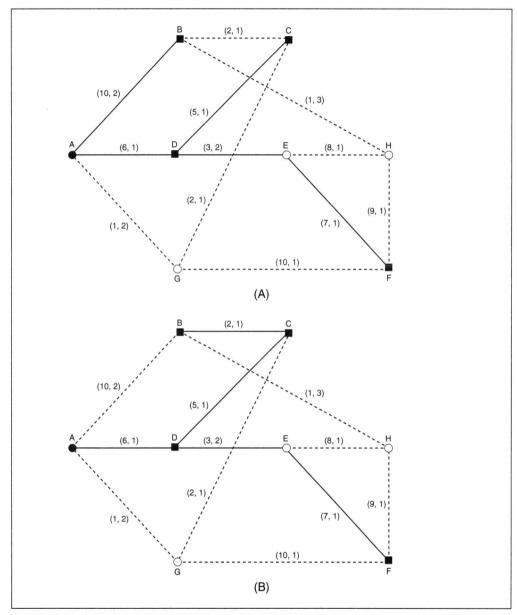

**Figure 4–2** Example of Path Replacing: Path AB in (A) is Replaced by Path BC in (B)

for the two end nodes of the path, are relay nodes. Each relay node connects exactly two tree edges. As shown in Figure 4–2, path DEF is one of these superpaths. Relay node E is connected by two tree edges DE and EF. In the algorithm, every superpath is a candidate path for path replacing.

To reduce the cost of $T_i$, DDBMA deletes a Superpath ($\phi$) in $T_i$, resulting in two subtrees $T_{i1}$ and $T_{i2}$, where $T_i = T_{i1} \cup T_{i2} \cup \phi$. The DDBMA then finds a delay-bounded shortest path ($\psi$) to reconnect $T_{i1}$ and $T_{i2}$. The new tree will be $T_{i+1} = T_{i1} \cup T_{i2} \cup \psi$. In the algorithm, the cost of every superpath is defined as the sum of the cost along the corresponding path in tree $T_i$.

The delay-bounded shortest path ($\psi$) between $T_{i1}$ and $T_{i2}$ is defined as the path with the smallest cost, subject to the constraint that the new tree $T_{i+1} = T_{i1} \cup T_{i2} \cup \psi$ is also a delay-bounded tree. To get the delay-bounded shortest path ($\psi$), the DDBMA tries to construct the first shortest path, using Dijkstra's shortest-path algorithm, between two subtrees $T_{i1}$ and $T_{i2}$. Since the first shortest path found may not satisfy the DB, second shortest path, third shortest path, and even $M$th shortest path have to be computed in order to find a feasible delay-bounded shortest path ($\psi$) for replacing the superpath ($\phi$) during $i$th tree refinement. If the first shortest path found satisfies the DB, the delay-bounded shortest path ($\psi$) equals the first shortest path. Otherwise, the DDBMA incrementally constructs the $M$th shortest path between two subtrees $T_{i1}$ and $T_{i2}$ until a feasible shortest path is found, and the delay-bounded shortest path ($\psi$) equals the $M$th shortest path. Using the $M$th shortest path algorithm, the construction of the $M$th shortest paths starts from the first shortest path that we have already found, then the second shortest path, then the third one and so on. Parameter $M$ is an unknown a priori, and is determined only after a delay-bounded shortest path ($\psi$) is found, which has resulted in a delay-bounded tree $T_{i+1}$. The $M$th shortest path algorithm incremental construction will stop when one of the following two conditions is satisfied: (1) The shortest path found does not result in the new tree violating the DB. (2) The shortest path found has equal path cost to the path just deleted.

To construct the first shortest path between two subtrees $T_{i1}$ and $T_{i2}$, we consecutively apply Dijkstra's shortest-path algorithm. The key step is extending Dijkstra's shortest-path algorithm to construct the first shortest path between two subtrees instead of two nodes. To do that, all possible paths between subtree $T_{i1}$ and $T_{i2}$ have to be computed, using Dijkstra's shortest-path algorithm, and the path with the smallest cost will be selected as the first shortest path.

Figure 4–3 is an example showing how to generate two subtrees $T_{i1}$ and $T_{i2}$ from tree $T_i$ and how to reconnect these two subtrees by the delay-bounded shortest path ($\psi$). In Figure 4–3(A), removing a superpath AB in $T_i$ results in two subtrees $T_{i1}$ and $T_{i2}$, where $T_i = T_{i1} \cup T_{i2} \cup AB$ with tree $T_{i1}$ containing only node B while tree $T_{i2}$ includes the node set (A, C, D, E, F) and corresponding edges (AD, CD, DE, EF). In Figure 4–3(B), a delay-bounded shortest path BC is found and used to reconnect two subtrees $T_{i1}$ and $T_{i2}$; this results in a new tree $T_{i+1}$, where $T_{i+1} = T_{i1} \cup T_{i2} \cup BC$. Adding path BC makes the delay bounds satisfied in the new tree $T_{i+1}$ as shown in Figure 4–3(B), where the maximum path delay for the new tree $T_{i+1}$ is 4 which satisfies DB = 5. The cost reduction from replacing path AB by path BC is 10 – 2 = 8.

The example in Figure 4–3 shows that the path replacing operation can reduce the tree cost and guarantee tree $T_{i+1}$ to be a delay-bounded tree. The path replacing operation is iteratively used to refine tree $T_i$ to tree $T_{i+1}$ until such operations can no longer reduce the total tree cost, which means that the final DMST is found.

Finding Minimum Cost Multicast Trees with Bounded Path Delay 153

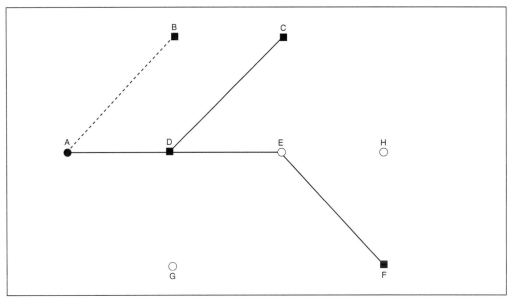

**Figure 4–3 (A)** Removing One of the Superpaths AB Results in Two Subtrees $T_{i1}$ and $T_{i2}$

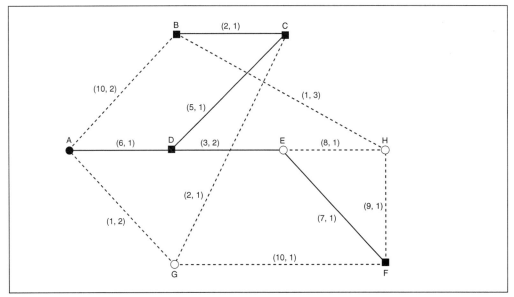

**Figure 4–3 (B)** The Shortest Path $\psi$ = BC Reconnects $T_{i1}$ and $T_{i2}$

Initially, the DDBMA constructs a minimum-delay Steiner tree ($T_0$) using Dijkstra's shortest-path algorithm. Tree ($T_0$) is an initial tree for path replacing and all superpaths in $T_0$ are initially unmarked. We define a set $\Phi$ in which are all unmarked superpaths. Among all unmarked superpaths in set $\Phi$, the superpath ($\phi$) with the highest path cost is selected and replaced by a delay-bounded shortest path ($\psi$) during $i$th tree refinement.

Instead of using the lexicography tie-breaking rule for multiple paths, tie-break choice rules are added for multiple same cost superpaths and multiple same cost delay-bounded shortest paths during the $i$th tree refinement. The superpath ($\phi$) chosen is the path with larger delay. The delay-bounded shortest path ($\psi$) chosen is the path with lower delay, so that the resulting tree will have a lesser total path delay.

Such a path replacing procedure guarantees that the resulting paths from source to destinations are all delay bounded. Set $\Phi$ is changed after each path replacing since the delay-bounded shortest path ($\psi$), which replaces the highest path cost superpath ($\phi$) will become another unmarked superpath available for the next path replacing. One of two cases must occur during the path-replacing procedure: (1) the delay-bounded shortest path ($\psi$) = superpath ($\phi$), or (2) the delay-bounded shortest path ($\psi$) $\neq$ superpath ($\phi$).

If $\psi = \phi$, the DDBMA marks $\phi$, which means the superpath ($\phi$) was examined, and continues to select another superpath with the highest path cost among the remaining unmarked superpaths available for path replacing in set $\Phi$. If $\psi \neq \phi$, the DDBMA unmarks all marked superpaths, and continues to select a superpath in set $\Phi$ with the highest path cost for path replacing. The DDBMA path replacing will terminate when all superpaths in set $\Phi$ are marked in the tree. That means all possible superpaths in the tree were examined for path replacing without success in reducing the total tree cost.

The purpose of dynamically reorganizing the existing delay-bounded minimum multicast tree ($T_m$) is to respond to changes in the destination set so that each individual destination node is allowed to join or leave the network during the life of connection. We assume that a node currently in the multicast tree may leave the tree after issuing a leave request. Similarly, a node that wishes to join this multicast tree will issue a join request. Under this scenario, it is necessary to dynamically update the existing multicast tree once a join or leave request is received.

To handle the dynamic problem, the DDBMA algorithm can be used to create a new multicast tree when there is a join or leave request. There is a certain overhead associated with this approach, including the computational cost of running the DDBMA, and the cost of the network resources involved in the transition from an existing multicast tree to the new tree. Because the new tree can be significantly different than the existing tree, the overheads can be high. Furthermore, such an approach may cause receivers totally unrelated to the destination nodes to be added or deleted, and to experience disruption in service. All these drawbacks make this approach inappropriate for real-time environments and applications where frequent changes in the destination set are anticipated.

Another approach attempts to minimize the cost incurred during the transition period as well as the disruption caused to the receivers. In this approach, the existing multicast tree will never be modified unless it is necessary to do so. Even then, the new tree is not computed from the beginning; instead, a feasible, delay-bounded minimum multicast tree is constructed by making incremental and localized changes to the existing tree. This approach is described as follows.

Let us first discuss the leave request, and assume that node $u$ in the tree issues a leave request to end its participation in the multicast session. If node $u$ is not an end destination node in the existing tree ($T_m$), no action needs to be taken. The new tree, $T_m'$, will be the same as $T_m$ (i.e., $T_m' = T_m$). The only difference is that node $u$ will stop forwarding the multicast packets to its local user. If $u$ is an end destination node of $T_m$, then to avoid wasting bandwidth, tree $T_m$ has to be pruned to exclude node $u$ and related relay nodes and links used in tree $T_m$, solely for forwarding packets to $u$.

There are two situations when a new node $v$ intends to join tree $T_m$: (1) The node $v$ that wants to join the tree is not the node in tree $T_m$. (2) The node $v$ that wants to join the tree is a relay node of tree $T_m$.

In case 1, a feasible path is expected to connect a new node $v$ and tree $T_m$ without violating the delay constraints. To do that, we first try to find the first shortest path, $P_1$, from $v$ to tree $T_m$ using Dijkstra's shortest-path algorithm. If the first shortest path ($P_1$) found satisfies the delay bound, the new tree $T_m' = T_m \cup P_1$. Otherwise, the $M$th shortest path algorithm will be used to find delay-bounded $M$th shortest path, $P_m$, between node $v$ and tree $T_m$ making new tree $T_m' = T_m \cup P_m$. Hence, the transition from the existing multicast tree $T_m$ to the new tree $T_m'$ involves only the establishment of a new path and does not affect any of the paths from the source-to-destination nodes already in the multicast tree $T_m$.

If case 2 occurs, supposing node $v$ belongs to tree $T_m$ (i.e., node $v$ is a relay node of $T_m$), then tree $T_m$ is a feasible tree (i.e., $T_m' = T_m$), and it can be used without any change other than having node $v$ forward the multicast packets to its user, in addition to forwarding them to downstream nodes.

Figure 4–4 shows a construction of a DMST using DDBMA. The construction includes six steps, and starts from the initial minimum-delay tree $T_0$ and stops at tree $T_5$ when the DDBMA terminates. From the initial tree $T_0$ to tree $T_5$, the total path cost of each tree consecutively decreases; accordingly, $COST(T_0) = 32$, $COST(T_1) = 31$, $COST(T_2) = 23$, $COST(T_3) = 21$, $COST(T_4) = 21$, and $COST(T_5) = 20$. Comparing this with the initial tree $T_0$ and tree $T_5$, the total path cost of the tree is reduced from 32 (the initial minimum-delay tree) to 20, and the maximal path delay of $T_5$ is 4 < DB ( = 5) that satisfies the DB. Tree $T_5$ is the final tree with the lowest total path cost, which is the DMST.

The minimum-cost tree constructed by the KMB algorithm is shown in Figure 4–5(A), the DMST constructed by the KPP algorithm is shown in Figure 4–5(B), and the DMST constructed by DDBMA is shown in Figure 4–5(C). Since the KPP algorithm and DDBMA are delay-constrained multicast routing algorithms, delay bounds are required for multicast tree construction, and we assume the DB = 5 in this example. The results of the total path cost and delay of trees constructed by three different algorithms are as follows:

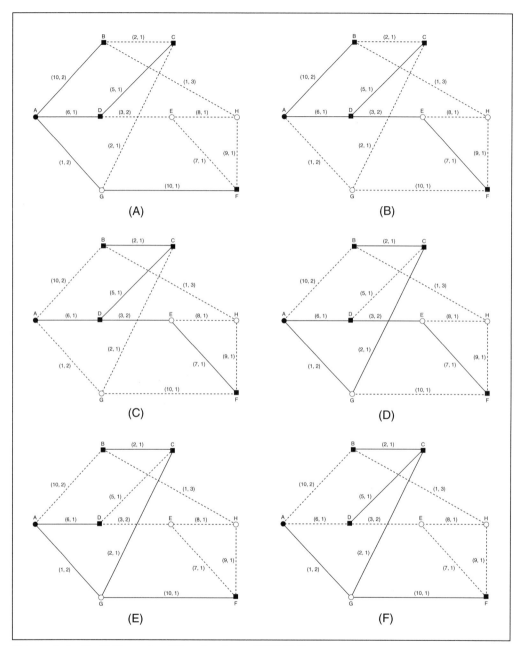

**Figure 4–4** (A–F) Example of Constructing a DDBMA Tree

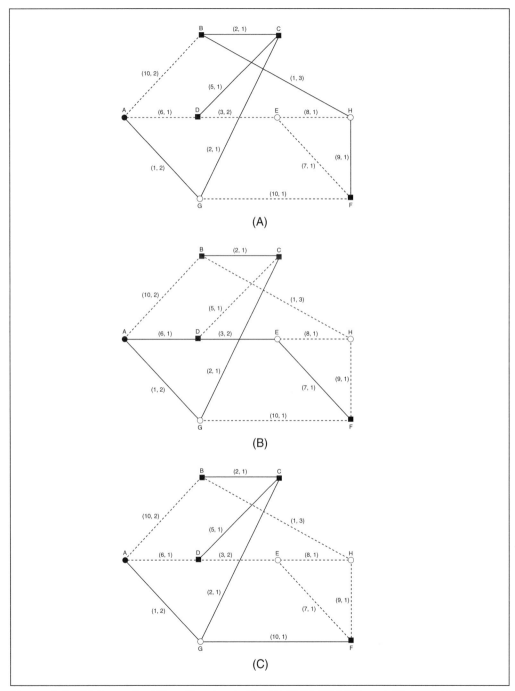

**Figure 4–5** Minimum Cost Tree Constructed by (A) KMB, (B) KPP, and (C) DDBMA

|  | KMB | KPP | DDBMA |
|---|---|---|---|
| Total Path Cost | 20 | 21 | 20 |
| Total Path Delay | 9 | 8 | 6 |
| Max. Path Delay | 8 | 4 | 4 |
| Delay Bounds | 5 | 5 | 5 |

The tree constructed by the KMB algorithm has the lowest, total path cost of 20, the largest, total path delay of 9, and the maximal path delay of 8, which is larger than the DB (DB = 5). The KMB tree is a minimum-cost tree, but it does not satisfy the delay constraint. Since the KPP algorithm and DDBMA are delay-bounded multicasting algorithms, the trees constructed by those algorithms are all delay-bounded trees. Both the KPP tree and DDBMA tree have the same maximal path delay of 4 that satisfy the DB (DB = 5). The KPP tree has a total path cost of 21 (slightly larger than of the KMB's), and the total path cost for the DDBMA tree is 20 (the same as that of the KMB's). The KPP tree has a total path delay of 8, and the total path delay for the DDBMA tree is only 6. DDBMA has much better performance than both the KPP and KMB algorithms.

According to Cayley's theorem, for an $n$ node network, there are $n^{n-2}$ possible spanning trees on $n$ nodes. Thus, the number of Steiner trees spanning is bounded by $n^{n-2}$. Let us consider a Markov chain of $n^{n-2}$ states, where each state corresponds to a spanning tree. We sort these states in a nonincreasing order from left to right with respect to the cost of the tree. We replace each state in the sorted list with $n$ copies of itself to get a total of $n^{n-1}$ states.

In this Markov chain, the transition from state $S_i$ goes only to a state to the right of $S_i$. Assume that each possible transition is equally likely. Thus the probability of a transition from $S_i$ to $S_j$ is

$$P_{ij} = \frac{1}{i-1} \quad (1 < j < i)$$

where $P_{11} = 1$.

Let $T_i$ be the number of transitions needed from state $i$ to state 1 and $Y$ be the random variable of the next state of the first transition. The expected value ($E[T_i]$) is

$$E[T_i] = E[E[T_i/Y]]$$

$$= \sum_y E[T_i/Y = y]P\{Y = y\}$$

$$= \sum_y E[T_i/Y = y]\frac{1}{i-1}$$

$$= \frac{1}{i-1}\sum_{y=1}^{i-1}(1 + E[T_y])$$

$$= 1 + \frac{1}{i-1} \sum_{y=1}^{i-1} E[T_y]$$

Since $E[T_1] = 0$, we have $E[T_i] = \sum_{y=1}^{i-1} \frac{1}{y} \approx \log(i)$.

Therefore, if the ideal algorithm starts in the most expensive state (i.e., $n^{n-1}$), then the expected number of transitions (i.e., iterations), is $O(log(n^{n-1})) = O(nlog(n))$. This is for an ideal algorithm to find the minimum-cost tree.

In the DDBMA algorithm, path replacing procedures correspond to a transition from state $S_i$ to $S_1$. Sometimes the DDBMA is likely to terminate earlier in finding the minimum (i.e., when the path replacing procedure fails consecutively for every superpath). Thus, the maximum expected number of path replacements for the DDBMA is $O(nlog(n))$. Path replacing primarily involves computing the $M$th shortest path. This can be done in $O(MS(n))$, where $S(n)$ is the time complexity of doing a single, shortest-path algorithm. Dijkstra's shortest-path algorithm is used in the DDBMA, and implementation of Dijkstra's shortest-path algorithm takes $O(n^2)$. The expected time complexity of DDBMA is

$$O(MS(n) \times nlog(n)) = O(Mn^2 \times nlog(n)) = O(Mn^3 log(n)).$$

The value of $M$ for the delay-bound shortest path construction depends on the delay bound. The value $M$ in the $M$th shortest path algorithm indicates that the $M$th shortest path has to be computed if the $(M-1)$th shortest path found violates the delay constraint. So a tight delay bound can result in more computation for the $M$th shortest path, and a relaxation of delay bound can result in less computation.

In comparison, the KMB algorithm has the worst case time complexity of $O(Dn^2)$, where $D$ is the number of destination nodes and $n$ is the number of nodes in the network. The analysis of the KPP algorithm shows that the majority of the time is spent on computing the lowest cost delay constrained path between all nodes. The implementation of the KPP algorithm takes $O(\Delta n^3)$, where $\Delta$ is the delay bound and $n$ is the number of nodes in the network. It is obvious that $O(Mn^3 log(n)) > O(Dn^2)$ is always true since $n > D$. So DDBMA takes longer to compute than the KMB algorithm. It is not clear which algorithm takes longer to compute between the DDBMA and KPP algorithms. Depending on the value of $M$ and $n$ in DDBMA, and value of $\Delta$ in the KPP algorithm, either algorithm can have a lower computation time.

## 4.2 Multicast and Other Routing Algorithms

Multimedia multiparty communication services are supported by networks having the capability to set up and/or modify the following five basic types of connections: point-to-point, point-to-multipoint (also called multicast), multipoint-to-point (also called concast), multipoint-to-multipoint, and point-to-allpoint (also called broadcast).

Many types of communication require transmission of certain information from the source to a selected set of destinations. This could be the cast of a multipoint video conference, the distribution of a document to a selected number of persons via a computer network, or the request for certain information from a distributed database.

When we consider the multicast routing method from the viewpoint of efficient resource utilization, a packet should be transmitted through the tree-shaped path, which is rooted in the source and all branches terminate at destinations. The multicast packets are copied in the branching nodes.

When using the number of hops in the network to decide on message routing, link-state routing supports the shortest-path multicast routing. Routers are considered as part of the state of a link, so that we know the set of groups that have members on that link. Whenever a new group appears or an old group disappears on a link, the routers attached to that link flood the new state to all other routers. Given full knowledge of what groups have members on what links, any router can compute the shortest-path multicast tree from any source to any group of destinations by using Dijkstra's algorithm. If the router performing the computation falls within the computed tree, it can determine which links it must use to forward copies of the multicast packets from the source to the destinations.

Another conventional multicast routing algorithm is based on finding the Steiner tree for the participating routers in a network when considering the total minimum costs as the criterion. Since the Steiner tree problem has been proven to be NP-complete, only heuristic algorithms are of practical interest. The heuristic algorithms are based on minimum spanning tree (MST) algorithms. Kou, Markowsky, and Berman take the subnet generated by applying the basic MST algorithm, and then prune those branches which do not contain any vertexes from destination set, providing their solution to the Steiner tree problem. These algorithms guarantee to output a Steiner tree whose cost (with respect to distance) is no more than $2 \times (1 - 1/L)$ times that of the optimal tree, where $L$ is the number of leaves in the optimal tree.

A number of algorithms that can be used in the LANs with the multicast destinations is presented by Deering and Cheriton. The single-spanning tree multicast routing is shown on the example of a bridged LAN. The distance-vector multicast routing uses a distance vector from the sender routing table. Some extensions to the reverse-path forwarding algorithm are proposed by using distance vector routing. The source-routing algorithm and its properties are described by Hagouel. These algorithms use the number of hops in the network to decide on message routing.

Yum and Chen present a multicast source routing scheme in packet-switched networks. A reverse path address code is designed that allows an individual destination node to retrieve the reverse path address without searching the topology database and invoking any route computation program.

The kernel implementation for high-performance multicast communication is described by Gait. The communication is allowed among shared memory multiprocessors using the Ethernet. The process communication is asynchronous and the overhead of the message management is masked by parallel computation. This implementation can be used in LANs.

A distributed algorithm for resource management in a hierarchical network is introduced by Hać. The algorithm uses dynamic information about the system state in LANs and WANs. The algorithm uses broadcast information to make decisions about process transfer and message routing, considering processor availability and system security. This algorithm can be implemented as a part of an operation system software to make decisions about process transfer and message routing in a hierarchical network.

A distributed multicast algorithm with congestion control and message routing is presented by Hać. It uses a source routing strategy to establish the path from the source to the multicasting destination. The algorithm considers congestion control and bandwidth management techniques by limiting trunks and switches occupancy. The routing algorithm uses a tree of trunks and switches from the source to the multicasting destination. The message transfer time is calculated and the path with the shortest transfer time is chosen.

### 4.2.1 Multicast Routing Algorithms Reducing Congestion

Algorithms solving the multicast problem by using the dynamic priority spanning tree are described in three versions. The first version is designed for the networks where all nodes have the multicast capability. The second version slightly modifies the first version for networks where some nodes do not have the multicast capability. The third version is designed for those networks that can be regarded as undirected graphs. Algorithms use the source-routing strategy to establish the path from the source to the multicasting destination, and find the path with fewer numbers of links and hops. The algorithms also prevent congestion caused by copy operations in a node. Network congestion is reduced by minimizing the total number of links in the tree.

Figure 4–6(A) is a nine-node network, with the destination set $D = \{a, c, i, m\}$, and a different capacity in each edge. We assume that all nodes in the network have the multicast capability. If the number of hops is used as the criterion, the shortest-path-spanning-tree (SPST) algorithm (Dijkstra's algorithm) is used as shown in Deering and Cheriton. Figures 4–6(B), and (C) illustrate the application of the SPST algorithm to this network. The SPST algorithm first finds the shortest path tree from the source to all other nodes in the network as shown in Figure 4–6(B). It then prunes those branches that do not contain any nodes from the destination set, and provides the final solution in Figure 4–6(C). We call the solution to the SPST algorithm the SPST multicast tree. Note that in this scenario, there are six links and seven hops used. Furthermore, if each link remains at 3.5Mb/s capacity in the network, and the multicast communication consumes 3Mb/s bandwidth, then the congestion occurs when additional transmissions between node $a$ and the other nodes require more bandwidth than 0.5 Mb/s. This also prevents large packet transmissions from node $a$ to the other nodes. Nodes can be regarded as routers, switches, or gateways in a wide-area packet switching network.

Hać and Wang propose a multicast-routing algorithm based on the dynamic priority search technique to build a multicast spanning tree, called the dynamic priority spanning tree

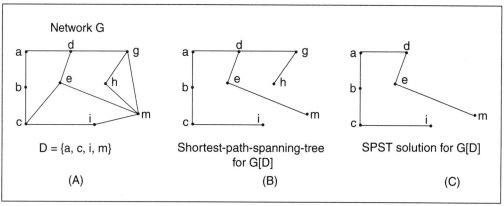

**Figure 4–6** Example Application of the SPST Algorithm in the Network

(DPST) algorithm. The DPST algorithm can reduce congestion and increase the path finding rate by using fewer links in the spanning tree than the other algorithms. For the network in Figure 4–6(A), using the DPST algorithm allows for fewer links and hops to be used than in the SPST algorithm. The solution of the DPST algorithm shows that only four links (path *a-b-c-i-m*) are needed to complete this task. In this case, the link *a–d* is still available for large packets transmission between node *a* and all other nodes. Thus, the problem that causes the congestion by large packet transmissions (over 0.5 Mb/s) from *a* to other nodes in the SPST algorithm does not exist if the DPST algorithm is used.

In the DPST algorithm, we define the links-used reducing rate as

$$1 - \frac{DPST(total\_links\_used)}{SPST(total\_links\_used)}.$$

The links-used reducing rate of DPST algorithm in this example is (1 – 4/6) = 1/3 in this simple network (Figure 4–6(A)). Thus, the consumed bandwidth for the transmission is saved.

The principle of the DPST algorithm is that instead of finding the shortest path between the source and the multiple destinations, it always chooses paths that can make the total number of links (or hops) in the multicast tree as few as possible, therefore leaving the bandwidth open for other transmissions. Intuitively, fewer links implies less bandwidth used and reduced congestion. The DPST algorithm also considers the balanced use of the network by limiting the duplication of multicast packets in the nodes to minimize congestion.

#### 4.2.1.1 The Algorithm

The DPST multicast routing algorithm uses the source-routing strategy to establish the tree of paths from the source to the multicasting destination. The algorithm allows for the transfer of packets from the source to the multicasting destination by using routing and packet duplication in network switches.

The multicast routing algorithm is distributed, and it can be executed in every source in the network. The algorithm checks the destination address of received packets, then builds the tree of paths from the source to the destination. If the tree is built, the requested message is sent to the multicast destinations. Otherwise, a message is sent to the source that the packets cannot be transmitted. The latter situation can occur if the network is congested.

The multicast tree is rooted in the source and all branches terminate at destinations. The packets are duplicated only when two branches of the path diverge. Postponing the duplication of packets in this way saves network bandwidth due to the transmission of fewer packets. If a single link leads to multiple destinations, only a single copy of the packet traverses the link; if necessary, the packet will be duplicated later.

The DPST is a heuristic multicast-routing algorithm. It can find the minimum-cost multicast path in a reasonable computation time. When we define the cost for each link as equal to one unit, the algorithm obtains the multicast path with the minimum links used.

The following notations are used in the algorithm:

| | |
|---|---|
| $G(V, E, C)$ | The network is defined as a given directed graph, with the vertex set $V$, the link set $E$, and the remaining capacity set $C$ for each link. |
| $D$ | The destination set. |
| $D_i$ | Subset of $D$. |
| $T(D)$ | The multicast tree $T(D)$ is rooted in source $s$, $s \in V$, with all branching nodes being multicast nodes, and with its leaf nodes in $D$ (destination set), $D \in V$. |
| $S(T)$ | The nodes in $T(D)$. |
| $P(u_1, u_2, ..., u_p)$ | A path from $u_1$ to $u_2$, ..., to $u_p$ in network $G = (V, E, C)$; $u_1, u_2, ..., u_p$, for all $k$, $1 \leq k < p$, $\{u_k, u_{k+1}\} \in E$, and $u_k \in V$. |
| $\text{Cost}(L\{u, h\})$ | Cost of the link from $u$ to $h$. We define the cost of the link as the unit, when we consider the fewer links consumed result. However, it can also be the function of the state of the link, such as: {*length* $(L\{u, h\})$}. This way, every link can have a different cost in the network. |
| $\text{Cost}(P(u_1, u_2, ..., u_p))$ | Equal to $\sum_{k=1}^{p-1} \text{cost}(L\{u_k, u_{k+1}\})$. If $\text{cost}(L\{u_k, u_{k+1}\})$ is defined as a unit, the total cost of the path is the sum of links in the path. |
| $SP(u_1, u_2, ..., u_p)$ | The shortest path from $u_1$ to $u_2$, ..., to $u_p$. $SP(u_1, u_2, ..., u_p)$ is the path from $u_1$ to $u_p$ whose cost is minimal among all the possible paths from $u_1$ to $u_p$. |
| $\text{Cost}(T_{\text{DPST}}(D))$ | The total cost of the DPST multicast tree $T_{\text{DPST}}(D)$. |
| $\text{Cost}(T_{\text{SPST}}(D))$ | The total cost of the SPST multicast tree $T_{\text{SPST}}(D)$. |
| $\text{Cost}(T_{\text{MST}}(D))$ | The total cost of the MST (minimum spanning tree) multicast tree $T_{\text{MST}}(D)$. |
| MAX | A bound of the number of hops in the search region. |

Branch-Bound       An integer bound of the number of allowed packet copies per node.

The DPST algorithm uses the source to find the first closest destination, and then it uses this established path to find the next destination close to this path. Thus, the algorithm forms an uncompleted multicast tree, which is used to find another close destination that has not been yet included in the tree. The uncompleted multicast tree is then updated and used to find the other destinations until all destinations have been included in the multicast tree. The procedure called the priority first search for the shortest path (PFSSP) is used to find the shortest path from the source, or path (or tree) to a destination that has not yet been included in the path (or tree).

The algorithm is efficient in transmitting the multicast packets across as few links as possible, and uses the following factors that may affect the routing decision.

1. To limit the search in a reasonable area, we set MAX as the bound of the number of hops in the search region; if the search depth falls beyond this region, the algorithm is terminated, and no solution is found. Thus, when the node's priority in the top of the heap is not greater than zero, it means that the search failed, and the algorithm should be terminated.
2. We set up a Branch-Bound as the bound of the number of packet copies per node. Too many copy operations of multicast packets in a particular switching node will require the larger hardware architecture in the node and cause congestion in the switch. For example, in the input buffer switches such as Batcher-Banyan switches, packets requiring too many copies will stay longer in the head of the input buffer. This causes head of the line (HOL) blocking for other packets which go through the input buffer. Thus, preventing concentrations of a packet copy operation in a switching node is the important issue in the multicast routing problem.
3. We can define the function to calculate the cost of each link according to the state of the link. However, since we focus on the number of links used, we define the cost of each link as the unit. Thus, in the PFSSP procedure, we have $t$'s priority $-$ cost($L(t, \text{node})$) = $t$'s priority $-$ 1, where $t$ is an arbitrary node in the network.

Here is the first version of the DPST routing algorithm for networks where all vertices have multicast capabilities. Let $T(D_1)$ denote an incomplete multicast tree, where $D_1$ is a subset of $D$. Let $S(T)$ denote the set of nodes in $T(D_1)$. The PFSSP procedure is described later. The DPST algorithm is as follows:

Algorithm input: network $G = (V, E, C)$, destinations $D$.
Required capacity: $C_{need}$ for the task.
Algorithm output: multicast tree $T(D_1)$, $(D_1 = D)$.

Step 1: Mark all destination nodes in $G$. Begin from the source $s$. Let $S(T) = \{s\}$; use the PFSSP procedure to find the closest destination $d_1$, $d_1 \in D$. This path is denoted by $P(s, u_1, u_2, ..., u_p, d_1)$.

Step 2: Set $D_1 = \{d_1\}$. Delete $d_1$ from $D$. Let multicast tree $T(D_1) = P(s, u_1, u_2, ..., u_p, d_1)$. The vertices in $T(D_1)$ are: $s, u_1, u_2, ..., u_p, d_1$. Then $S(T) = \{s, u_1, u_2, ..., u_p, d_1\}$.
Step 3: while $D$ is not empty do
Step 4: Start from vertices in $S(T)$; use the PFSSP procedure to find the shortest path from $S(T)$ to the current closest destination $d_n$, which is $P(u_k, u_{k+1}, ..., d_n)$, ($u_k$ is a vertex in $T(D_1)$, $k \in T(D_1)$). Set $D_1 = D_1 + \{d_n\}$, delete $d_n$ from $D$.
Step 5: Update $T(D_1)$. ($T(D_1) = T(D_1) + P(u_k, u_{k+1}, ..., d_n)$). Update $S(T)$, any vertices in $T(D_1)$ must occur only once in $S(T)$.
   endwhile

The PFSSP procedure used in the DPST algorithm is similar to the first priority search for the shortest path algorithm. This algorithm starts with the source node. The PFSSP procedure used here can start with several nodes. The major difference is how we define the priority. The principle of dynamic priority in the procedure is that the vertex in $S(T)$ always has the highest priority, and the priority decreases along the search. The search will always start from the nodes in $S(T)$, and stop when the new destination is found.

The heap structure is used in the program as the priority queue. The purpose of the function "update the heap" is to ensure that the given node appears in the queue with a given priority. If the node is not in the heap, an "insert" operation is done. If the node is there but has a low priority, then a change operation is used to increase the priority. To forbid a loop in the tree, each node can have only one parent node, and several children nodes. Let $C(t, t_i)$ denote the remaining capacity of the link from $t$ to $t_i$, and $C_{need}$ denote the required capacity for transmission. The PFSSP is as follows.

Procedure Input: network $G = (V, E, C)$, node set $S(T)$, destinations $D$.
Required capacity: $C_{need}$ for the task.
Procedure output: the shortest path from $S(T)$ to destination.
[1]  initialize all nodes' priority to zero
[2]  set all nodes' priority in $S(T)$ to MAX, put all nodes in $S(T)$ to heap.
[3]  while heap is not empty, do
[4]    pump a node from the heap;
[5]    mark the node visited;
[6]    if the node's priority < 0,
[7]    then search failed, exit;
[8]    endif
[9]    let $t$ denote the node;
[10]   if $t$ is destination, $t \in D$,
[11]   /* The first found destination is the current closest destination to $T(D_1)$. */
[12]     then return path $P(s, ..., t)$;
[13]   else if

[14]        the number of branches in *t* > Branch-Bound,
[15]        then discard *t*;
[16]        else
[17]            for (all the nodes connected to *t*), do
[18]                if node is visited or $C(t, \text{node}) < C_{\text{need}}$,
[19]                    then discard it;
[20]                else if
[21]                    the node's priority < *t*'s priority − cost($L(t, \text{node})$),
[22]                    then
[23]                        set node's priority = *t*'s priority − cost($L(t, \text{node})$);
[24]                        set the node's parent node to *t*;
[25]                        update the heap;
[26]                    endif
[27]                endif
[28]            endfor
[29]        endif
[30]    endif
[31] endwhile

Figure 4–7 explains the algorithm step by step. The algorithm begins from source *a*, thus $S(T) = \{a\}$, and it uses the PFSSP procedure to find the closest destination, which is *c*. $P(a, b, c)$ is the shortest path from *a* to *c*. Let $S(T) = \{a, b, c\}$, and Steps 1 and 2 in the DPST algorithm are completed as shown in Figure 4–7(A). In Step 3 of the DPST algorithm, $D = \{i, m\}$ is not empty, and it uses the PFSSP procedure to find the next closest destination *i*, as shown in Figure 4–7(B). The algorithm prunes branches that do not contain any vertexes in *D*, and the result of Step 5 of the DPST algorithm is shown in Figure 4–7(C). This way, by repeating Steps 3, 4, and 5, the final solution for network *G* is shown in Figure 4–7(E).

In the DPST algorithm, the nodes in $T(D_1)$ (any incomplete multicast tree) have a certain higher priority MAX, therefore the nodes close to $T(D_1)$ are more likely to be visited (have the higher priority) and included in the tree. For example, in Figure 4–7(D), from *i* to *m* the cost is only one link (*i* is in $T(D_1)$), thus, node *m* is included in the path $P(a, b, c, i, m)$ immediately without waiting for the path $P(a, d, e, m)$ to be explored.

For the networks with both multicast and nonmulticast vertices, we modify the DPST routing algorithm. In this version, we avoid multicast tree branching in the nodes that do not have the multicast capability.

We designate nodes that do not have the multicast capability as no-multicast nodes. If the node is already in $S(T)$, and is a no-multicast, we discard it hereafter. Thus, these nodes have only one child as it is already in $S(T)$, and no more nodes will use this node as a parent node or append it to the path. If the node is not included in $S(T)$, it will participate in the search. Thus, only one path will include this node.

# Multicast and Other Routing Algorithms

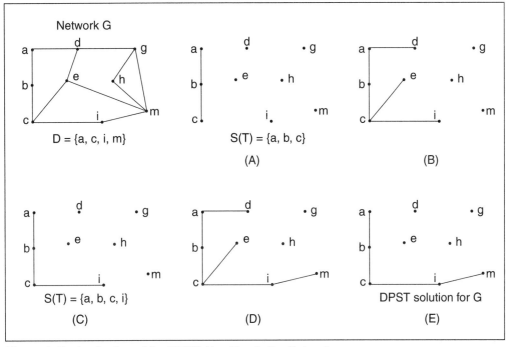

**Figure 4–7** Application of the DPST Algorithm in the Network

The second version of the DPST algorithm is changed slightly inside the PFSSP procedure. The second version of the PFSSP algorithm is as follows.

Procedure Input: network $G = (V, E, C)$, node set $S(T)$, destinations $D$.
Required capacity: $C_{need}$ for the task.
Procedure output: the shortest path from $S(T)$ to destination.

[1] initialize all nodes' priority to zero
[2] set all nodes' priority in $S(T)$ to MAX, put all nodes in $S(T)$ to heap.
[3] while heap is not empty, do
[4]     pump a node from the heap;
[5]     mark the node visited;
[6]     if the node's priority < 0,
[7]         then search fail, exit;
[8]     endif;
[9]     let $t$ denote the node;
[10]     if $t$ is a destination, $t \in D$
[11]         then return path $P(s, ..., t)$;
[12]     else if
[13]         $t$ is in $S(T)$ and is no-multicast,

| [14] | then discard $t$; |
| --- | --- |
| [15] | else if |
| [16] | the number of branches in $t >$ Branch-Bound, |
| [17] | then discard $t$; |
| [18] | else |
| [19] | for (all the nodes connected with $t$), do |
| [20] | if node is visited or $C(t, \text{node}) < C_{\text{need}}$, |
| [21] | then discard it; |
| [22] | else if |
| [23] | the node's priority $<$ $t$'s priority $-$ cost($L(t, \text{node})$), |
| [24] | then |
| [25] | set node's priority $=$ $t$'s priority $-$ cost($L(t, \text{node})$); |
| [26] | set the node's parent node to $t$; |
| [27] | update the heap; |
| [28] | endif |
| [29] | endif |
| [30] | endfor |
| [31] | endif |
| [32] | endif |
| [33] | endif |
| [34] | endwhile |

In the MST algorithm, the network is regarded as an undirected graph, and all nodes have multicast capability. The purpose of the third version of the DPST algorithm is to modify the first version slightly to obtain an efficient result in this networking environment. In this version, we add Step 6 to the first version of the DPST algorithm. The third version of the DPST is as follows.

> Algorithm input: network $G = (V, E, C)$, destinations $D$.
> Required capacity: $C_{\text{need}}$ for the task.
> Algorithm output: $T_{\text{DPST}}(D)$.

Step 1: Mark all destination nodes in $G$. Begin from the source $s$, let $S(T) = \{s\}$; use the PFSSP procedure to find the closest destination $d_1$, $d_1 \in D$. This path is denoted by $P(s, u_1, u_2, ..., u_p, d_1)$.

Step 2: Set $D_1 = \{d_1\}$. Delete $d_1$ from $D$. Let the multicast tree $T(D_1) = P(s, u_1, u_2, ..., u_p, d_1)$. The vertices in $T(D_1)$ are: $s, u_1, u_2, ..., u_p, d_1$. Then $S(T) = \{s, u_1, u_2, ..., u_p, d_1\}$.

Step 3: While $D$ is not empty, do

Step 4: start from vertices in $S(T)$; use the PFSSP procedure to find the shortest path from $S(T)$ to the current closest destination $d_n$, which is $P(u_k, u_{k+1}, ..., d_n)$ ($u_k$ is a vertex in $T(D_1)$, $k \in T(D_1)$). Set $D_1 = D_1 + \{d_n\}$, delete $d_n$ from $D$.

Step 5:    update $T(D_1)$. ($T(D_1) = T(D_1) + P(u_k, u_{k+1}, ..., d_n)$). Update $S(T)$, any vertices in $T(D_1)$ must only occur once in $S(T)$.

   endwhile

Step 6: Convert $S(T)$ in $G$ to form a new small graph $G_S(D)$, which contains only the nodes in $S(T)$ and links between them. Find the minimum spanning tree $T'(D)$ of $G_S(D)$, then prune those branches that do not contain any nodes from destination set. Output this tree $T_{DPST}(D)$.

The PFSSP procedure is the same as in the first version, but we remove the test for the Branch-Bound in Lines 13, 14, and 15, since we are not concerned about it in the third version.

In this version of the DPST algorithm, the multicast tree $T(D)$ is built in the first five steps of the DPST algorithm (as it is in the first version), and a new graph $G_S(V', E', C')$ is converted from the graph $G(V, E, C)$. In this new graph only $V'$ contains the nodes in $T(D)$, and $E'$ contains the links directly connecting those nodes. Then, the minimum-spanning tree algorithm is applied to $G_S(V', E', C')$, and a minimum-spanning tree is found which has the minimum sum of costs among all trees collected. This makes the cost of the DPST output multicast tree as low as possible. Thus, $cost(T_{DPST}(D)) \leq cost(T(D))$, where $T_{DPST}(D)$ denotes the output of the DPST (third version), and $T(D)$, which represents the tree set up by the first five steps ($T(D)$ is also the result of the first and second versions).

The third version of the DPST algorithm is only applicable to undirected networks where all nodes must have multicast capability, and the network is not concerned about the congestion caused by too many copy operations occurring in a node.

### 4.2.1.2  Cost and Performance Analysis

There are basically three costs incurred in large network routing. These costs include the storage required in each network node, the computation required to calculate optimal paths, and the bandwidth consumed by the path update message.

The cost of storage in the DPST algorithm is approximately the same as the other algorithms, for instance, the link-state multicast algorithm. We will discuss the computation cost of the DPST routing algorithm.

In a sparse network such as most existing internetworks, the computation time of the DPST algorithm in all versions is $O(M(E + V)\log V)$, where $E$ is the number of links in the network, $M$ is the number of destinations, and $V$ is the number of vertices.

The cost of computation time for the PFSSP procedure is $O((E + V)\log V)$. In the second version of DPST for networks, which have no-multicast nodes, the change inside the PFSSP procedure needs $O(1)$ computation time, so the complexity for this procedure is equal to $O((E + V)\log V)$.

Since the major operation in DPST Step 1 is the PFSSP procedure, it costs $O((E + V)\log V)$. DPST Step 2 needs maximum $M$ times searching through the destination list, so the total computation cost for Step 2 is $M$. Because the major operation in DPST Step 4 is a PFSSP procedure, and DPST Step 5 costs $O(V)$, DPST Step 3 runs $(M - 1)$ times, then the computation cost of Steps 3, 4, and 5 in DPST is $O((M-1)(E+V)\log V) + O(V) \approx O((M-1)(E+V)\log V)$. Thus, the total computation cost for the first version of the DPST algorithm is: $O((E+V)\log V) + M + O((M-1)(E+V)\log V) \approx O(M(E+V)\log V)$. The computation time for the second version is the same as for the first version.

The computation time for Step 6 in the third version of the DPST algorithm is $O((E'+V')\log V')$ ($E' \leq E$, $V' \leq V$). Thus, the total computation cost for the DPST algorithm (third version) is: $O((E+V)\log V) + M + O((M-1)(E+V)\log V) + O((E'+V')\log V') \approx O(M(E+V)\log V)$.

*4.2.1.2.1   Performance Comparison of the DPST and SPST Algorithms*

We can prove that the cost of generating a spanning tree in the DPST algorithm (first and second versions) is not higher than in the SPST algorithm.

*Lemma 1*:   If the path $P(\text{source}, ..., i, .., d)$ from the source to destination $d$ in the DPST multicast tree is not the same as the shortest-path $SP(\text{source}, ..., d)$; $i$ is a node in the multicast tree $T(D_i)(D_i \subset (D-d))$, in which $d$ is included in the DPST multicast tree $T(D_{i+1})(D_i = D_i + d)$ by path $SP(i, ..., d)$; then the cost of path $P(\text{source}, ..., i, .., d)$ is

$$\text{cost}(P(\text{source}, ..., i, .., d)) = \text{cost}(P(\text{source}, ..., i)) + \text{cost}(SP(i, ..., d)),$$

where $\text{cost}(SP(i, ..., d)) < \text{cost}(SP(\text{source}, ..., d))$.

*Proof*:   Let us assume that Lemma 1 is not true, and $\text{cost}(SP(i, ..., d)) > \text{cost}(SP(\text{source}, ..., d))$.

We know that before $d$ is appended to the DPST multicast tree, both the source and $i$ are in the DPST multicast tree $T(D_i)(D_i \subset (D-d))$. When using $T(D_i)$ as the input in the PFSSP procedure to find destination $d$, $d$ should be appended to the shortest path from $T(D_i)$ to $d$. If $\text{cost}(SP(\text{source}, ..., d)) < \text{cost}(SP(i, ..., d))$, $d$ should follow the path $SP(\text{source}, ..., d)$ instead of the path $SP(i, ..., d)$. This contradicts the condition that $d$ is not in $SP(\text{source},..., d)$. Therefore, the assumption $\text{cost}(SP(i, ..., d)) > \text{cost}(SP(\text{source}, ..., d))$ is not true, and by contradiction, Lemma 1 holds.

*Lemma 2*:   If there is a path $P(\text{source}, ..., i, ..., d)$ from the source to destination $d$ in the DPST multicast tree that is not the same as the shortest path $SP(\text{source}, ..., d)$, then the DPST algorithm guarantees a lower cost than the cost in the SPST algorithm.

*Proof*:   Let $T_{\text{DPST}}(D_i)$ and $(D_i = D - d)$ denote the DPST multicast tree before $d$ is found.

Because $P(\text{source}, ..., i)$ is in $T_{\text{DPST}}(D_i)$, then $\text{cost}(P(\text{source}, ..., i))$ is already included in $\text{cost}(T_{\text{DPST}}(D_i))$. Thus, the cost for the DPST multicast tree is $\text{cost}(T_{\text{DPST}}(D)) = \text{cost}(T_{\text{DPST}}(D_i)) + \text{cost}(SP(i, ..., d))$. On the other hand, the cost in the SPST multicast tree is $\text{cost}(T_{\text{SPST}}(D)) = \text{cost}(T_{\text{SPST}}(D_i)) + \text{cost}(SP(\text{source}, ..., i, ..., d))$. Since $T_{\text{DPST}}(D_i))$ is the same as $T_{\text{SPST}}(D_i)$ from the condition that only the path from the source to $d$ is not the same as the shortest path, and in Lemma 1, we proved $\text{cost}(SP(i, ..., d)) < \text{cost}(SP(\text{source}, ..., i, ..., d))$, thus, $\text{cost}(T_{\text{DPST}}(D)) < \text{cost}(T_{\text{SPST}}(D))$. Lemma 2 holds.

*Theorem 1*: The cost of generating a spanning tree in the DPST algorithm is not higher than in the SPST algorithm.

*Proof*: From Lemmas 1 and 2, the DPST algorithm generates either the spanning tree different than that in the SPST algorithm, which makes the total cost of the DPST multicast tree lower than the total cost of the SPST multicast tree; or the spanning tree is the same as the SPST multicast tree, which is the worst case for the DPST algorithm. This guarantees that the DPST algorithm will not cost more than the SPST algorithm. Theorem 1 holds.

The DPST algorithm achieves the goal of fewer links used when the cost of each link is the same, a unit. The lower cost of the multicast tree means fewer links used. Thus, Lemmas 1, 2 and Theorem 1 also prove that the DPST algorithm can decrease the number of links used. Because fewer links transmit the multicast packets, the unused links are available for other transmissions. This way, the congestion is reduced.

The DPST algorithm selects paths reducing network use, and causes little overhead in computation time. The DPST algorithm does not always choose the shortest-path from the source to some destinations, thus, the link transmission delay for a pair of source and destination may sometimes be longer than in the SPST algorithm. When the network is lightly loaded, and the bandwidth is not critical, the DPST algorithm will not perform better than the SPST algorithm. In the network with a heavy load when the bandwidth is critical, especially when several large-size video conferences simultaneously occur in the network, the delay caused by network congestion is the major factor affecting transmission speed. The network congestion may even cause the path finding to fail. In this case, the DPST algorithm shows an advantage over the other algorithms, including the SPST algorithm.

### 4.2.1.2.2 Performance Comparison of the DPST and MST Algorithms

Finding the minimum cost multicast path is similar to solving the Steiner tree problem, and the heuristic MST algorithm can find close to the optimal worst case bound. We prove that the DPST algorithm (first and second versions) has the same worst case bound as the MST algorithm.

In our comparison, the network is an undirected graph, and all nodes have multicast capability. The new terms used in the MST algorithm are defined as follows:

$G'[D]$          the complete graph $G'$, where $cost(PG'(u, v))$ is the length of the shortest path from $u$ to $v$ in G; $cost(PG'(u, v)) = cost(SP(u, ..., v))$

$T'(D)$          the minimum spanning tree in $G'$.

Before we compare the DPST and MST algorithms, we briefly describe the MST algorithm. The MST algorithm is as follows.

Algorithm input: graph $G$, destinations $D$.
Algorithm output: MST multicast tree.

Step 1: Construct a complete graph $G'$, where the cost($PG'(u, v)$) is the length of the shortest path from $u$ to $v$ in $G$;
Step 2: Construct the minimum spanning tree $T'(D)$ in $G'$;
Step 3: Convert the edges in $T'(D)$ to paths in $G$ to form solution $T_1$.

Figure 4–8 illustrates the application of the MST algorithm with set $D = \{a, d, e, g\}$ and the cost of each link indicated, the complete graph $G[D]$, the minimum spanning tree in $G[D]$, and the final solution.

*Lemma 3*: The total cost of the DPST multicast tree is not higher than the total cost of the minimum-spanning tree of complete graph in Step 2 of the MST algorithm.

*Proof*: The first path $SP$(source, ..., $d_1$) included in the DPST multicast tree is the shortest path from source to destination $d_1$. This path should be the same as the first path chosen in the minimum spanning tree in $G'[D]$ in Step 2 of the MST algorithm.

The second path (or branch path) in the minimum spanning tree in $G'[D]$ in Step 2 of the MST algorithm is selected from $PG'$(source, $d_2$) and $PG'(d_1, d_2)$, and the shortest path is chosen. On the other hand, the second path (or branch path) included in the DPST multicast tree is chosen as the shortest path from the first path $SP$(source, ..., $d_1$) to destination $d_2$, which means it selects the shortest path from SPG(source, ..., $d_2$), SPG($d_1$, ..., $d_2$), and SPG($V_i$, ..., $d_2$), where $V_i$ is the node in path $SP$(source, ..., $d_1$) ($i = 1, 2, ...$). Thus, the second path in the DPST algorithm is either (a) the same as the minimum spanning tree in $G'[D]$, or (b) different than that in the minimum-spanning tree in $G'[D]$. In case (a), the second path is chosen from SPG(source, ..., $d_2$), or SPG($d_1$, ..., $d_2$), and the costs of the multicast trees in both the DPST and MST algorithms are the same, as explained in the notation that cost($PG'$(source, $d_2$)) = cost(SPG(source, ..., $d_2$)), and cost($PG'(d_1, d_2)$) = cost(SPG($d_1$, ..., $d_2$)). In case (b), the second path is chosen from $PG'(V_i, d_2)$ ($i = 1, 2, ...$), and the cost of the DPST multicast tree is lower than the cost of the MST multicast tree, since the DPST algorithm chooses only the shortest path from the paths: SPG(source, ..., $d_2$), SPG($d_1$, ..., $d_2$), and SPG($V_i$, ..., $d_2$). Continuing on this way, we can compare the paths selected in the DPST algorithm and in the minimum spanning tree in $G'[D]$. We find that all paths chosen are in one of the above cases (a) and (b), thus, the cost of branch paths in the DPST algorithm is either the same as the cost of the minimum-spanning tree in $G'[D]$, or lower than the cost of the minimum-spanning tree in $G'[D]$. Lemma 3 holds.

*Theorem 2*: The DPST algorithm has the same worst case bound as the MST algorithm.

*Proof*: Kou, Markowsky, and Berman proved that the total cost of the minimum-spanning tree in $G'[D]$ has a close to optimal worst case bound. From Lemma 3, the total cost of the DPST multicast tree is lower than or equal to the cost of the spanning tree in the complete graph $G'[D]$ in the MST algorithm. Thus, the DPST algorithm has the same worst case bound as the MST algorithm. Theorem 2 holds.

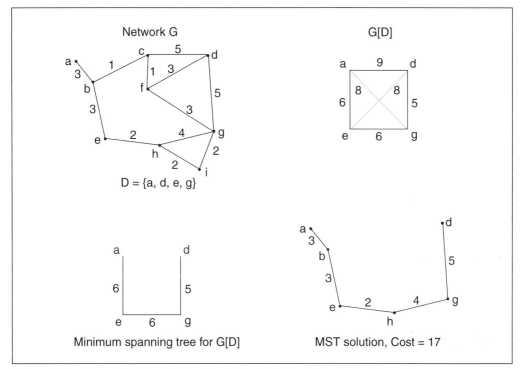

**Figure 4–8** Example of the Application of the MST Algorithm in the Network

The solution given by the MST algorithm is not always optimal as shown in Figure 4–8. The DPST algorithm used in the same example and shown in Figure 4–9 yields the tree with the total cost lower than the cost in the MST algorithm. In this case, the DPST algorithm performs better, and with the cost closer to the optimal than the MST algorithm. The computation time for the MST algorithm is $O(M(E + V)\log V)$ in sparse network. This computation time is the same as for the DPST algorithm.

The multimedia multicast connections are dynamic and complex. The multicast routing algorithm should solve the multicast connection in the networks where some nodes do not have the multicast capability. The algorithm should also restrict the concentration of packet duplication in some nodes to avoid the congestion caused by too many copy operations in a node. The MST algorithm cannot solve these problems in a simple and inexpensive way as does the DPST algorithm.

The average results for a 100-node network with multicast vertices, where the link costs for networks are generated by randomly selecting a number from one to nine are as follows:

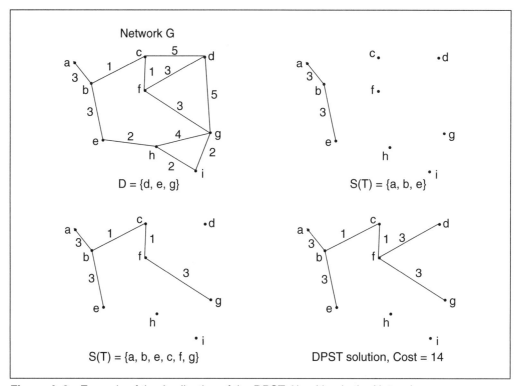

**Figure 4–9** Example of the Application of the DPST Algorithm in the Network

| Destination | DPST cost | MST cost | SPST cost |
|---|---|---|---|
| 4 | 34.87 | 35.01 | 45.96 |
| 5 | 35.96 | 36.02 | 46.33 |
| 8 | 47.86 | 48.13 | 61.74 |
| 10 | 53.38 | 53.64 | 79.03 |
| 15 | 66.23 | 66.73 | 84.56 |
| 20 | 77.07 | 77.87 | 99.00 |
| 25 | 88.96 | 89.29 | 109.36 |

The DPST algorithm uses fewer links, thus reduces the cost of the multicast path in comparison with the SPST algorithm, especially in a large size of destination groups. The DPST algorithm also performs better than the MST algorithm in finding the minimum-cost multicast path in an average case.

#### 4.2.1.3 Applying the Multicast Routing Algorithm to a Multipoint Connection in Broadband Networks

Convergence cast, also known as concast, is a connection from multipoint to a point. Concast merges traffic along the convergence path. As it approaches the destination, more bandwidth will be required for the links. An application monitors the system collecting information sent from a cluster of sensors that are geographically isolated.

The concast path is a multicast path with reverse direction links requiring a different bandwidth. The heuristic multicast routing algorithm can be modified to traverse the link in a reverse direction, as well as to increase the link bandwidth required along the path.

Video conference service with a centralized presentation control can be attended by participants from all over the world. Each participant can view the same image as all the others. It could be the composite image of all the participants, a particular speaker, or a viewgraph. To fulfill the preceding requirements, this service needs to be set up from a multipoint-to-multipoint connection. The service needs a conference bridge located optimally in the network to minimize the cost of all connected paths. The paths, those from the participants to the video conference bridge and vice versa, must be the shortest.

In a multiparty video call, each party may select a subset of the video sources and compose them according to the presentation description into a single video stream for a user display device. One method is to use a bridge to gather all the video sources and produce customized images according to each user's presentation description. The other method is to merge some of the video sources along the way to the particular user. Although the latter approach requires more than one bridge, it saves bandwidth and can be implemented with many simple modules. The trade-off depends on the cost ratio between the bridges and links.

An information provider may have several centers within the network that have identical copies of certain pieces of information. This results in the network distribution, and the multiple multicast connection, while distributing information to the subscriber. The remedy is to implement a combination of the network partition and multicast routing algorithm. However, the most difficult task for each of the distribution centers is to search for the optimal partition of local exchanges. The multicast routing algorithm can play a role in calculating the multicast path cost for each proposed partition, and in setting up and modifying the multicast paths that distribute the information.

## 4.3 Routing in Wireless Personal Networks

Routing is an indispensable control function that connects a call from the origin to the destination. Such control functions in wireless personal communication networks use a general routing procedure, a rerouting procedure, and a handoff. Along with the features of wireless communication, the user mobility control function tracking locations of networks subscribers should be associated with routing schemes during the communication connection. Network routing mechanisms in traditional wireless communications networks use a centralized approach which requires additional signaling channels. In wireless communications networks supported by

B-ISDN, the network topology is established by virtual paths. Virtual paths are logical direct radio links between all switch nodes. The bandwidth of a virtual path can consist of a number of virtual channels. Because of the features of wireless communications networks, the network topology is highly dynamic. Hence, the traditional centralized approach for wireless communications networks is hard to implement on such a highly dynamic topology since network control mechanisms must be dynamically modified to match the current topology. The bandwidth of wireless communications networks is limited, the traffic increases quickly, and it is hard to schedule incoming traffic on time in the centralized approaches. Obviously, the centralized method is not efficient when the network size increases and the network services are enhanced.

The subscribers in wireless communications networks roam. To create connections among all communication parties to deliver incoming and outgoing calls, the first consideration is the current location of mobile users and hosts. In wireless communications networks, the key service for providing seamless connectivity to MHs is the creation and maintenance of a message forwarding path between two known locations of MHs that call and are called. A routing decision in wireless communications networks is made not only using the states of paths and internal switching nodes, but also using the location of available information.

The geographical area covered by wireless communication networks is partitioned into a set of cells. A routing path may be inefficient while an MH hands off to another cell coverage area. The connection paths need to be reestablished each time to continue communication. As a result, the network call processor can become involved many times during the lifetime of the mobile connection. When wireless communications networks move toward smaller size cells to accommodate more MHs or to provide higher capacity, the handoff becomes a more frequent part of communications. Conventional routing procedures for connecting MHs fail due to a frequent handoff when the network call processor becomes a bottleneck. Hence, routing efficiency in wireless communications networks has a critical dependence on the propagation of location information into the network. However, excessive information propagation can waste network resources, while insufficient location information leads to inefficient routing.

Wireless communications networks can provide different personal communications services, which have different transmission time delay requirements. For cellular telephone communications, the shortest time delay or most strict time delay to transmit a voice message is required. In portable computer communications or other data communications, the requirements of transmitted time delay are not very strict. Transmitted data can be stored in buffers and be transmitted later when channels are available. However, to provide a high QoS, the transmission time delay is an important factor in the wireless communications networks. Routing procedure in the wireless communications networks should depend on different requirements of transmission time delay, and on how to balance the transmission load and find minimum-cost transmission paths.

Several routing schemes in wireless networks have been proposed. Yuan proposes a concept called a "friend network" for building the access history into the routing scheme. This concept provides only partial solutions to the location and routing problem. Perkins proposes continuous network access to provide routes for MHs, no matter how often MHs' location changes. This scheme does not improve routing efficiency, and requires more memory space because the data structure of the information is long. Nakajima proposes automatic routing for mobile communication networks. This routing scheme provides only a method that finds routes for MHs based on HLR and VLR information. There is no method to handle handoff when MHs cross different radio frequency zones. This scheme has the potential to interrupt communication when communications are longer and MHs roam with high mobility. Pollini proposes a scheme for routing information between interconnected cellular mobile switching centers, and discusses the switch time performance for interswitch handoff from the old path to the new path. This scheme does not consider congestion. When user density increases in switch coverage, gateways between interswitches can become congested. In these situations, the method of interswitch handoff cannot be used. Communication between MHs will be blocked or interrupted. These proposals for routing in wireless communication networks focus on efficiently tracking the MHs' locations. The proposals do not consider how to handle the situations to find an efficient and low cost rerouting path during MHs handoffs. Neither do they consider congestion when MHs communicate with each other.

Hać and Zhu propose an extended shortest path (ESP) routing algorithm, and a dynamic state-dependent shortest path (DSDSP) routing algorithm to improve performances of the ESP routing algorithm for wireless communications networks in the call block rate, handoff drop rate, routing distance, and rerouting distance. In both algorithms, each switching center can control the same number of BSs, and each BS can serve the same number of mobile hosts.

### 4.3.1 The Architecture and Model of a Wireless Communications Network

Wireless communications networks consist of a set of MHs, mobile base stations, mobile service switching centers, communication channels, and radio links. MHs can move from one BS to another and communicate with other MHs by using communication channels. Each MH is assigned a permanent ID. MHs access the wireless communications networks by using radio links.

Each radio cell has a BS that is fixed in the wireless communications network. Mobile BSs control the directions of message (upstream or downstream). The number of BSs are connected to a fixed subnetwork, which is controlled by a switching center in a cluster. A mobile BS can serve several MHs in its service call by using radio links. When an MH connects to a mobile BS, the MH finds and then registers with this BS. If the MH is initially registered with this BS, this BS is referred to as the MH's home station, otherwise, this is a foreign station. The mobile BS provides a wireless interface between MHs and the rest of the wired networks. Each mobile BS has its own address in the network. This address can be represented as an MH's current location.

Each cluster has a switching center. These switching centers in the wireless communications networks are connected to a higher level subnetwork. MSCs are connected to each other by communication channels. These switching centers are supported by B-ISDN and SS7 which provide service profiles and locations of MHs. Each switching center has a unique ID number. Based on the capability of a mobile service switching center, each switching center controls a

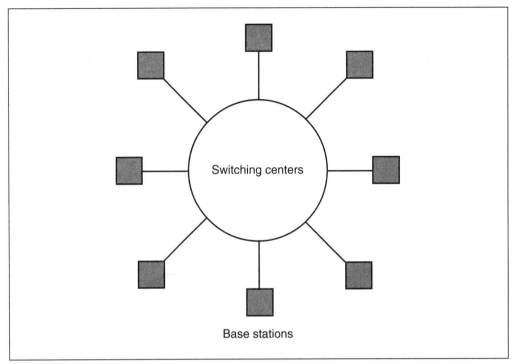

**Figure 4–10 (A)** Vertical Section View of Star Topology

certain number of BSs. Each switching center and its BSs are connected by a wired LAN. Several switching centers and their base stations are connected by a wired MAN. Routing decisions are made by the switching centers.

We divide this architecture into a wired subnetwork and a number of MHs. A wired subnetwork consists of a number of switching centers and BSs. This subnetwork is the central part of the wireless communications network. The topology of this subnetwork depends on the distribution of switching centers. The topology of switching centers can be a mesh, a ring, or other topology. The topology of each switching center and its BSs is a star topology in a vertical section view and a two-level tree topology in a longitudinal section view, as shown in Figure 4–10 (A) and (B).

There are a number of MHs in a wireless communications network. The wired subnetwork is surrounded by MHs. This is done by the radio links that are between BSs and MHs. The switching centers, BSs, and MHs are connected by using communication channels and radio links. A virtual topology of the wireless communications network in a longitudinal view and vertical view is shown in Figure 4–11 (A) and (B).

Graph $G(V, E)$ represents a topology of wireless communications network. Graph $G$ consists of two sets: a finite set $V$ of vertices and a finite set $E$ of edges. The network model can be defined as follows.

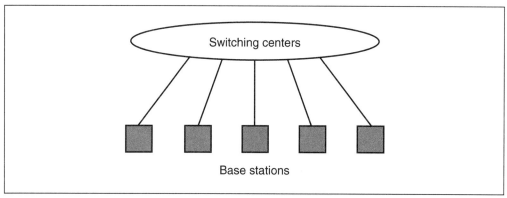

**Figure 4–10 (B)** Longitudinal Section View of Two-Level Tree Topology

The topology of the higher level of a wired subnetwork is represented by a directed subgraph $G_1 = (V_1, E_1)$, where each edge $e \in E_1$ represents the communication channels and each node in $V_1$ represents a switching center. $G_1 \subset G$, $V_1 \subset V$, and $E_1 \subset E$.

The subtopology of the BSs and their related switching centers are represented by an undirected subgraph $G_2 = (V_2, E_2)$, where each edge $e \in E_2$ represents the channels between the BS and the switching center that are directly connected, and each node in $V_2$ represents the BS connected to a switching center. $G_2 \subset G$, $V_2 \subset V$, and $E_2 \subset E$.

The relationships of $G_1$, $G_2$, and $G$, etc., are as follows.

$$G = G_1 \cup G_2, \quad V = V_1 \cup V_2, \quad E = E_1 \cup E_2.$$
$$G_1 \cap G_2 = 0, \quad V_1 \cap V_2 = 0, \quad E_1 \cap E_2 = 0.$$

Each edge $e_i$ in $G_1$ has a limited capacity to carry a number of calls. Each call occupies one unit. The capacity of each edge is defined as $C_1(e_i)$. The edges have the same capacity. Similarly, each edge $e_i$ in $G_2$ has a limited capacity to carry a number of calls, and each call occupies one unit. The capacity of each edge is defined as $C_2(e_i)$. The edges have the same capacity. The capacity relationship between $C_1(e_i)$ and $C_2(e_i)$ can be different or can be the same.

Two different BSs can establish a channel connection by allocating a channel path or a sequence of paths, possibly across several switching centers.

A general communication call request is denoted by $r_i = (s_1, s_2, d_1, d_2)$. This call request consists of four elements: $s_1$ and $s_2$ are the source switching center and source BS, respectively, and $d_1$ and $d_2$ are the destination switching center and destination BS, respectively.

For each edge $e_i$, which is between the switching center and the BS, the total number of channels allocated for a set of call requests $R_2(r_1, r_2, ..., r_n)$ cannot exceed the capacity of the edge. The set of call requests $R_2$ consists of call requests $r_i$ from different BSs of the switching center. That is

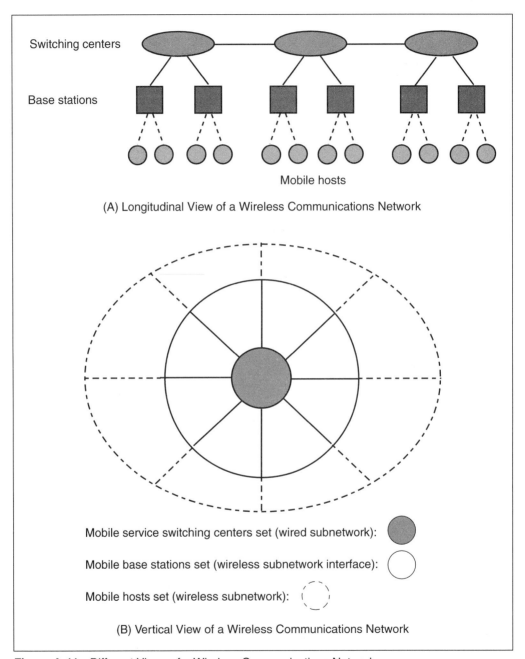

**Figure 4–11** Different Views of a Wireless Communications Network

$$\sum_{i \in E_i} R_2(r_i) \leq C_2(e_i) \quad \text{for all } i, \ i \in E_2.$$

For each edge $e_i$, that is between switching centers, the total number of channels allocated for a set of call requests $R_1(r_1, r_2, ..., r_n)$ that have arrived in the switching centers from their BSs cannot exceed the capacity of the edge between switching centers. That is

$$\sum_{i \in E_i} R_1(r_i) \leq C_1(e_i) \quad \text{for all } i, \ i \in E_1.$$

Let Idle $(r_1(e_i))$ denote the available number of channels $e_i$ between switching centers and Idle $(r_2(e_i))$ denote the available number of channels $e_i$ between the switching center and its BSs. A call request $r_i = (s_1, s_2, d_1, d_2)$ is rejected if Idle $(r_1(e_i)) < R_1(r_i)$ or Idle $(r_2(e_i)) < R_2(r_i)$.

A handoff call request denoted by $H_i = (s_1, s_2, d_1, d_2, h_1, h_2)$ consists of three node pairs: a source switching center and its BS, a destination switching center and its BS, and a handoff switching center and its BS. The $h_1$ and $h_2$ are a handoff switching center and its handoff BS, respectively. A handoff call request is a special call request. A general call request is $r_i = (s_1, s_2, d_1, d_2)$, and the handoff request options are $h_1 = 0$, and $h_2 = 0$. When a call request is involved in a handoff request, the handoff request options become $h_1 \neq 0$, and $h_2 \neq 0$.

If a handoff is requested in the same cluster, the handoff switching center and the old origin switching center are the same ($h_1 = s_1$), and the handoff base station and the old origin BS are different ($h_2 \neq s_2$). This handoff is a noncluster handoff request. If a handoff is requested in different clusters, the handoff switching center is different from its old origin switching center ($h_1 \neq s_1$), and the handoff BS is different from its old origin BS ($h_2 \neq s_2$). This handoff is a cluster handoff request.

A handoff request is a special call request. A call request which involves a handoff request has a higher priority. The priority of a call request that is the cluster handoff request is the highest, the noncluster handoff request is in the middle, and the general call request is the lowest.

Any MH can access the network directly via a BS.

The time length of communication calls and the mobility of different MHs can be different. There are three situations that can arise during mobile communications.

1. The communication time is very short and both MHs move very slowly. When the communication is completed, the caller and callee are still in the original radio cells. This communication is similar to those in wired networks. There is no handoff call request involved.
2. The communication time is not only long but also the MH caller has high mobility. During this communication, the mobile caller can leave the original radio cell and move to a new neighbor radio cell which is served by the same cluster. A noncluster handoff call request is involved in this situation.

3. If one of the MH callers is located in the boundary of the original cluster, this MH caller has high mobility. During this communication, the MH caller can cross different clusters and different radio cells. In this situation, several cluster handoff call requests are involved.

### 4.3.2 The ESP and DSDSP Routing Algorithms

The decision to select call requests and routing paths is made by the switching center. Since each switching center connects a number of BSs, there can be several call requests arriving in a switching center. If call requests arrive at different times at the switching centers, the call request is selected from different BSs based on a first-come-first-served policy. If call requests arrive at the switching center at the same time, a call request is selected based on the policy of the higher priority first. However, if call requests have the same priority and the same arrival time, a call request with the smallest ID in the BS is selected. Two main parameters, transmission delay and network throughput, affect the decision about routing paths selection, and lead to the shortest path and the maximum flow problems.

The shortest path problem is to find the shortest path between a given source-destination node pair in a given topology:

$$\sum_{e_i \in E} e_i = \text{minimum}.$$

Here $e_i \in E$, $e_i$ is the edge between the source and destination nodes $(s, d)$, and $E$ is the set of edges in network topology, for call request $r = (s, d)$ in the communications network.

The maximum flow problem is to allocate the maximum traffic flow at the destination node:

$$\text{maximum} \sum_{e_i \in E} r_i(s, d) \leq C(e_i).$$

Here $e_i \in E$, $e_i$ is the edge between the source and destination nodes $(s, d)$, $E$ is the set of edges in the network topology, and $C(e_i)$ is the capacity of edge $e_i$, for call requests $r_i(s, d)$.

Since transmission time delay is based on the distance of message propagation, the shortest path guarantees the shortest transmission time delay. However, if the capacities of the channels are known, an efficient schedule of traffic leads to the maximum throughput. Hence, these two variables are used to measure the performance of communications networks. Since the current wireless communications networks are real-time systems, the possibility of maximum flow under the condition of the shortest transmission time delay must be considered. Dijkstra's shortest-path algorithm is the shortest path searching algorithm for a given source node to the other destination nodes in a given topology of the network. This algorithm builds the minimum spanning tree that starts in the root of the source node. Any path between the source node (root) and

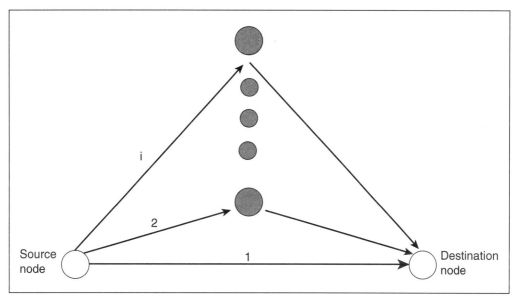

**Figure 4–12** State Dependent Shortest Path

other nodes (leaves) in the MST is the shortest. This shortest path is the only path crossing several nodes in the tree.

If one channel capacity of all channels in the network is full, this channel cannot accept any other calls from call requests, until the channel releases some calls. The original topology changes with the channel capacity. The shortest path for a given node pair can also change, since the states of the topology change. The shortest possible path between a given node pair for the topology of different states is shown in Figure 4–12. When a call request $r = (s, d)$ arrives at the source node, all capacities of channels between the source and destination nodes are idle. The state-dependent shortest path is channel 1, which directly connects these two nodes. If the capacity of channel 1 is not idle, the capacities of other channels are idle. The state-dependent shortest path will become channel 2 or another channel that indirectly connects these two nodes. The decision of searching the shortest path routing is made based on the current traffic states of each channel in the network topology. This is the traffic state dependent shortest path. The traffic state dependent shortest path for a source and destination node pair is the only path at a certain time and in a certain traffic state of the network. This traffic state dependent shortest path is different at different times and in different traffic conditions.

*Lemma*: The shortest path between a given node pair is the only path, where the routing distance $D(s, d) = \sum_{i \in E} e_i = $ minimum, where $i \in E$, for a connection request $r = (s, d)$ in the network.

*Proposition*: Let $D = (D_1, D_2, ..., D_n)$ be a vector satisfying

$$D_j \leq D_i + \alpha_{ij}, \quad \text{for all } (i, j) \in E \tag{4-1}$$

and let $P$ be a path starting at node $i_1$ and ending at node $i_k$. If

$$D_j = D_i + \alpha_{ij}, \quad \text{for all edges } (i, j) \text{ of } P, \tag{4-2}$$

then $P$ is the shortest path from $i_1$ to $i_k$.

*Proof:* By adding the equation (4–2) over the edges of $P$, we can see that the length of $P$ is $(D_{ik} - D_{i1})$. By adding the equation (4–1) over the edges of any other path $P'$ starting at $i_1$ and ending at $i_k$, we can see that the length of $P'$ must be at least equal to $(D_{ik} - D_{i1})$. Therefore, $P$ is the shortest path.

A handoff occurs whenever the signal strength measured at the BS falls below a threshold value. The number of handoffs depends on the following four factors: (1) the size of the radio cells, (2) the size of a cluster, (3) the mobility of MHs, and (4) the communication time length. Once there is a handoff call request at a switching center, this handoff switching center becomes a new source switching center and its BS becomes a new source BS. The original source BS and its switching center become the older one. A decision of rerouting the path between the older source node pair and the new source node pair is made in the older source switching center. There is a strict time limit for making rerouting decisions. If the path searching time is too long, the MH that gives the handoff call request finds that the communication is not seamless from one radio cell to another. If the time is too short, the rerouting path cannot be easily found. This call is dropped before it is completed. Handoff is an important factor affecting the quality and completion of communication in wireless communications networks. Hence, it is important not only to consider the selection of the routing path but also the seamless shifting of the BS and/or switching center to complete the communication call when a handoff call request occurs.

#### 4.3.2.1 The ESP Routing Algorithm

The basic idea of the ESP routing algorithm is as follows. If a call request does not involve a handoff request, the state dependent shortest path is searched to respond to this call. If a call request involves a handoff request, the previous paths are reserved. The rerouting paths are extended from the old source switching center and its BS to the new switching center and its BS in the shortest path searching time to guarantee a seamless continuation of communications. We assume that the previous paths are the shortest between the original source and destination switching centers.

A call request is accepted while the routing paths between the source and destination are all available at call request time. There are three different situations in wireless communications, as described above, and the routing algorithm must consider all of them. Different situations will lead to different routing.

In the first situation, MHs are in the same radio cell during communication. Handoff does not occur, so there is no radio cell shifted or cluster shifted in this communication. The MHs can be modeled as fixed hosts. This is the simplest situation in wireless communications, and is sim-

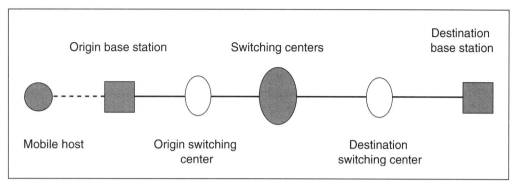

**Figure 4–13** A Nonhandoff Situation in a Wireless Communications Network

ilar to wired communications. Based on the idea of state-dependent shortest path, the routing decision is made only by using three kinds of channel traffic status: (1) the status of the channels between the source and destination switching centers, (2) the status of the channel between destination BS and its switching center, and (3) the status of the channel between source BS and its switching center. Figure 4–13 shows a nonhandoff situation.

In the second situation, MHs are in different radio cells, which are served by the same cluster during communication. Noncluster handoff call requests occur since the original radio cell is shifted to its closest neighbor radio cell. However, the shifted radio cell is served by the same cluster. After this noncluster handoff has occurred, the original source switching center remains the same as before this handoff. Only the original source BS is changed. The rerouting path searching procedure is keeping and searching. The keeping procedure keeps the previous two-fragment routing paths: (1) between the source and destination switching centers, and (2) between destination BS and its switching center. The searching procedure searches the extended routing paths between the original source switching center and the new source BS. The condition of continuing the communication occurs only while this extended routing path is available. Figure 4–14 shows the noncluster handoff situation.

In the third situation, MHs are not only in different radio cells but also in different clusters during communication. This is a cluster handoff case. To continue communication, the original source switching center and its BS are changed to a new source switching center and its BS. Two new fragment extending paths, which are between the new and old source switching center, and between the new source base station and its switching center, should be found to accept this cluster handoff request. The rerouting path searching procedure is also keeping and searching. The keeping procedure keeps the previous routing paths: (1) between the old source and destination switching centers, and (2) between the destination switching center and the destination BS. The searching procedure searches two fragment extending paths, one is between the old and new source switching center, the other is between the new source BS and its switching center. Figure 4–15 shows a cluster handoff situation.

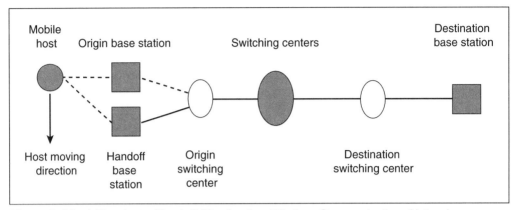

**Figure 4–14** A Noncluster Handoff Situation in a Wireless Communications Network

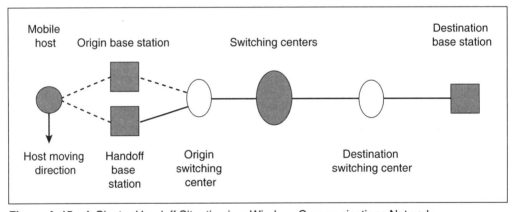

**Figure 4–15** A Cluster Handoff Situation in a Wireless Communications Network

The time between a call request and a call-accepted response should be short. For a call request $r_i = (s_1, s_2, d_1, d_2)$, the source and destination switching centers $(s_1, d_1)$ are directly connected by the state-dependent shortest path. The path searching only needs to check the status of the edge between $(s_1, d_1)$. The path searching time is very short. If the source and destination switching centers are not directly connected, the path searching is not easy. This searching procedure needs to search all of the other switching centers in the network. If the size of the network is small, the searching time is short. Otherwise, the searching time is long.

The completed routing paths consist of three fragments: (1) between the source switching center and its BS $(s_1, d_1)$, (2) between the destination switching center and its BS $(d_1, d_2)$, and (3) between the source and destination switching centers. Since the paths between switching centers and their BSs are directly connected node pairs, these paths do not have to be searched. Hence, it is clear that the searching time of the routing paths depends on the network size and the distance between the source and destination switching centers.

In wireless communications systems, the time to seamlessly shift the radio cells is very strict. To reduce the path searching time, we separate the source and destination switching centers into four categories for a call request: In the first category, a source and the destination switching centers, and their BSs, are the same. In the second, a source and the destination switching centers are the same, but a source and destination BSs are different. In the third, a source and the destination switching centers are directly connected. In the fourth category, a source and the destination switching centers are not directly connected. Since each category of routing paths consists of three segments, each segment has two states: full and idle. Each category in each communication situation can have several kinds of combination states of these paths.

For a nonhandoff situation, the completed routing paths have three fragments. The combinations of the states of these possible routing path fragments between (1) the destination and source switching centers, (2) the destination switching center and its BS, and (3) the source switching center and its BS, total at least 14 in all three categories. The possible path states combination is $2^1$, $2^2$, $2^3$, respectively, under the states of the first, second and third category of call requests, totalling $2^1 + 2^2 + 2^3 = 14$. In addition, the possible path states combination in the fourth category of call requests is uncertain. The total number of combinations of possible routing paths is at least 14 in a nonhandoff situation.

For a noncluster handoff situation, the continuing routing paths consist of two fragments: (1) previous routing paths and (2) a rerouting path that is between the new BS and the original switching center. Since the previous routing paths are already reserved, the states of these completed routing paths are the states of the rerouting path. The total number of combinations of possible rerouting path states between new BSs and the original switching center is 2.

For a cluster handoff situation, the continuing routing paths consist of three fragments: (1) previous routing paths, (2) a rerouting path between the new switching centers and the new BS, and (3) rerouting paths between the new and the old switching centers. The previous routing paths are already reserved, hence the states of this fragment do not need to be considered. The rerouting path between the new switching center and the new BS is directly connected, so there are only two states in this rerouting path. The rerouting paths between the new and the old switching centers can be directly connected or not. There are at least two kinds of states in this fragment. Thus, the total number of combinations of states of possible rerouting paths is at least 4.

When a call request arrives at a BS, a routing decision is made to match one of the combinations in one of the communication situations. If the paths between the destination and source MHs are available, the origin switching center sends beacons to reserve the paths for call requests. After the communication is terminated, these paths are released. This matching procedure saves the path searching time and improves the qualities of communications and services. If the possible routing and rerouting paths do not match any of the path combinations, the routing or rerouting path needs to be searched. The path searching procedure can require a certain amount of time.

Hać and Zhu propose an ESP routing algorithm for a wireless communications network. The main idea of the ESP routing algorithm is that the routing paths are created by building the MST, and the rerouting paths are created by the simplest matching steps to extend the previous routing paths from the origin switching center to the new switching center when handoff occurs. This routing path searching and rerouting path matching are decided by the origin switching center. The ESP routing algorithm consists of the following algorithms: the call initialization algorithm, the path searching algorithm, and the network control algorithm.

The call initialization algorithm is executed in every origin switching center in the wireless network. The basic process of call request selection is first-come-first-served. While several call requests arrive at the origin switching center at the same time, the origin switching center checks the type of call requests and selects the call request with the highest priority first. There are three classes of call requests. The highest priority is the cluster handoff request, a lower priority is the noncluster handoff request, and the lowest priority is the nonhandoff request. If all call requests have the same priority and the same arrival time, a call request with the smallest ID of the BS is selected.

The path searching algorithm is executed in every origin switching center in the wireless network. The main processes of this algorithm are the routing path search for nonhandoff call requests and the rerouting path search for handoff call requests. The routing path search builds the MST in the origin switching center. If the tree is built, the call request is sent to the destination and the routing path is reserved for this call. If the tree is not built, the call request cannot be sent to the destination and the call request is rejected. The rerouting path search builds an extended path tree in the origin switching center. The previous acknowledgment of the routing path and this extended rerouting path in the tree consist of paths completed during communication. If the extended path tree is built, the call continues to the destination switching center from the previous routing path and the extended path of the tree. This extended path of rerouting path is reserved for this continuing call. If the extended tree is not built, the call is terminated. A call request is rejected or a call is terminated if the network is congested. The channel occupancy of call requests is updated.

The network control algorithm is executed in every origin switching center in the wireless network. The algorithm receives messages from the source BS, source switching center, and destination switching center. Upon receipt of a message from the source BS, the message is sent to the source switching center and the call initialization algorithm is invoked. A call request message is sent to the destination switching center by the path searching algorithm. The network control algorithm checks the status of the calls in the network. If a call is finished, the channels of the call are released. The network status is updated.

4.3.2.2    The DSDSP Routing Algorithm

In the DSDSP routing algorithm, if a call request does not involve a handoff request, the state-dependent shortest paths are searched to respond to this call. If a call request involves a handoff request, a new state-dependent shortest routing path is searched. If the new path is available, this path substitutes for the previous paths. The previous routing path is released, and can be used for other call requests. The new path is between the new switching center and the destination switching center. The new path is the shortest at that moment. This algorithm allows for lower channel occupancy, a lower call drop rate, and a higher network throughput.

In the ESP routing algorithm, the routing paths are always the shortest paths in nonhandoff call request situations. For a nonhandoff call request $r_i = (s_1, s_2, d_1, d_2)$ the routing paths, which are between source switching center ($s_1$) and destination switching center ($d_1$) belong to the MST whose root is in $s_1$. In the noncluster handoff call request situations, where MHs do not move out of their original source switching centers, the only change involves their source BSs. Network topologies for the original and noncluster handoff successful call request are not changed. Hence, rerouting paths after the handoff call request are still the shortest paths that were previously searched in the noncluster handoff call request situations.

In the cluster handoff request situations, the rerouting paths that are between the new source switching center and destination switching center are not guaranteed, after handoff, to be the shortest paths in the ESP routing algorithm. The previous path is the shortest path, which is a part of the MST with the root in the source switching center built before the cluster handoff call request occurred. The extended path is also a part of the MST with the root in the original source switching center built after the handoff call request occurred. These two trees were built at different times, and after the network status changed. Thus, these two trees may not be the same, and the extended path and the previous path consist of a path, which is not the shortest for a handoff call request since a loop is generated after handoff call request is accepted.

In the cluster handoff call request situations in the ESP routing algorithm, the extended rerouting paths, which are between the new source and destination switching centers, are extended from previous shortest paths to the new source switching centers after the handoff is accepted. If the new source switching center is directly connected to both the destination and older source switching centers, the message flow along this rerouting path goes backwards from the new source switching center to the old switching center and then forward to the destination switching center. These rerouting paths after the cluster handoff call request are not the shortest paths because the new source switching center is not directly connected to the destination switching centers. A triangle loop is created by these three switching centers as shown in Figure 4–16. Additional communication channels are used in this communication. If the size of clusters and radio cells decrease and the mobility increases, the number of cluster handoff call requests increases. Thus, the wireless communications networks become congested, call drop rate increases, and the network performance decreases.

The time to search the rerouting path affects seamless communication. If we search the state-dependent shortest path in a reasonably short time, the quality of communications and services is not affected. A state-dependent shortest path search involves both factors, the time and path length. The search builds the MST with the root in the new source switching center. If the spanning tree is available, the path between the root and the destination switching center is the shortest path at that moment.

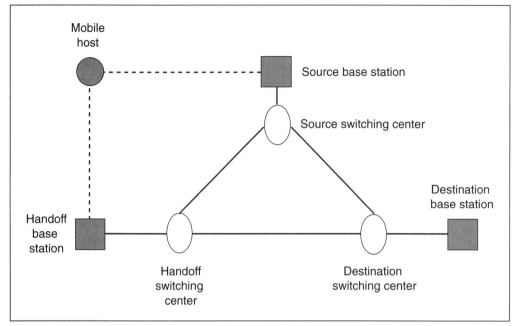

**Figure 4–16** A Loop Example in the ESP Routing Algorithm

The ESP routing algorithm finds the extended path more easily than the DSDSP routing algorithm. For a handoff call request, the DSDSP routing algorithm is involved in the extended rerouting path search procedure, if the new shortest path is not available. If the new shortest path and the extended rerouting path are both available at the same moment, the new shortest path is selected first. The previous shortest paths are released. If the new shortest path is not available and the extended rerouting path is available, this extended path is reserved for the completion of the communication. The DSDSP routing algorithm decreases the call drop rate, transmission time delay, and network congestion. The algorithm also improves network performance, and the QoS since this algorithm has more options.

Hać and Zhu propose the DSDSP routing algorithm for a wireless communication network. The main idea of the DSDSP routing algorithm is that the routing paths are searched by building the MST, and the rerouting paths can be created by a possible MST at the new origin switching center in the best case, or by the extended paths between the new and the old switching centers in the worst case. The state-dependent shortest path search is completed in a negligible amount of time. The extended path searching is also completed. If the state-dependent shortest path is not available, the extended path can be used as the rerouting path while the extended path is available. If all the paths are available, the state-dependent shortest path is chosen as the new routing path to continue communication. Previous routing paths and the extended path are released.

The DSDSP routing algorithm involves three subalgorithms: the call initialization algorithm, the option path searching algorithm, and the network control algorithm.

The call initialization algorithm is executed in every source switching center in the wireless network. The basic process of call request selection is first-come-first-served. If several call requests arrive at the source switching center at the same time, the source switching center checks the type of call requests and selects the call request with the highest priority first. There are three classes of call requests. The highest priority is the cluster handoff request, a lower priority is the noncluster handoff request, and the lowest priority is the nonhandoff request. If all call requests have the same priority and same arrival time, the call request with the smallest BS ID is selected.

The option path searching algorithm is executed in the original switching center in a wireless network. The main processes of the algorithm are the routing path search for nonhandoff call requests, and the rerouting path search for handoff call requests. The routing path search builds the MST in the origin switching center. If the tree is built, the call request is sent to the destination and the routing path is reserved. If the tree is not built, the call request cannot be sent to the destination and the call request is rejected. The rerouting path search has two strategies. The first strategy is to build the MST in the handoff switching center. The MST creates the shortest path between the new switching center and the destination switching center. The shortest path is reserved for the continuing call and the previous routing path is released. The call continues through the shortest path to the destination. The second strategy is based on the previous routing path acknowledgment of the extended rerouting path availability from the origin switching center to the handoff switching center. If the tree is built and the extended rerouting path is available, the tree is selected, and the extended rerouting path is released. If the tree is not built, and the extended rerouting path is available, the call continues through the extended rerouting path to the destination. If neither the tree nor the extended rerouting path are available, the call is terminated. A call request can be rejected or a call can be terminated if the network is congested. The channel occupancy for the call requests is updated.

The network control algorithm is executed in every source switching center in the wireless network. The algorithm receives messages from the source BS, source switching center, and destination switching center. Upon receipt of a message from the source BS, the message is sent to the source switching center and the call initialization algorithm is invoked. The call request message is sent to the destination switching center by the path searching algorithm. Meanwhile, the network algorithm checks the status of calls in the network. If a call is completed, the channels for the call are released. The network status is updated.

### 4.3.2.3  A Comparison between the ESP and DSDSP Routing Algorithms

The routing path searching procedure for both the ESP and DSDSP routing algorithms builds the MST.

The rerouting path searching strategies for the ESP routing algorithm are matching and extending the path searching methods. If the rerouting paths match one kind of the extended rerouting paths, the rerouting path searching time is the shortest. If the rerouting path does not match any possible rerouting path combinations, the rerouting path extended search procedure searches the extended rerouting path from the old origin switching center to the new switching center. This rerouting path search occurs in part of the network. The number of available rerouting paths are smaller than the number of paths in the entire network. Meanwhile, the completed communication routing paths consist of both rerouting paths and previous routing paths. These communication routing paths do not guarantee that the paths are the shortest. The ESP routing algorithm emphasizes the shortest path searching time.

The rerouting path searching strategies of the DSDSP routing algorithm are to build the MST at the new switching center and search the extended paths at the old switching center. This rerouting path searching occurs in the entire network. The rerouting path searching time can be longer than the searching time of the ESP routing algorithm. If the network size is very large, the rerouting path search is expensive. The distance of the rerouting paths can be shorter than the distance found by using the ESP routing algorithm. Since the rerouting path searching occurs in two phases at the same time, the rerouting paths have more options. The handoff drop rate is smaller than by using the ESP routing algorithm. The DSDSP routing algorithm is more efficient due to lower handoff drop rates and shorter rerouting paths. The DSDSP uses a slightly longer path search time than the ESP algorithm.

Both the ESP and DSDSP routing algorithms are sensitive to network topology and the distribution of call requests. The complexity of both algorithms is $O(N^2)$ in the worst case. Both the ESP and DSDSP routing algorithms consist of three subalgorithms. The computation in two subalgorithms, which are call initialization and network control requires $N$ steps, where $N$ is the number of nodes in the network. Thus, the complexity of these two subalgorithms is $O(N)$. The path searching subalgorithm in the ESP and option path searching subalgorithm in the DSDSP build the MST in the network. Each step in both subalgorithms requires a number of computations proportional to the number of nodes, $N$, and the steps are repeated $N-1$ times, thus, the complexity of computation in the worst case is $O(N^2)$. Combining all complexities of these three subalgorithms in the ESP and DSDSP routing algorithms, results in $O(N^2)$ for both the ESP and DSDSP routing algorithms.

The ESP and DSDSP algorithms use the state-dependent path searching method to establish the routing path from the origin switching center to the destination switching center. The routing path is the state-dependent shortest path by using the MST built at the origin switching center. The rerouting path can be the state-dependent shortest path by using the MST that is rebuilt at the new origin switching center and releases the old MST that was built at the old origin switching center in the DSDSP algorithm. The rerouting path searching time of the ESP algorithm is shorter than the searching time of the DSDSP algorithm. The time length of the rerouting path searching time in the DSDSP algorithm strongly supports seamless communications. The DSDSP algorithm has a lower handoff drop rate and call block rate than the ESP algorithm. The performances of both algorithms are sensitive to the network topology and distribution of call requests.

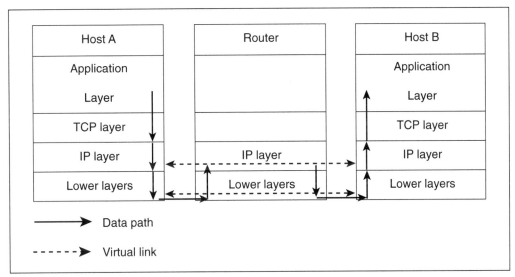

**Figure 4–17**  Virtual Link of Hosts

## 4.4 MH Protocols for the Internet

In TCP/IP, an application is connected with another application through a router as shown in Figure 4–17. Each host in the Internet has a unique address. An IP address is a 32-bit binary number that can also be assigned in dotted decimal notation. IP addresses contain two parts: a network address that identifies the network to which this host is attached and a local address that identifies the host. A local address can be separated into two parts: subnet address and local address.

The hierarchical address makes routing simple. A host who wants to send packets to another host only needs to send packets to the network to which the target host is attached. The host does not need to know the inside of the network. However, a computer's IP address cannot be changed during the connection and communication. If the user wants to move the computer to another area while using it, this will be difficult because the physical IP address of the computer must be changed in a different subnet. To solve this problem, a number of protocols have been proposed: Virtual IP (VIP), loose source routing IP (LSRIP) and Internet engineering task force mobile IP (IETF-MIP).

### 4.4.1  VIP

A VIP uses two 32-bits IP style addresses to identify MHs: a VIP address and a temporary IP address (TIP). A VIP address is the IP address that the MHs get from their home network. MHs always use a VIP as their source address inside the IP packet. When MHs move to another network, they get another IP address from the foreign network, which is a TIP address. Each VIP packet contains information to combine VIP and TIP, so the packet target to the VIP can be routed through the general Internet to its temporary network by reading its TIP. The VIP uses additional space inside the packet to carry this information; a new IP option to identify the VIP while original address fields carry the TIP.

When an MH moves to a foreign network, the information about the MH's current location is sent to the MH's home network. During the transmission, each intermediate router that supports the VIP protocol can receive this information and update this router's cache. This cache is a database that stores information about MHs' current location.

If a host wants to send a packet to an MH, it only knows the VIP address of the MH and does not know its TIP, the current location. The packet will be sent to the MH's home area. If any intermediate router that supports the VIP protocol receives this packet, it will modify this packet according to its cache so that new packet will include information about the TIP that can be routed to the MH's current location. If the packet does not reach any intermediate router that supports the VIP and it has cache information about the MH, the packet will be routed to MH's home network. The gateway in the MH's home network can modify this packet according to this gateway's cache and route the packet to the MH's current network. This gateway always has the MH's current location, because each time the MH moves to a new network, the MH notifies its home network of its current location. The optimized path in VIP is mainly based on the number of intermediate VIP routers. If many intermediate routers support the VIP protocol, the optimized routing path should be obtained. The option the VIP uses to carry information of VIP is an option of IP, and not all of the routers will support this option. Some computers even discard all of the options the IP packets carry. The VIP protocol also needs several extra IP addresses for foreign networks to assign to the MHs.

### 4.4.2 LSRIP

LSRIP uses one of the IP options to cause the IP packets to be routed through a series of intermediate routers to the destination. This IP option is the loose source and record route (LSRR). For example, if host A wants to send a packet to host C with an LSRR option, the packet can reach host B through general IP routing. Then host B replaces the destination IP address in the IP header with the first IP address, C, in the LSRR option and routes the packet to a new destination, C. Also, host B points to the second IP address in the LSRR option. Host C can perform the same procedure: replace the destination IP address with next IP address in the LSRR option, increase the pointer to the next IP address in the LSRR option, and reroute the packet to a new destination. Not until the pointer points to the last IP address in the LSRR option, is the packet to its destination address. The LSRIP uses the LSRR option to carry the information of an MH.

When an MH moves to a foreign network, information about the MH's current location is sent to the MH's home network. During the transmission, each intermediate router that supports LSRIP protocol can receive this information and update this router's cache.

If a host wants to send a packet to an MH, which is not in its home area, the packet will first be sent to the MH's home network. During the path to the home network, if an intermediate router has the cache of the MH, this router can put the LSRIP option into the packet and cause it to be routed to the current network of the MH. The gateway of the current network reads information from the LSRIP option and can determine that the MH is this packet's destination. Thus, the gateway can route the packet to the MH that is connected to the current network.

If the packet does not reach any intermediate router that supports LSRIP and has cache information about the MH, the packet will be routed to its home network. The gateway in the MH's home network can add the LSRIP option to this packet according to the gateway's cache, and route the packet to the MH's current network. This gateway always has the MH's current location because each time the MH moves to a new network, the MH notifies its home network of its current location.

The LSRIP needs more intermediate routers to achieve an optimized routing path. Also, the option used by the LSRIP is not compatible with current routers. We can test this by sending a packet using a trace route, a tool that checks the path through which one packet has passed. Trace route uses the LSRR option to record the path of the packet and the transmission time to each intermediate host. If any intermediate host does not support the LSRR option, the trace route bypasses this host. After sending a packet, we can find from the messages sent back that some sites are not displayed correctly, shown as "*", which means that these sites do not support the LSRR option.

### 4.4.3 IETF-MIP

The IETF-MIP is the most usable protocol for mobile IP in the Internet. The basic idea is to use two agents to handle the job related to the MH. When the MH moves to other networks, it will notify the foreign network's agent, the foreign agent (FA), and its home agent (HA) about its current location. Then when the packet to the MH is sent to the HA using general IP, the HA will modify the header of the IP packet, change the destination address to the FA's address, and add some fields to the packet including the MH's permanent address. When the FA receives this packet, it will know it is for an MH that is now at its location, the FA will modify this packet again, and send it directly to the MH through the local network. In IETF-MIP protocol, there are cache agents to optimize performance. A cache agent is a host that can maintain a database that stores the MH's current location. This database can be changed as the location of the MH changes.

In this protocol it is difficult to achieve the optimized routing path, especially in a WAN. For example, a user in London wants to send a packet to an MH in London, whose home network is in New York. First, this packet will be sent to New York, modified by the home agent, and then sent back. This takes about one circle of the whole earth. The optimized path can only be about 100 feet. Thus, the only solution for IETF-MIP is to set up as many cache agents as possible in the entire Internet. When the cache agents receive a packet, they can modify the packet instead of sending it to the HA of the MH.

### 4.4.4 A Comparison between Mobile IPs

The mobile IP should be compatible with current IPs, which means the current protocols and applications do not need to be changed. Mobile IP also needs to have an optimized routing path, which means the protocol should be efficient in routing packets. VIP and LSRIP use the option of IP to carry information about MHs. But some of the current hosts do not support an IP option. When these hosts receive a packet that includes options, they discard the options of the IP packet. Meanwhile, IETF-MIP only uses the basic IP header and packet, and does not use any IP option. From the compatibility point, IETF-MIP is the best protocol out of the three. From the point of the optimized routing path, all three protocols depend on intermediate cache hosts. If there are enough intermediate cache hosts inside the Internet, the three protocols can find an optimized routing path. The IETF-MIP is the best out of the three mobile IPs, if we consider both the compatibility and optimized routing path. The IETF-MIP is also the mobile IP used in the Internet.

### 4.4.5 Scalable Mobile Host IP (SMIP) Protocol

Hać and Guo propose a mobile IP protocol, SMIP, that solves the disadvantages of IETF-MIP protocol in a WAN, and considers the concept of a hierarchical structure of the Internet. In this protocol, we propose a packet format, packet routing path, MH registration procedure, MH notification procedure, registration message format, notification message format, and authentication procedure. The SMIP protocol can be used in a hierarchical network, such as networks configured using an open shortest path first (OSPF) protocol. By using a hierarchical structure, most of the networks can be considered as being connected to the backbone through a border router (BR), physically or virtually. All of the packets sent out will pass through these BRs, as shown in Figure 4–18.

In SMIP protocol, each MH has a permanent IP address that is received from its home network, and the protocol will use this address to send and receive IP packets. The home agent is the router or a host inside the MH's home network. The home agent will arrange the location information of MHs that belong to this subnet. Each home agent has a list to trace the current location of all of the MHs that belong to this home agent. The HA also has a function to modify the general IP packet so that this packet can be routed to the current location of the MH. The FA is a router or a host in another subnet, which will handle the registration and location information of MHs from other subnets. Each FA has a list so it can remember all of the MHs visiting it. Also, each FA has a function by which it can recognize the packet sent by the HA or other agents, and send it to the MH. The border router (BR) is a router for the entire area. It holds the information about all MHs inside this area. Each BR also has a function to modify the packet sent by the HA and then to route it to the correct FA inside its area. All of the MHs, HAs, FAs, and BRs, have a cache function to achieve the optimized routing. This cache function will modify and route the packet to the correct destination according to the unexpired cache for this MH. Each time this cache function receives a notification message regarding the new location of an

# MH Protocols for the Internet

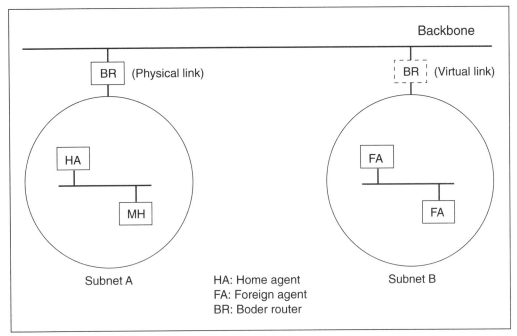

**Figure 4–18** Hierarchical Network

MH, it can update the cache information about this MH. The cache information is stored in a database.

The basic operation of the SMIP is as follows. When an MH moves from its HA inside area A to a new FA inside a new area B, the MH can register to this new FA. This new FA sends the registration message to the HA of the MH through the BRs of areas A and B. Then the new FA puts the MH onto the new FA's visiting list. The HA, BR of area A, and BR of area B, have the current location information of this MH. The difference for the HA and BRs is that the current location address of the MH in the HA and BR of area A, is the address of the BR in area B. The current location address of the MH in the BR of area B, is the address of the new FA. At this time, if host S wants to send a packet to the MH, the packet can be sent to the MH's home network because S does not know the MH's current location. The packet received by the HA can be modified and then routed to the current location address of the MH. The modification procedure is to make a copy of the old IP header to keep the original information, change the destination address of the packet to the current location address of the MH, change the source address of the packet to the HA's address, and change the packet protocol to SMIP from IP. After the modification, the packet can be routed as a general IP packet to the target, which is the BR of area B. The BR of area B reads the information about the packet and knows that this is a SMIP packet. The BR also knows the current location address of the MH, which is the address of the new FA inside area B. The BR modifies the packet again by changing the destination address of the packet to the address of the new FA, and by changing the source address of the packet to the address of the

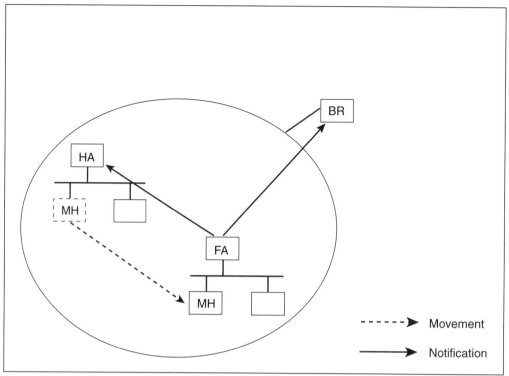

**Figure 4–19** The MH Moves Inside the Same Area

BR. Then the packet will be routed to the new FA as a general IP packet again. When the new FA receives this packet, the FA knows that this is a SMIP packet for the MH, which is connected to the local network of the FA. The FA modifies the packet by restoring the original source address and destination address. This information comes from the SMIP part in the packet header. After modification, the packet can be sent to the MH through the local network.

We describe SMIP by using different scenarios. Each scenario includes actions like movement of the MH; registration to the FA, BR and HA; updating the message sent to previous FAs, HAs and BRs, and finally, the routing of the packets.

#### 4.4.5.1   HA to FA Inside the Same Area

In Figure 4–19, the MH moves from its HA to an FA inside the same area. This MH will first register with the FA, and the FA will notify the MH's HA and BR. Then the HA and BR will have the location information of the MH. Meanwhile, the HA will broadcast the address resolution protocol (ARP) update information inside its local network so that the host in the network will not try to send a packet directly to the MH, which is not yet in the same local network. The FA also broadcasts the ARP update information inside its local network so that each host in the network will recognize the MH and can send packets directly to the MH.

MH Protocols for the Internet                                                                                                            199

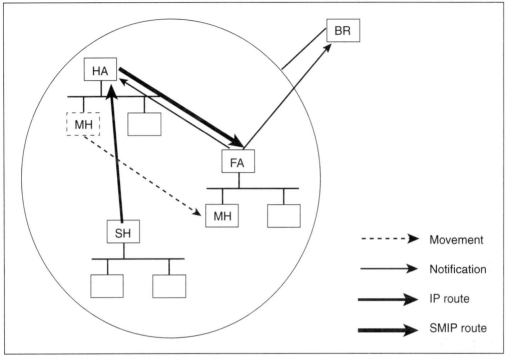

**Figure 4–20**  A Packet Sent Inside the SH

If a host inside the same area, the SH, wants to send a packet to the MH, as shown in Figure 4–20, the IP packet will be routed to the HA as a general IP packet. The HA will modify this packet by changing the destination address to the FA, and adding more fields to store the actual address of the MH. Then the HA routes the packet to the FA by using a general IP routing. Then, the HA will send a message to the SH to notify it of the correct location of the MH. If the SH supports SMIP, it will cache this information for later use. In this situation, the packet goes through the same path as the IETF-MIP because the packet is routed inside a small area.

When a host in the other area wants to send a packet to an MH, as shown in Figure 4–21, the packet will pass through the BR first. The BR has information about the MH, so it will modify the packet and route it to the FA directly. In this situation, the packet transfer time will be lower than for IETF-MIP, because there is no need to send the packet to the HA again. In IETF-MIP, the packet will first be sent to the HA, and then routed to the FA because the BR does not have the current location information of the MH.

### 4.4.5.2     HA to FA in the Other Area

If the MH moves to the other area, as shown in Figure 4–22, the MH will register to the FA in area B. This FA will notify the previous FA and previous BR in area A regarding the new location of the MH. Also this new FA in area B will notify the HA of the MH, and then the BR in

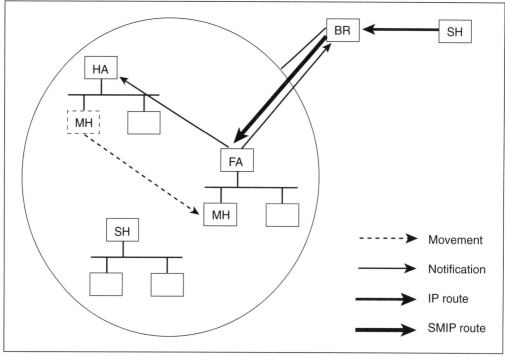

**Figure 4-21** A Packet Sent Outside the SH

area B about the information of the MH. The new location of the MH reported to the previous FA, previous BR and HA will be the BR's address of area B. The location of the MH that reported to the BR in area B will be the address of the new FA in area B. The transfer of packets in this scenario can be described in three situations. First, the packet is sent from the host inside area A, which is the home area of the MH. Second, the packet is sent from area B, which is the current area of the MH. Third, the packet is sent from other areas, such as area C.

In the first situation, the SH in area A sends the packet to the MH's HA if it does not have the current location information of the MH. The HA then modifies and routes the packet to area B's BR according to its cache. (Also, the HA will send notification message to the SH.) After receiving the packet, the BR in area B will modify and route the packet to the FA in area B. Finally, the FA in area B removes the additional information the BR added and passes the packet to the MH using local network access. The path optimization in this situation is the same as in IETF-MIP, but one additional process in the BR is added. For IETF-MIP, when the MH moves to the FA in area B, the current location of the MH reported to the HA is the address of the FA in area B. The packet sent to the MH is modified by the HA so that the destination address is the FA in area B. This modified packet will pass the BR in area B as a general IP packet because the destination is the FA. In this situation, an additional process is added by the BR in area B for SMIP.

MH Protocols for the Internet                                                                201

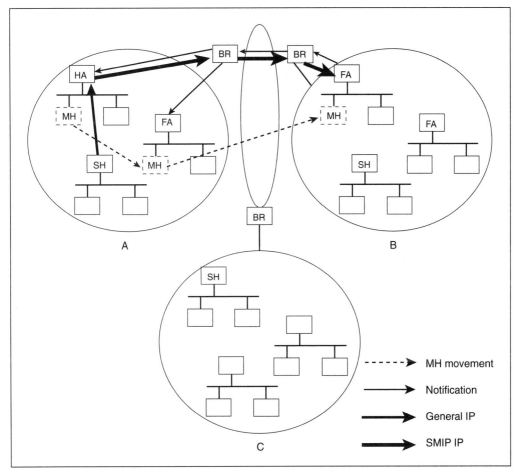

**Figure 4–22**  Home Area to the Other Area

In the second situation, shown in Figure 4–23, the SH in area B wants to send a packet to the MH. Because the SH does not know the location of the MH, it will send the packet to the MH's HA. This packet will go through area B's BR, and this BR knows the current location of the MH. This packet will be modified by the BR and routed directly to the correct FA inside area B. Also, the BR will notify the SH in area B that current location of the MH is the FA. In this situation, a significant amount of time has been saved because the packet does not need to travel to the MH's HA and be bounced back.

In the third situation, when the SH in area C sends a packet to the MH, the packet will first go to the MH's HA if the SH does not have the cache information. The BR in area A will receive this packet and modify it (because it has the cache information of the MH), then route the packet to the BR in area B. Then the BR in area A can send a notification message to the BR in area C

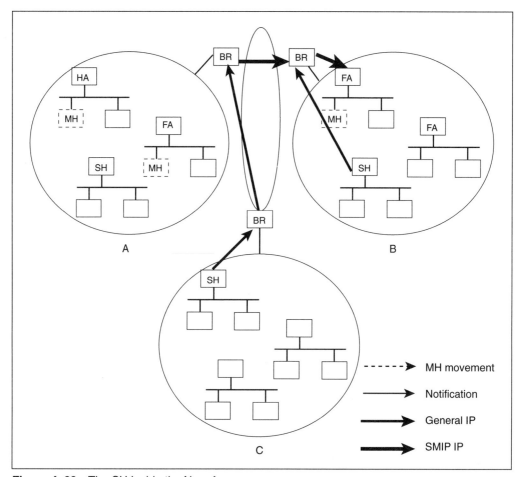

**Figure 4–23** The SH Inside the New Area

to indicate that the current location of the MH is the BR of area B. The BR in area B will modify the packet again after receiving it and route it to the correct FA in area B.

4.4.5.3   Moving Inside the Other Area

Suppose the MH changes its location inside area B, as shown in Figure 4–24, to a new FA. The MH needs to register to this FA. The new FA will determine the previous FA's location. If previous FA is in the same area, the new FA will only notify the previous FA and BR in area B. There is no need to notify the HA of the MH, because the binding address of the MH for the HA is the same as before.

In this scenario, we consider three situations. First, the SH in area A sends a packet to the MH. If the SH does not support SMIP, at this time, nothing changes. The packet follows the same path as in the previous scenario, from the HA, then to the BR of area A, then to the BR of

MH Protocols for the Internet 203

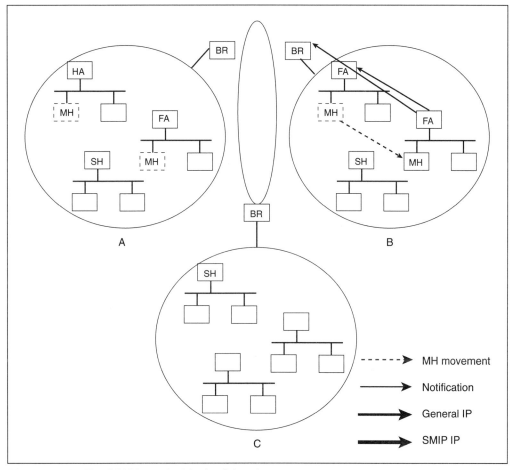

**Figure 4-24** The MH Moves Inside the Other Area

area B, and finally gets to the new FA. If the SH supports SMIP, then the last time the SH sent a packet to the MH, the SH can get cache update information from the HA. The packet moves to the BR of area A, then to the BR of area B, and finally gets to the new FA in area B. Except for IETF-MIP, this packet is first sent to the old FA in area B, and then is modified by the old FA, and finally is sent to the new FA. In comparison, SMIP saves the packet transmission time from the old FA to the new FA.

In the second situation, if the SH supported SMIP the last time the SH sent a packet to the MH, the SH would get a cache update notification from the BR of area B. Then the SH tries to send a packet to the old FA. The old FA then forwards the packet to the new FA according to its own cache information of the MH. If the SH does not support SMIP, the path through which the packet is routed will be the same as before.

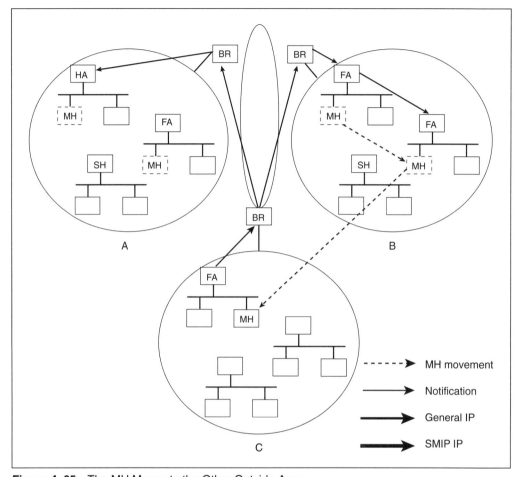

**Figure 4–25** The MH Moves to the Other Outside Area

In the third situation, either the BR of area C or the SH supporting SMIP causes the packet to be routed to the BR in area B directly. This is almost twice as fast as before, because the packet does not need to be routed to the BR in area A. If both the BR and SH do not support SMIP, the packet follows the same path as before.

### 4.4.5.4  Moving from the Outside Area to a New Outside Area

In a scenario shown in Figure 4–25, the MH moves from area B to C. The MH registers with the new FA. The FA, after determining that the MH has come from the other area, will notify the previous FA, the BR of area B, the HA of the MH, and the BR of area C. The binding information sent to the previous FA, the BR of area B, and the HA, will be the address of the BR in area C. The binding information sent to the BR of area C is the address of the new FA in area C.

**MH Protocols for the Internet**   205

In this scenario, we consider three situations. In the first situation, the SH in area A wants to send a packet to the MH. The packet is first sent to the BR in area A. The BR in area A has the current location information of the MH because this BR received the information when the MH registered to area C. The BR in area A can modify this packet and route it to area C. The BR in area C can modify the packet again when this BR receives it. Then the packet is routed to the FA in area C to which the MH is attached.

In the second situation, if the SH wants to send a packet to the MH, the SH first sends the packet to the old FA in area B, according to the SH's cache information about the MH. The old FA in area B can modify this packet and route it to the correct destination, area C. Also, the old FA can send a notification message to the SH to update SH's outdated cache information about the MH.

In the third situation, the packet sent by the SH to the MH must pass the BR in area C. The BR in area C has the current location information of the MH. The BR in area C can modify this packet, and route the packet to the current FA in area C.

### 4.4.5.5  SMIP Packet Format

The SMIP packet is shown in Figure 4–26. When a general IP packet comes to an agent, which has cached the MH's location, the agent will first copy the entire IP header, then add it to the top of the original IP packet, change the source address to the agent's address, the destination address to the address of the FA to which the MH is being connected, and the IP to the SMIP. After modification of the original IP packet, the agent will send it again like a general IP packet.

The entire IP header is copied for the purpose of compatibility with the current Internet routing. Only by copying the entire IP header, some IP packets which were fragmented can be assembled correctly when they reach the target. When the MH is attached to area B as shown in Figure 4–22, and the HA receives a packet to the MH, the HA will put the BR's address into the destination field, the HA's address into the source field, and change the IP to SMIP. When the BR in area B receives the packet, the BR knows that this is the SMIP packet, and the destination is for the MH inside its area. The BR looks for the cache information about the MH, and then puts the current FA address into the destination field, making the BR address the source address, and then routes the packet to the current FA. The field of previous agent addresses is used for sending the notification message more efficiently. An example is shown in Figure 4–27. The first location of the MH is FA1, then after some time the MH moves to a new location FA2, and after some time this MH finally moves again to another new location, FA3. Now only FA2 knows that the current location of the MH is FA3. If the SH wants to send a packet to the MH, this packet will be sent to FA1. FA1 has cache information about the MH's current location, which is FA2. FA1 got this cache information when the MH moved from FA1 to FA2. FA2 sent the notification message to FA1 when the MH registered to FA2. FA1 will modify the packet and forward the packet to FA2 according to FA1's cache. After receiving this packet, FA2 checks the cache for the MH and forwards the packet to FA3. FA2's cache information about the MH is different from FA1. In FA2, the current location of the MH is FA3. From the packet that FA2 receives, FA2 knows that this is the SMIP packet. FA2 also knows that the agent that modified and sent the

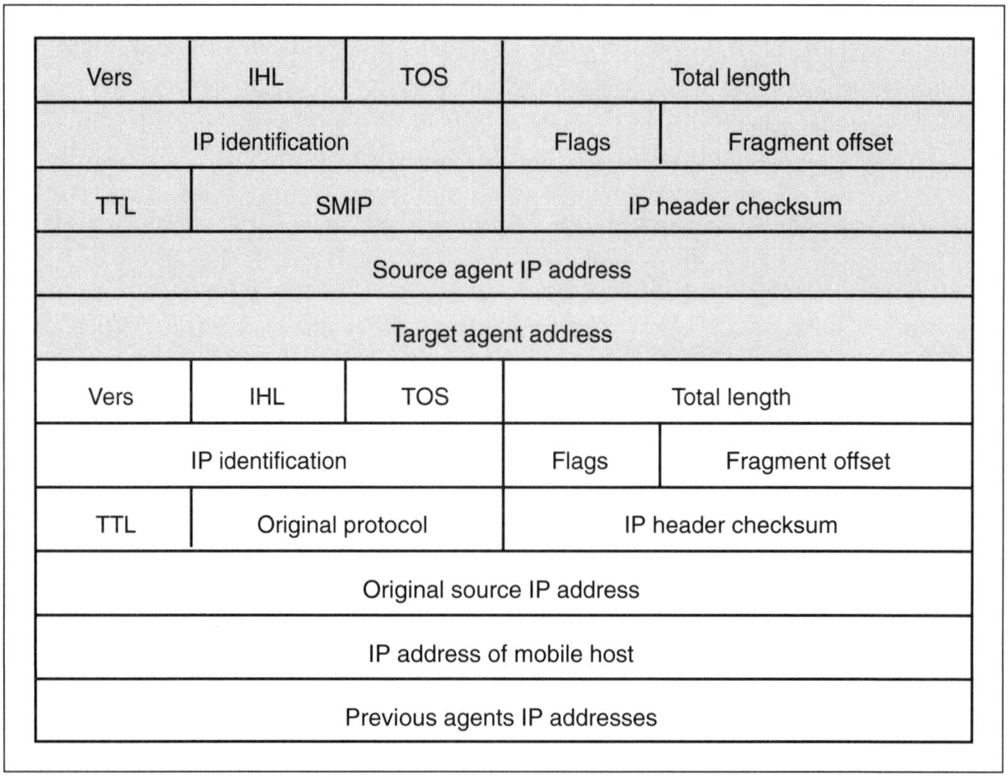

Vers
IHL
TOS
Total length
IP identification
Flags
Fragment offset
TTL
SMIP
IP header checksum
Source agent IP address
Target agent address
Vers
IHL
TOS
Total length
IP identification
Flags
Fragment offset
TTL
Original protocol
IP header checksum
Original source IP address
IP address of mobile host
Previous agents IP addresses

**Figure 4–26** SMIP Packet

FA1
FA2
SH
MH
MH
FA3
MH
MH movement
Notification
General IP
SMIP IP

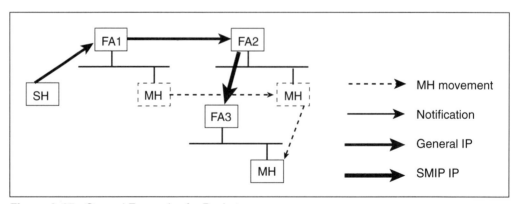

**Figure 4–27** Several Forwards of a Packet

packet has outdated cache information about the MH. During the modification, FA2 puts FA2's address as the agent source IP address, and puts FA1's address, which is the old agent source IP address, into the previous agents addresses field. Then, when FA3 receives this packet, FA3 will

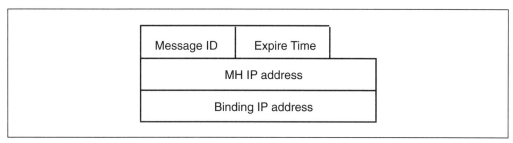

**Figure 4–28** SMIP Notification Message

know that FA2, which is in the agent source address field, has the current location of the MH. But FA1, which is in previous agents addresses field, does not have the correct location information of the MH. FA3 will send a notification to FA1 to update FA1's cache for the MH.

#### 4.4.5.6 SMIP Notification Message Format

The SMIP notification message is shown in Figure 4–28. The message ID field represents the type of message. There are three kinds of messages that can be sent: warning, binding, and request, represented by 1, 2, and 3, respectively. When an agent receives a message whose message ID is 2, that means this message is a binding message. If the message ID inside a message packet is 1, it is a warning message. If the message ID inside a message packet is 3, it is a request message. The host that receives a message packet can act according to the message ID. If the message ID is 1, the host will search the cache database for the MH inside the message packet and delete the cache information for this MH. If the message ID is 2, the host will update the cache information for this MH or create a new cache entry for it if there is no cache information about it. If the message ID is 3, the host will send back a new message to the required host. This new message ID is 2, and includes the current location of the MH. When agent A's cache for an MH has expired, and it still receives a packet from agent B, then agent B has an outdated cache. Agent A will then send warning message to agent B. The received packet will be sent to the MH's HA. After receiving the warning message, agent B will delete the cache information for this MH. Then agent A and agent B need to send a request message to the MH's HA. If the message ID is 3 and the binding field is left blank, the HA of the MH will send back a binding message to the agent who sent the required message. The binding message includes new binding address of the MH and the expired time of this binding. Another situation of sending binding messages occurs when the MH moves to the other area. The current FA will send the binding message to the previous FA, HA, or BR.

#### 4.4.5.7 SMIP Authentication

SMIP considers an authentication problem in the Internet. There are two general situations: registration period and message sending period.

The preferred structure of SMIP allows each organization to have its own security key, and different BRs can set up a shared security key. When the MH moves inside its HA, it can use the security key to encrypt registration information and register with the HA and other FAs. When

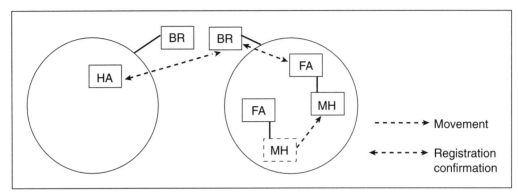

**Figure 4–29** MH Registration to a New Area

the MH moves to another area, as shown in Figure 4–29, only the BR needs to confirm the registration of the MH to the HA the first time the MH enters. All other FAs inside the same area will accept the MH's registration. Using this method, a large number of registration confirmation messages can be saved. This organization only needs to take charge of their own security key.

The notification in SMIP can also take advantage of this structure. When an MH moves inside the same area, every binding message will be encrypted using the security key from the area, which will make all updating messages more stable and trustable.

If the network cannot use the same security key inside one subnet, SMIP will use a simple authentication method to make sure the registration and notification messages are secure. When the MH registers to an FA, A, which does not have a shared security key with the HA, this FA, A, will forward the registration message to the HA, then get the reply from the HA, and the registration will succeed. Now the FA, A, can create a random number for this MH as the temporary security key for future use. The next time the MH registers to the other FA, B, the MH can send this random number with the registration message to the FA, B. The FA, B, can check the registration for previous FA, A, by sending the random number to A, and if B gets a reply from A with the same random number, that means the MH has the correct registration message. As a confirmation, the FA, B, can reconfirm the MH's registration with the MH's HA using the same method. The important advantage of this method is that the MH can receive fast service and does not affect the continuity of communication.

### 4.4.5.8 The Popup Method

When the MH moves to an area where there is no FA, the MH will try to get an IP address from the local network and use this IP address as its FA IP address. That means the MH will be its own FA. This is called the popup method. During popup, all packets sent to the MH will be sent to this temporary IP address first, then the MH will remove the additional header before processing it. For example, in Figure 4–30, the MH has a permanent IP address, such as 1.1. When this MH moves to a network whose router does not support SMIP, the MH can get a temporary IP address from local network, such as 2.1. When the SH wants to send a packet to the MH, this

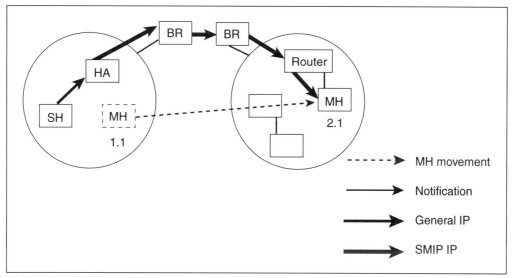

**Figure 4-30** The Popup Method

packet can be first sent to the HA. The HA modifies this packet so that the destination of the packet is the current location of the MH, which is now 2.1. Then the packet can be routed as a general IP packet and finally reach address 2.1. The host in this address is the MH itself. But before accepting this packet, the MH will process the packet as an FA. The MH can modify the packet and restore the original information from the SMIP part inside the packet header, and then accept the packet and pass it to the upper layer.

## 4.5 Location Update and Routing Schemes for the Mobile Computing Environment

MH protocols support host mobility in conventional internet works. The major difference between these protocols is the way in which the location information is propagated into the network to trace an MH, and how packets are routed to destination MHs. The main problems of mobile computing are how to distribute location information efficiently into the network, and how to utilize the location information to efficiently route packets to MHs.

Host migration results in a change of its network address, and network addresses returned by name servers cannot identify migrating hosts. Therefore, the packets bound for the host changing its point of network attachment must be delivered with an extra redirection support without losing its ability to communicate. A number of MH routing architectures supporting host mobility within conventional environments have been proposed. The architectures can be different, and they all provide migration transparency services to support host mobility. The basic ideas for the architectures are very similar. A permanent (home) address is assigned to the MH, which is a logical identifier of the MH. The forwarding (current) care of address associated with its current physical location is obtained by the MH dynamically when it moves around and

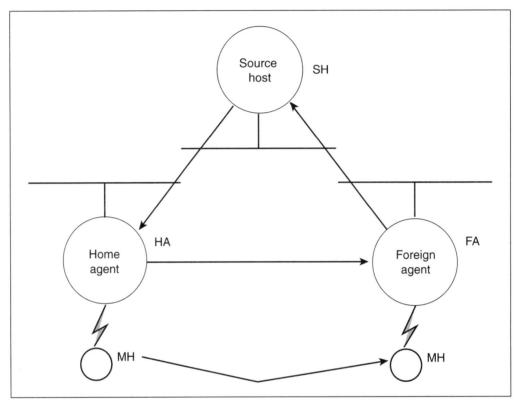

**Figure 4–31** Triangle Routing

registers with a new mobility agent. The association between the MH's home address and its care of address is known as its "mobility binding," which can be used to resolve the MH's current physical location.

The base mobile IP allows any MH to move around, changing its point of network attachment to the Internet, while being addressed by the same home address. Source hosts sending packets to an MH, address them to the MH's home address in the same way as to any stationary destination.

While an MH is connected to the network away from its home network, all packets addressed to the MH are routed using the routing mechanisms to the MH's home network, where they are intercepted by the MH's HA, which then tunnels each packet to the MH's current care of address. The outgoing packets from the MH use its current foreign agent as a default router but require no other special redirection routing.

The base mobile IP presented by Perkins does not provide route optimization. Without route optimization, all packets destined to the MH must be routed through that MH's HA, which then tunnels each packet to the MH's current location. This results in a "triangle routing" (Figure 4–31) for every packet addressed to the MH even when the source is in the same network as the

MH. In Figure 4–31, suppose the MH changes its point of network attachment, and registers with an FA. An SH somewhere in the network sends packets to the MH. The SH addresses packets to MH's home address. Packets are routed to MH's home network and intercepted by MH's HA. The HA tunnels packets to the FA currently serving the MH. The FA locally delivers packets to the MH. The outgoing packets from the MH use the FA as the default router. Therefore, the packets going between the SH and MH form triangle routes.

Intermediate routers are introduced by Myles and Skeller, Perkins, et al. and Johnson to achieve an optimal or suboptimal route bypassing the HA. However, location information is arbitrarily given to intermediate routers along the way to the MH. Thus, efficiency achieved by introduction of intermediate routers greatly depends on the intermediate routers location in the network. Another drawback is that the location cache size is proportional to the number of MHs in the network, which can be very large with the growing population of mobile users. An important feature of the protocols is the introduction of a special tunnel packet to avoid routing loops. A special tunnel packet is a tunnel packet whose encapsulation destination address is the same as the original destination address, which is always routed to the MH's HA without redirection.

Mobile host protocols (MHP) are compared by Myles. The Columbia MHP has fewer compatibility problems with existing networks. The Columbia MHP uses the definition of a virtual mobile subnet created by placing a small number of cooperating mobile subnet routers (MSRs) wherever MHs can be connected to the network. The MSRs have similar function as the FAs, however, the MHs do not have their respective HAs. When an MH migrates, it detects and registers with an MSR at its new location. The MH informs its previous MSR, which still caches its location information. When the MSR receives a packet addressed to an MH, it looks for the MH's current location in the cache table. The appropriate entry can usually be found in the cache. If no cache entry exists for the MH, the MSR sends a message to the other MSRs asking for the MH's current location. The MSR caches a returned answer for future use and forwards the packet using encapsulation. This procedure is only appropriate for a small number of MSRs, thus, the MH's mobility is limited. In the LAN and small campus environments, the Columbia MHP offers a better solution. It provides close to optimum routing to and from MHs in the local area in a manner that is robust and makes minimal demands on the existing network. However, in WANs, the Columbia MHP offers a workable but unsatisfactory solution.

Extensions made to the Columbia MHP provide better scaling properties to WANs and large campus environments. A special form of router, called the Mobility Support Border Router (MSBR), is introduced. Special mobility functions are introduced into the OSPF area BRs. The MSBR is essentially the same as an OSPF area BR. The MSBR serves as the second level of mobile routing and tracking hierarchy. Path-reversal techniques are used to retrieve the OSPF area BR address at the other end of the active communication path, thereby minimizing the expensive wide area traffic due to the mobile-related location information propagation. Update and propagation of location information are accomplished in a distributed and hierarchical manner.

Location update schemes proposed by Aziz, Myles et al., Perkins et al., and Johnson do not consider routing efficiency. Location notification is sent when one mobility entity determines that another entity may have an incorrect location binding for an MH. The memory size to cache the location information of MHs is proportional to the number of MHs in the network, and the network is flooded with notifications.

In Cho and Marshall, two concepts, "local region" and "patron hosts," are developed to use the locality properties of an MH's pattern of movements and access history. For each MH, a local region is a set of designated networks within which the MH often moves, and the patron hosts are the hosts from which the majority of traffic for the MH originates. Introduction of these two concepts provides efficient location update and routing. A location update is limited to the local region and to those hosts which are the most likely to communicate with the MH.

The problem of the scheme proposed by Cho and Marshall is the difficulty of local region maintenance for a mobile host. Each time the MH changes its interest-moving regions, the new local region has to be set up with the help of network administrators. Also, since the patron service is invoked only when the MH crosses in and out of its local region, the scheme is only suitable for the boring professor mobility model, but not for the traveling salesman mobility model or the popup mobility model. In the boring professor mobility model, the MHs tend to have relatively unchanging mobility behaviors. Mobile users choose some locations that they visit more often than others. In the traveling salesman mobility model, the users have an epicenter in their movement patterns. The zone of movement is specified by the number of hops from the epicenter. The popup mobility model corresponds to the case when the mobile user has a tendency to popup at different points in the network. The new location has no relation to the previous movements of the MH. This mobility model can be considered as an extension of the traveling salesman model.

Another location update scheme different from triangle routing is the static update scheme. This is an improved strategy proposed for high call-to-mobility ratios, where the MH actively informs a set of remote sources, those most likely to communicate with the MH, of its current address. Performance of the static update scheme at high call-to-mobility ratios can serve as a benchmark.

The concept of patron host is introduced in Cho and Marshall to take advantage of MH traffic pattern locality. Hać and Huang further define the concept of the patron area. The patron area is the OSPF area where one or more patron hosts reside. We place redirection agents, similar to intermediate routers, at the OSPF area BRs. We update the location information in redirection agents of the patron areas whenever the MH moves in a nonhierarchical architecture, or the MH crosses the OSPF area boundaries in the hierarchical architecture. Thus, SHs within the patron areas, whether they are patron hosts of the MH or not, all benefit from the location update.

Hać and Huang use the definition of the concept patron area and the update of redirection agents in those patron areas instead of an individual patron host. We call the OSPF routing area, where the MH's home network is located, the MH home area, which is unrelated to the mobility models. Instead of using the arbitrary update of intermediate routers along the way to the MH, we limit the location update to a small set of patron areas from where the communication requests to the MH most likely originate. A special tunnel packet introduced by Myles et al. is used. We compute the redirection agent address in another OSPF area from a node's network address in that area by using the path-reversal technique described in Aziz.

This scheme has the advantage of limiting location updates to a set of designated patron areas, and provides nearly optimal routing for most communication, while increasing network and host scalability. The scheme also has the advantage of releasing the patron host from storing the calling list, and has no restriction on the model's mobility. Since the number of patron areas is smaller than the number of patron hosts, we reduce the storage space required to store a patron list in each MH. Though more routers are involved, the scheme provides a more suitable architecture for the mobile network, such as temporarily setting up a local area network and an island network. Hierarchically distributed storage of location information is also a preferred design.

By introducing the mobile functionality into the OSPF Area BRs, we can bypass the MH's home network and HAs whenever it is possible, which results in an optimal or nearly optimal routing for most of the communication.

The location information of the mobile user is provided by a profile and stored in the network BSs hierarchically. The environment of mobile computing is similar to a PCS network. To reduce delay in delivering packets to MHs, and consumed network resource, we propose a hierarchical update scheme. The location information of an MH is propagated into the network and stored in a four-level hierarchical architecture (Figure 4–32). For some packet routing, the local and home level are the same, and the hierarchy reduces to three levels.

### 4.5.1 Locality Traffic Pattern

An analysis of wide area TCP traffic collected at the University of Southern California, University of California, Berkeley (UCB), and Bellcore™, presented at the conversation level shows that locality of reference exists between host pairs and network pairs in wide area internetworks. One of the observations in the study of WAN locality was that 50% of the telnet conversations at UCB are directed to just 10 sites. These observations were made at the network level. The experiments made at the user level also show the locality of the user traffic pattern. Because of the property of locality in hosts and networks with which the MH communicates, the concept of patron hosts was used in Cho and Marshall to refer to the set of hosts that communicate with the MH most often. A potential set of sources that can communicate with the MH is large, but the set of patron hosts is relatively small.

The locality of a traffic pattern of an MH exists not only on the host basis, but also on the network basis. That is, the set of networks that communicate most often with the MH is relatively stationary and small. Therefore, we introduce the concept of a patron area, which is the OSPF area where one or more patron hosts reside, and from where connection requests to the MH are generated most often to further reduce update cost and enhance routing efficiency. Each patron area has at least one patron host of the MH, that is, the number of patron areas cannot be

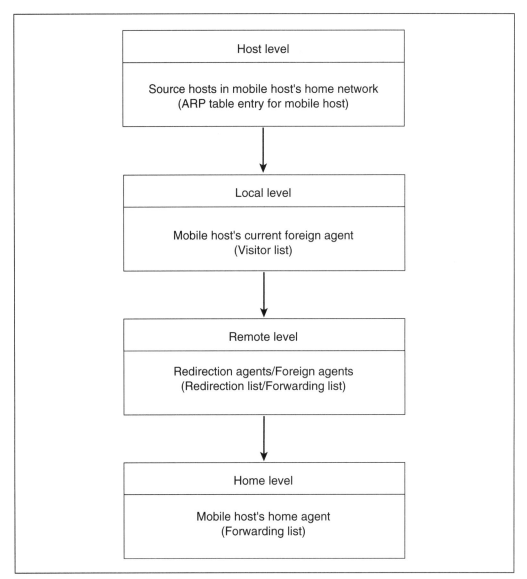

**Figure 4–32** MH's Location Information Hierarchy

greater than the number of patron hosts. Thus, the number of messages exchanged per location update is reduced. The location information update is only propagated to the small selected set of patron areas, and avoids unnecessary use of wide area communication links. Patron areas reflect the traffic pattern of an MH. The set of patron areas is relatively stationary, and can be configured by a mobile user.

### 4.5.2 Scheme Description

An MH must initially register with a mobility agent, which is its HA in MH's home network. A mobile host can communicate with the rest of the network using either the wired network or wireless link.

FAs must send out agent advertisements periodically to indicate their existence to the MHs visiting their control areas. Upon receiving the agent advertisement, an MH can go under the control of a new mobility agent. An MH sends a registration packet, which includes the addresses of its HA, previous FA (if applicable), and redirection agent in its home area, to its current FA. If the registration procedure goes through normally, the MH receives a registration ACK from the FA confirming registration and permission to send or receive data packets. If there was a fault in the registration procedure, the MH will retry the procedure until it times out.

As stated by Hać and Huang, the location information is propagated to related mobility agents hierarchically. When an MH moves to a new region, it first identifies a new current FA and registers with it. The FA then updates the MH's HA, redirection agents in both the MH's home area and its own OSPF area, and other related mobility agents. The redirection agent in the current area updates the redirection agent in the MH's last visited area if the MH crossed the area boundary.

The OSPF is a link state routing protocol and is designed to be run internally to a single autonomous system. Each OSPF router maintains an identical database describing the autonomous system's topology. From this database, a routing table is calculated by constructing the shortest-path tree. OSPF allows collections of contiguous networks and hosts to be grouped together into an area. Each area runs a separate copy of the basic link state routing algorithm, which means that each area has its own topological database and corresponding graph. Having a complete topological database, the redirection agent (the OSPF area BR) is able to learn about other area redirection agent's addresses. This is accomplished by computing the packet reverse path until it reaches a redirection agent in another area. This also explains why it is the responsibility of the redirection agent to forward the registration packet to the MH's last visited area. The router not participating in the OSPF backbone routing algorithm simply does not know how to reverse the path to compute the redirection agent address in the MH's last visited area from the address of the MH's previous FA in that area.

The efficiency of location update and routing schemes is measured by the total cost to route a packet to its destination MH. This cost includes the update cost and routing cost. The update cost consists of both the registration cost and patron update cost. The routing cost consists of the search cost and the hop counts between the nodes, in which the MH's location binding is found, and the destination MH.

Our approach is to reduce the total cost to route a packet to the destination MH. Because the number of patron areas cannot be greater than the number of patron hosts, we can reduce the number of messages needed to update the location information for the MH. The update is sent to redirection agents at the border of each patron area instead of the individual patron host within those areas. It costs less to update the redirection agent at the area border than to update the indi-

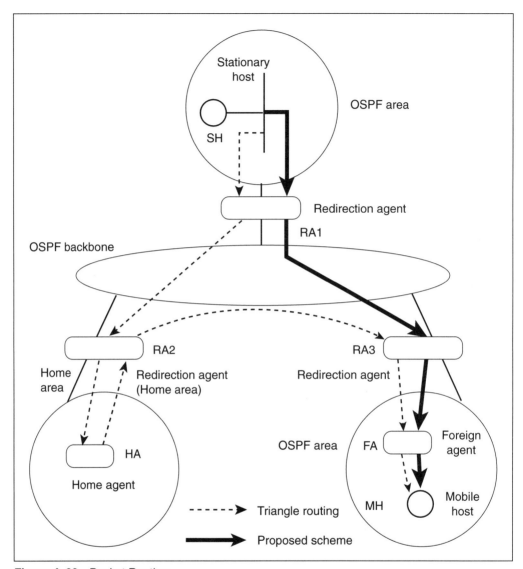

**Figure 4–33** Packet Routing

vidual patron host within that area. Therefore, the cost of location update associated with the MH's movement and network architecture is reduced. Further, because we update the mobility agents at the border of the OSPF areas, all the SHs in those areas, either patron hosts of the MH or not, benefit from location updates. As a result, traffic from SHs within those patron areas, either the hosts communicating with the MH on a frequent basis or infrequently, can always achieve nearly optimal routing even if the hosts are located far away from the MH. Figure 4–33 shows packet routing using both the triangle routing scheme and ours. In triangle routing, the packet sent by a stationary host to an MH traverses path SH-RA1-RA2-HA-RA2-RA3-FA-MH. In our scheme, the same packet traverses path SH-RA1-RA3-FA-MH. In triangle routing, the packet traverses twice the diameter of the OSPF backbone and twice the diameter of the area. In our routing scheme the packet traverses only once the diameter of the OSPF backbone and once the diameter of the area.

### 4.5.3 Architecture

The location update and routing scheme consists of a set of functional entities: MHs, mobility agents, and redirection agents. Mobility agents include HAs and FAs, whose functions are similar to the HLRs and VLRs in a cellular telephone network. Redirection agents are the same as the OSPF area BRs.

An MH can use either the wired network or a wireless link to communicate with the rest of the network. Because of the asymmetric wireless communication medium, the transmission of messages from an MH consumes more power than the receipt of messages. We assume an architecture where much of the responsibility for the location information propagation happens in the mobility agents, but not in the MH. When registering with a current FA, the MH informs the FA about the addresses of its HA, the redirection agent in the home area, and previous FA, if applicable. The FA tries to forward a registration packet to each of the corresponding mobility agents on behalf of the MH. The FA only passes a registration acknowledgment packet indicating the registration success to the MH.

#### 4.5.3.1 Mobile Host (MH)

An MH has additional software that allows the host to move within the network. The host movement is transparent to the user and to the protocol layers above the network layer within the host. An MH is assigned a constant, unique home address that belongs to a home network, in the same way as any other stationary hosts. The permanent (home) network address remains fixed regardless of the host attachment, thus acting as the MH logical identifier.

An MH must initially register with its home network mobility agent as its HA. The MH connecting to its home network is just another stationary host in the network. The care of address in the MH provided by its HA is set to its own home address. The MH answers the ARP requests from the local network. All packets destined for the MH use regular routing algorithms without packet redirection. Additional software in the MH supporting the mobility is not invoked.

When an MH is out of home range, connecting to networks other than its home network, it must be able to determine that it has moved to a new network, and can identify the new, current FA connected to the new local network with which to register. The FA is the default router for all outgoing packets from the MH. A care of address is provided to the MH by a current FA. The care of address and its fixed home address form a location binding of the MH.

Each MH maintains a patron list, which contains the addresses of redirection agents in MH's patron areas from where the connection requests are most often made. The addresses of redirection agents in its patron areas are obtained when the MH initially registers with its redirection agent in the home area. The redirection agent computes the addresses, and returns them to the MH. When the patron host list changes, the MH has to obtain the new set of addresses of the redirection agents in its patron areas by sending an initial registration packet to its redirection agent in the home area. Whenever an MH crosses the OSPF area boundary, the redirection agent in the MH's current visiting area notifies it about this movement. The MH then sends a patron service packet, including all addresses of redirection agents in its OSPF patron areas, to its current foreign agent, which in turn forwards the patron service update to each of redirection agents.

#### 4.5.3.2   Foreign Agent (FA)

During the registration process, an FA provides the MH with a care of address, which is usually its own network address. The care of address indicates the current point of network attachment of the MH. The association of an MH's home address and its care of address is known as "mobility binding," which can be used to resolve the physical location of the MH in the network. The FAs must periodically send out agent advertisements to indicate their existence to the MHs visiting their control area to recognize them and register with them as their current FAs. Upon receiving registration packets from the MHs, the FA creates and/or updates a visitor list entry for these MHs, and sends a registration acknowledgment packet to the MH if the registrations passed through smoothly. The FA, upon completion of the registration process for an MH, has the responsibility to forward the registration packet to that MH's HA, the redirection agent in its home area, and any other related mobility entities depending on its movement.

If the registration packet is forwarded to a redirection agent in the FA's own OSPF area, the FA includes the address of the previous FA of the MH (if applicable) in the forwarding registration packet. It is the responsibility of the redirection agent in the MH's current visiting OSPF area to notify the MH about crossing area boundary movements.

The visitor list maintained by each FA identifies all MHs currently registered. When an FA tries to deliver packets addressed to MHs locally in its visitor list, the scenario is the same as the case of local delivery of packets to a stationary host. MAC addresses are used to distinguish hosts in the local network. The MAC address of the MH can be provided during the registration process with an FA.

When an FA receives a registration packet addressed to an MH in its visitor list from another mobility agent, the MH has already moved out of the FA's control area and changed its point of network attachment. The FA creates a forwarding list entry, deletes the visitor list entry for the MH, and finally sends a registration acknowledgment packet to the mobility agent. The forwarding list is an array of pointers to FAs currently serving MHs.

With a forwarding list caching, the FA can also work as pointer to another FA currently serving as an MH. When an FA receives a data packet addressed to an MH in the forwarding list, which the FA maintains, it tunnels the packet to the address indicated by the forwarding list entry, which is generally the FA currently serving the MH.

When an FA receives a patron service packet from an MH in its visitor list, it sends the patron service update to each of the redirection agents in the OSPF patron areas listed in the patron service packet. These redirection agents' redirection list entries for the MH are updated by the address of its current visiting OSPF area redirection agent. Upon receiving all acknowledgment packets from the patron areas, an acknowledgment packet is sent to the MH.

### 4.5.3.3 Home Agent (HA)

An MH must initially register with its HA, which is a mobility agent in its home network. When a home agent receives the initial registration packet from an MH, it forwards the registration packet to the redirection agent in the home area for the MH. Upon receiving the registration acknowledgment packet from the redirection agent, the HA records the MH in its home list, which identifies all MHs the HA is configured to serve, and then sends a registration acknowledgment packet to the MH.

Suppose the MH moves back to its home area, and connects to its home network. The registration process proceeds in the same way as it does when the MH registers with a new current FA. When the HA receives the registration packet directly from the MH that initially registers it in HA's home list, the HA knows that the MH has returned. The HA then deletes the entry for the MH in its forwarding list if it exists, and returns the home address of the MH as its new care of address. The HA forwards the registration packet to the redirection agent in its OSPF area, along with address of the MH's previous FA. The HA further notifies the redirection agent of the MH's return by sending a notification. With the MH's previous FA address, the redirection agent can determine whether the MH has crossed its home area. When receiving notification of the MH's return, the redirection agent deletes the redirection list entry for the MH. The redirection efforts are no longer needed to route the packets destined for the MH's return.

After registering with a new FA when moving to a new location area, an MH must always register with its HA through its current FA. When the HA receives the registration packet from the FA addressed to MH in the home list maintained by the HA, the HA creates a forwarding list entry (if there is no entry for that MH), or updates the forwarding list entry for that MH (if there is an entry), and finally sends the registration acknowledgment packet to that FA. The HA always caches the most recently updated location bindings for the MHs initially registered with it as their HA. The HA must act as an FA for other MHs visiting its control area, and for these initially registered with it.

When an HA receives a data packet addressed to an MH in its home list and forwarding list, the HA uses the forwarding list entry to tunnel the packet to the FA currently serving the MH which is out of its home range. If the MH is in its home list but not in the forwarding list, it means the MH is at home, no action is needed at the HA, and the packet is delivered the same way as the other packets.

The HA must also provide the proxy ARP for those MHs that initially register with the HA and are out of home. When the MH is connected to its home network, it answers the ARP requests from local network. All packets destined for the MH use routing algorithms without any efforts to redirect. When an MH is not at home, connecting to networks other than its home network, the MH's HA must broadcast messages to all of the SHs in the local network to clear the

ARP entries they held for the MH while it was out of its home range. The HA also takes the responsibility for answering the ARP requests from the local network on behalf of the MH.

#### 4.5.3.4 Redirection Agent (OSPF Area BR)

The redirection agent is a mobility router that has a special function to redirect (tunnel) packets to or from an FA. We place redirection agents on the border of an OSPF area. The redirection agent is essentially an OSPF Area BR with extra functionality to support tracking and delivery functions for intradomain mobility scenarios. Each OSPF area has at least one redirection agent.

When the redirection agent receives a registration packet forwarded from a mobility agent, it updates the corresponding redirection list entry for the MH, and sends the registration acknowledgment packet to that agent. With the address of the MH's previous FA provided in the registration packet, the redirection agent can determine whether the MH has just crossed the OSPF area boundary. If the redirection agent (which is in the MH's current visiting OSPF area) notices a crossing OSPF area boundary movement of the MH, a notice packet is sent to that MH to notify it of the boundary crossing. The current redirection agent forwards the registration packet to the redirection agent in the MH's last visited OSPF area, whose redirection list entry for the MH is updated with the address of current redirection agent.

When the MH is back in the home network, the HA sends a notification to the redirection agent in the home area. When the redirection agent receives a notification of the MH's return, it clears the entry in its redirection list for the MH, and routing mechanisms are used to route packets to the MH.

When the redirection agent receives a patron service update forwarded from a mobility agent, the redirection agent updates the corresponding redirection list entry for the MH, and sends the registration acknowledgment packet back to that mobility agent.

The redirection list preserves location information for MHs. The redirection list works as a pointer array, pointing to the addresses of the FA's that currently serve the MHs. In a nonhierarchical architecture, when an MH moves, it must create and/or update the mobility binding with the redirection agents of its patron areas, including its HA and previous FA (if the MH had one). In the hierarchical architecture, every move of the MH causes a registration with its HA and previous FA, but only the movements across the OSPF area boundaries incur an update of mobility binding with redirection agents in the MH's OSPF patron areas. When the redirection agent receives a packet addressed to an MH in its redirection list, it uses the redirection list entry to tunnel the packet to the mobility agent that currently serves the MH.

### 4.5.4 Packet Routing

Figure 4–32 shows that location information about MHs is distributed and stored in a four-level hierarchical architecture. With all the mobility agents working together, and the location information cached at each level, the packets are routed directly to the destination MH whenever it is possible, without them being routed through that MH's home network or HA. To illustrate packet routing (tunneling), we describe how the packet traverses the MH's location information hierarchy and reaches its destination (Figure 4–34). The term packet refers to an untunneled

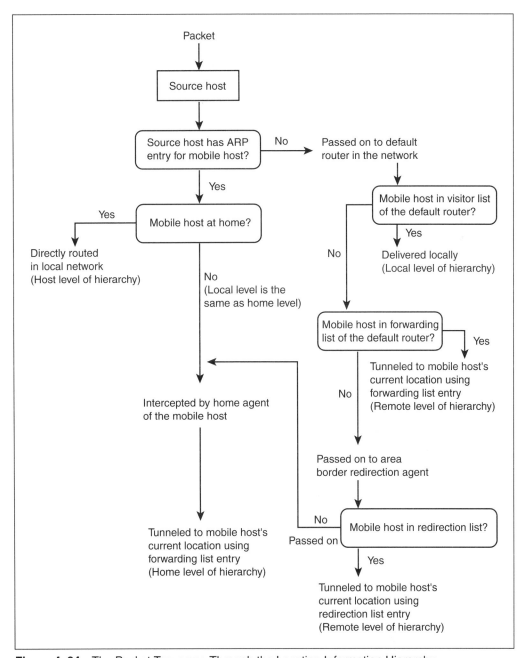

**Figure 4–34** The Packet Traverses Through the Location Information Hierarchy

packet or the inner packet carried by the tunnel. A special tunnel packet (a special case of packet forwarding (tunneling)) applies to all the mobility agents other than the HA.

A special case in packet forwarding, used to avoid possible routing loops, occurs for a tunneled packet in which the destination of the tunnel is the same as the original destination of the packet. The tunneled packet is called a special tunnel packet, and it is always forwarded to the destination MH's home network without redirection. No location cache entries or visitor list entries can be used in routing a special tunnel packet; the packet must be routed using only regular packet routing, and it reaches the destination MH's home network where it is intercepted by its HA if the MH is away from home. Tunneling protocol must be designed so that a special tunnel can be detected after packet fragmentation.

When a tunneled packet traversing through the second or third level of hierarchy is addressed directly to the node, and the network node (FA or redirection agent) is unable to provide location information, the special tunnel is used to tunnel the packet to the MH's home network. The special tunnel ensures that any loops can be easily and quickly detected and broken without having to rely only on the time-to-live field in the packet header.

SHs in the same network as the MH's home network usually have an entry for the MH in their ARP tables. When these hosts send packets to an MH in the same network, and the MH is at home, the packet is directly routed within the local network. An exception to this is when the MH is not at home, in which case the local level is the same as the home level, and the ARP entry for the MH is actually its HA's MAC address. The packet is directly routed to the HA, and tunneled to the MH's current location using the forwarding list entry.

The SHs in the networks different from the MH's home network do not have entries for the MH in their ARP tables. The packet is addressed to the MH's home address and passed on to the default router in the network using packet routing.

If the MH is visiting the network and registers with the network's default router as its current FA, the current FA has a visitor list entry for the MH. The packet destined for the MH is delivered locally. If the FA is serving the MH at the remote level of the information hierarchy, the FA has an entry in the forwarding list for the MH. The packet is tunneled farther to the destination. If no location bindings for the MH are cached, the packet is passed on using normal packet routing, and reaches the OSPF area BR, which is also the redirection agent for that area and the remote level of the information hierarchy.

If the redirection agent at the area border has the location binding for the MH, that is, the MH is in the redirection agent's redirection list, the packet is tunneled to the MH's current location using the redirection list entry. However, if no location bindings are found, the packet is passed on to the MH's home network using packet routing.

If no location binding is cached for the MH along the way to its HA, or a special tunnel packet is formed by the second or third level mobility agent, the packet gets to the MH's HA. The MH's HA always caches the MH's most up-to-date location information. If the MH is at home, the packet is delivered there, otherwise, the HA tunnels the packet to the MH's current location and the triangle route is formed.

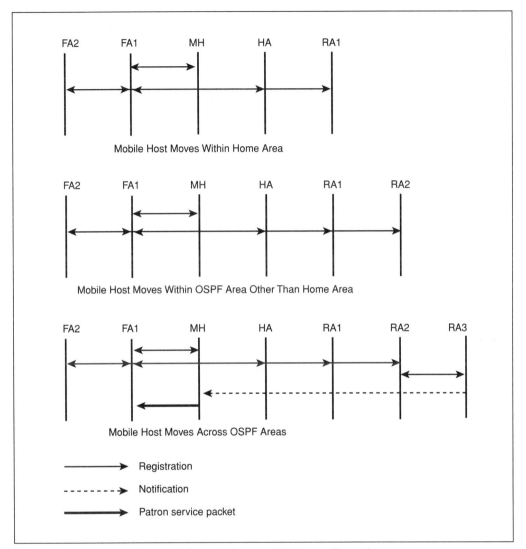

Figure 4–35  Registration and Informal Message Exchange Examples

### 4.5.5 MH Movements

Examples of registration and informal message exchange are shown in Figure 4–35. An MH moves in the network. The MH currently registers with an FA, FA1. The MH's previous FA was FA2. The HA and the RA are the MH's HA and redirection agent in the home area, respectively. RA2 and RA3 are redirection agents in the MH's current visiting area and the last area visited, respectively, when applicable. The registration and informal message exchange flows are different and depend on the MH's movement.

In the hierarchical architecture, an MH does not need to declare its movements within an OSPF area, that is, movements within an OSPF area do not have to be known to the outside world. This greatly facilitates mobility in the OSPF area. If the MH continues to move within the OSPF area, then all that is needed are local transfers of control information.

If the OSPF area is the MH's own OSPF area (we call it the home area of the MH), and the MH is away from home, then after identifying its new FA, FA1, the MH registers with FA1. FA1 which is currently serving the MH propagates the registration packet to the MH's HA, the previous FA, FA2 (if the MH has one), and the redirection agent, RA1, in the MH's home area.

If the OSPF area is an area other than the MH's home area, whenever the MH moves, it registers with a new FA, FA1. The registration packets are propagated to the MH's HA, previous FA, FA2, and both the redirection agent, RA1, in the MH's home area and the redirection agent, RA2, in the present visiting OSPF area.

When an MH leaves its home area and joins a mobility agent in another OSPF area, the MH registers with a new current FA, FA1, in that OSPF area. The current FA, FA1, sends the registration packet to the MH's HA, the previous FA, FA2 (which must be in an OSPF area other than the OSPF area, in which the MH is, since we discussed the situation when the MH crossed the OSPF area boundary), and the redirection agents in the MH's home area, RA1, and present OSPF area, RA2. Upon receiving the registration packet, with the address of the MH's previous FA, FA2 provided in the registration packet, the redirection agent, RA2, in MH's current visiting OSPF area can determine that the MH has just crossed the OSPF area boundary, and the redirection agent, RA2, informs the MH about the movement by sending the MH notification packet. The redirection agent, RA2, also propagates a registration packet to the redirection agent, RA3, in the MH's last visited area. Upon receiving the notification packet, the MH sends the patron service packet to its current FA, FA1, which in turn forwards the patron service update to each of the redirection agents in the patron areas listed in its service packet. The individual patron host of the MH is not aware of the movement.

The MH registers with a mobility agent within its home area as its current FA. The FA sends registration packets to the HA and the previous FA of the MH, and redirection agents within home area. The redirection agent in the home area notices the MH crossing the home area boundary. The redirection agent forwards the registration packet to the redirection agent in the MH's last visited OSPF area, and notifies the MH. Upon receiving the notification, the MH sends out the patron service packet.

CHAPTER 5

# Resource Allocation in Multimedia Wireless Networks

## 5.1 Dynamic Channel Assignment in Cellular Mobile Communications Networks

A service area of a wireless network is divided into cells as shown in Figure 5–1. Each cell has an assigned group of discrete channels from the available spectrum. The BS is established in each cell. Subscribers in a cell use channels assigned to that cell. Every MS or other terminal in the cell communicates through the BS via a channel or channels. A region initially served by a BS in a conventional mobile system is split into several cells, each of which has its own BS, served by a certain number of channels. A system can grow geographically by adding new cells.

The primary role of a BS is to terminate the radio link in the network side of the user-to-network interface. This involves an element of control as well as radio frequency (RF) and base-band processing. In a small, microcell system, a distributed channel assignment function can be implemented in a BS. BSs are connected to a switching network through wired lines.

Spectral efficiency is defined as the maximum number of calls which can be served in a given area. The cellular concept is about achieving high spectral efficiency. The basic concept of the cellular system is frequency reuse. If a channel of a certain frequency covers an area of radius R, then the same frequency can be reused to cover another area. Each area constitutes a cell. Cells using the same carrier frequency are called cocells. They are positioned sufficiently apart from each other so that cochannel interference can be within tolerable limits. With this concept, a region initially served by a BS in a conventional mobile system is split into several cells, each of which has its own BS served by a set of channels.

As illustrated in Figure 5–2, frequency reuse allows discrete channels, assigned to a specific cell A, be used again in any cell such as cell B and C that are separated from cell A by enough distance to prevent cochannel interference from deteriorating the QoS.

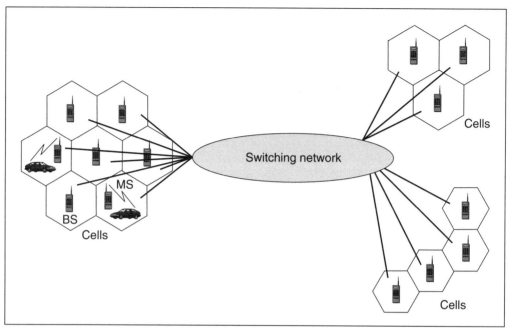

**Figure 5–1** Wireless Network Architecture

The major requirements for future mobile communication systems are high network capacity and flexibility to accommodate time-varying communication traffic. The limited availability of the radio frequency spectrum is the reason that future cellular systems must efficiently employ channel assignment methods to increase network capacity and adapt to time-varying traffic.

The wireless cellular system design maximizes the number of channels in the space and spectrum under the constraint that the impairments due to interference are lower than the prescribed limit. The capacity of a wireless cellular system is directly related to the number of channels available, the area in which these channels can be reused, and the minimum cell size that can be implemented in a given environment. The large subscriber capability requirement demands that the system be capable of serving many thousands of mobile users within a local serving area with a fixed number of hundreds of channels. Furthermore, the spectrum utilization requirement demands that the multiple use of the same channels in cells with geographical separation ensures that the radio spectrum is used efficiently. By using the cellular system, both the subscriber capability and spectrum utilization efficiency can be increased. A wireless system should be adaptable to traffic density. Since the traffic density differs from one point in a cellular coverage area to another, the capability to cope with different traffic demands must be designed as an integral feature of a wireless system.

The same frequency channel is reused in several cells simultaneously if they are separated by a distance long enough to avoid cochannel interference. The problem is to find a channel

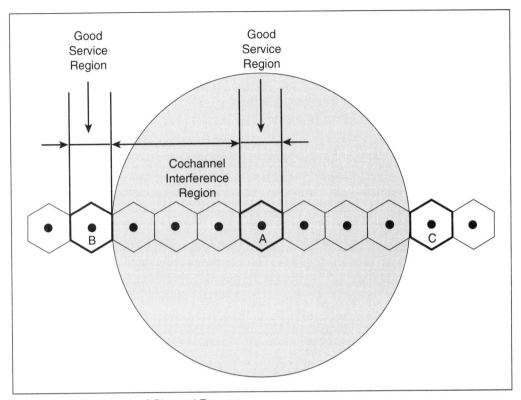

**Figure 5–2**  Illustration of Channel Reuse

assignment within each cell that avoids cochannel interference. When a call is requested, the wireless network goes through the channel assignment process. If a free channel is available, the process allocates the channel to the user. The strategies to assign channels to cells are fixed channel assignment (FCA) and dynamic channel assignment (DCA).

FCA allocates channels to each cell in advance according to the estimated traffic intensity in a cell. Channel assignments and handoffs are managed by the central switch. FCA requires a moderate amount of BS radio equipment and uses a simple control algorithm. FCA is not efficient in total channel use over the entire service area if traffic varies dynamically from cell to cell.

In DCA, channels are assigned dynamically over the cells in the service area to meet the traffic requirements. The channels are kept in a central pool, and any channel can be used in any cell. However, the channel used in one cell can only be simultaneously reassigned to another cell if the separation between two cells exceeds the minimum distance required for circumventing cochannel interference. DCA algorithms can be subdivided into three categories: traffic adaptive schemes, interference adaptive schemes, and channel reusability. The difference among these strategies is how to choose the optimal channel from the pool of free channels in response to a specific request.

Dynamic channel allocation is regarded as a means to cope with time varying traffic, where the standard fixed channel allocation method has some obvious disadvantages. In a microcell environment, the cells are packed closely enough together to make DCA feasible. Small-cell systems are interference limited. To maintain high utilization of the available radio channels in this environment, DCA becomes essential. DCA has the ability to assign radio channels to different cells, within the reuse constraints, to match radio BS capacity to the real-time traffic load.

Generally, to examine whether it is possible to assign a channel to a new call is the equivalent of the coloring problem in the graph theory. The solution to channel assignment is an NP-complete problem. Approximate algorithms must be used to find the solution in a practical time frame when DCA is used in a wireless system.

Centralized DCA algorithms often require a control center to handle channel allocation or handoff among all cells and mobiles. The control center must obtain local status information from all mobiles in a given cellular system. The current occupancy and interference level of channels and the propagation condition of mobiles within each cell can be measured and transferred to the control center periodically. Based on the actual conditions of the entire system, the decision is made to assign a certain channel to a mobile or to carry out a handoff.

By using computation over complete observation of cells states, centralized methods can achieve the optimum performance in the entire system: the maximum capacity or minimum blocking probability with certain interference level. However, as the number of cells increase, the centralized computation can become mathematically intractable. The method is also impractical under highly varying traffic because of the difficulty in measuring the actual conditions. In addition, the measurement and transmission of current local status information and the control instruction can occupy some spectrum resource and decrease the system capacity.

In the distributed DCA algorithm, each cell can handle the channel assignment or handoff autonomously. The allocation or handoff decision is based on local status information. The exchange of status information with relevant cells is needed to reduce mutual interference. This channel assignment scheme is self-organizing. Cells with busy channels, near full capacity, acquire new channels while cells with idle capacity release assigned channels. The specific channels acquired and released are selected on the basis of use patterns in the surrounding cells. In keeping with the distributed network control paradigm, the BS, rather than the central controller, makes the channel acquisition and release decisions.

Compared with the centralized scheme, the distributed algorithms can only achieve suboptimum allocation since the decision is based on the local and not on the global status of the system. These algorithms are more flexible and can accommodate an increase in traffic density, and adapt to the growth of traffic. In addition, the distributed algorithms are easier to implement and can avoid allocation delay or handoff delay that frequently occurs in centralized schemes, especially under heavy and nonuniform traffic. However, distributed channel access control may cause unpredictable interference. A cell may allocate a channel that, based on the local situation, seems to be unused, but it can introduce an unacceptable interference level to another mobile within other cells, thus resulting in a handoff. A timid algorithm tries to avoid channel allocation causing such interference, whereas an aggressive algorithm disregards it, allowing the handoff to solve the problem.

The most efficient method to improve network capacity is the introduction of microcells. A third generation wireless network serves a large user population by means of a dense grid of microcells. However, with shrinking cell sizes the radio network planning process becomes more complex because it depends highly on the "microscopic" structure of cells. A dense network of small cells will exhibit extreme temporal and geographical variations in traffic density. To overcome this problem, future DCA methods should be able to automatically adapt to instantaneous interference and traffic changes. To make efficient use of the limited spectrum, networks will have to rearrange transmission resources dynamically to meet the rapidly changing demand for communication channels. A solution to the network management problems created by the use of microcells is DCA with distributed control. Using the adaptive DCA method, the network capacity can be increased by the introduction of new cells, as necessary, but without the need for frequency planning. The network control will be distributed among many dispersed processors rather than concentrated in the MSC.

In third generation networks carrying multimedia traffic, channel assignment is more complex since multimedia traffic can require various bandwidth assignments. Users of different types of multimedia traffic must be assigned channels. In the integrated services wireless network, the allocation of channels to cells is similar to the allocation of system capacity in a conventional integrated services network. Different services require different amounts of some basic unit of system resource. For example, in a TDMA system, the basic unit is one slot per TDMA frame. A speech call requires the use of one slot per frame, while a high-speed data service can require 8 slots per frame. Therefore, a call can require the use of more than one radio channel (TDM slot) simultaneously. These radio channels must be allocated subject to the normal reuse constraints for radio frequencies. Many ways exist to allocate the total system capacity between the different services. At one extreme, the capacity can be rigidly partitioned so that calls of a particular type can use only a preassigned portion of the system capacity, for example, slots 1 to 8 in a TDM frame. At the other extreme, calls use any part of the system capacity, and partitioning is completely dynamic. Channel sharing techniques for multichannel traffic in a small cell environment are described by Everitt and Manfield.

Two major interference problems in wireless systems are adjacent-channel interference and cochannel interference. Adjacent-channel interference is basically caused by equipment limitations such as frequency instability, receiver bandwidth, and filtering. Although the equipment is designed for maximum interference performance of the system, a combination of factors, such as cellular architecture and random signal fluctuation, usually causes deterioration of a received signal, primarily due to the interference of adjacent channels. Cochannel interference is a complication that exists in mobile systems using cellular architecture. Channels are used simultaneously in as many cells as possible, with the minimum acceptable separation, to increase the

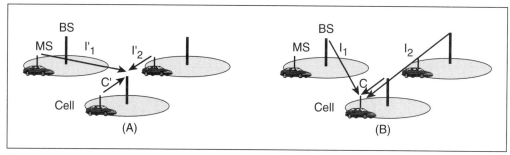

**Figure 5–3** Illustration of Interference Scenario: (A) Uplink (Mobile-to-Base) at Desired BS; (B) Downlink (Base-to-Mobile) at Desired MS

reuse efficiency. A BS receiving a wanted signal from an MS within its cell can also receive unwanted interference signals from mobiles within other cell clusters using the same channel. The determination of frequency reuse distance, and hence, the cell repeat pattern, has a direct influence on cochannel interference levels. Longer reuse distance implies smaller cochannel interference. However, this also implies a larger number of cells per cluster resulting in a smaller reuse efficiency. Therefore, a trade-off between interference and efficiency must be negotiated. Cochannel interference is the main factor affecting frequency reuse efficiency in channel assignments.

When a channel is assigned to a new call, cochannel interference must be considered. In the traditional DCA, the same channel is assigned with a certain separation called a reuse distance so that the radio wave interference is sufficiently small. A new cochannel interference model improves the spectral efficiency in the cellular mobile communication system. Lee introduced the channel assignment method using the carrier-to-interference ratio (CIR). The channels are assigned with a separation smaller than required for reuse as long as the required CIR is satisfied, which is based on the measurement of radio wave interference data. Tamura formalized the new model, where a channel can be assigned in a cell only if the CIR of the channel is equal to a threshold or greater than CIR in the cell. The same channels can be reused in cells even if they are located within the minimum separation distance, which is prohibited in the basic model.

The channel assignment is made by considering the cochannel interference level determined by the CIR requirement. The CIR is a random variable affected by random phenomena such as the location of the mobile, Rayleigh fading, log normal shadowing, antenna characteristics, and cell site location. There are two interference scenarios. The uplink (mobile-to-base) scenario is characterized by all the cochannel MS's interference with the desired cell's BSs as shown in Figure 5–3(A). The downlink (base-to-mobile) scenario is characterized by all the cochannel cell BS's interference with the desired MS as shown in Figure 5–3(B). The interference level for each link is the sum of the interference signals arriving in the MS or the BS. The availability of candidate channels depends on whether the required CIR is satisfied at both the BSs and MSs.

If we neglect the local noise, the CIR can be expressed as

$$\text{CIR} = \frac{C}{\sum_{k=1}^{n} I_k} \qquad (5\text{--}1)$$

where $C$ is the received carrier power in a desired cell, $n$ is the number of cochannel cells and $I_k$ is the interference (either uplink or downlink) from one of $n$ cochannel cells. If we set the CIR threshold to $\text{CIR}_{\min}$, then the same channel can be reused if the CIR level for the specified channel in the desired cell is equal to or larger than the specified threshold level $\text{CIR}_{\min}$.

An intracell handover or channel reassignment improves cochannel reuse. An intracell handover assigns a channel to a new call by reassigning new channels to calls already in progress. The channels are rearranged so that the blocking probability can be minimized or reduced. An intracell handover is a requirement for the DCA to adapt effectively to interference and variations in traffic. Using intracell handovers, the existing calls in a cell can be moved to new channels in the same cell, as necessary, to maximize channel reuse. If intracell handovers are not permitted, the DCA schemes follow inefficient reuse patterns as dictated by the specific pattern of call arrivals, call completions, and intercell handovers. When intracell handovers are permitted, a complete reassignment of calls to channels can be done systemwide, as often as necessary, so that the optimal assignment is achieved. This is called the maximum packing (MP) strategy.

The MP dynamic channel allocation approach is an idealization of the DCA. This approach assumes that a new call will be blocked only if there is no possible reassignment of channels to calls (including reallocation of the calls in progress), which results in the call being carried. When a new call arrives, and the system is in a state where the call would normally be blocked, the system finds it possible to reconfigure the radio channels in use to accept a new call, subject to reuse constraints. Under these conditions, the system reconfiguration will be done and the new call will be accepted. This can involve reallocating of calls in progress to other radio channels. In practical networks, such a strategy is not cost effective to implement from the control point of view since it requires system-wide information about channels in use in every cell, but it does have the properties of being analytically tractable, and provides a bound on the performance of the system. Maximal traffic adaptability is bounded by the MP strategy to carry the call.

A system can organize itself with little loss in capacity by using simple algorithms, where decisions are based solely on local observations. Call blocking can be reduced, in some cases, with only a small increase in the number of reconfigurations required (i.e., by seizing a channel even though there is interference nearby).

The assignment of intended channels relies on local information about the signal levels and interference present at individual channels in the cell or mobile site. These methods use periodic local measurements of CIR to make decisions about channel assignments. This CIR information can be measured in the BS, handset, or both. Algorithms can be used to autonomously select a channel with an acceptable signal-to-interference level such as those with the highest CIR. When a new call enters the system or when the status of the portables is checked, the cochannel interference is updated both at the BSs and the portables across the entire system.

Since BSs make autonomous decisions about channel acquisition and the reassignment of calls, temporal and spatial variations in interference require rapid intracell handovers and reassignments. Intercell handovers are requested whenever a BS with a stronger signal is detected at the portable. Intracell handovers are requested whenever the CIR drops below its threshold. Because channel assignment is derived from local interference measurements, this method does not require frequency planning. For large cellular systems, this is a very attractive feature.

The basic idea of DCA is that each BS strives to maintain a carrier use pattern resembling FCA, as long as this is compatible with the offered traffic pattern. The requirements of a good DCA strategy are two fold. In order to maximize carrier use, the DCA strategy should keep reuse of various carriers to the maximum. During the dynamic system evolution, the DCA strategy should maintain, as much as possible, the maximum packing of carriers. An ideal DCA is one which, at any time, succeeds in satisfying any carrier-to-cell request pattern, provided that the pattern is compatible with the frequency reuse constraints. Such an ideal DCA cannot be practically implemented since, in general, it would require a real-time reconfiguration of the carrier-to-cell assignment through the entire cellular network and hence, considerable signaling overhead. On the other hand, the DCA strategy implementation should require the minimum information exchange among BSs to minimize signaling overhead. Extending the resource assignment to the more general case of bandwidth or capacity assignment for multimedia traffic is more complex because each traffic type can require a different bandwidth assignment.

### 5.1.1 Cochannel Information Based Dynamic Channel Assignment (CDCA) Strategy

Hać and Mo propose a distributed DCA strategy for small cell and microcell systems. The channel assignment function is made by the BSs locally, according to the knowledge about their neighboring information, especially, the number of cochannels in the neighboring area. We call this CDCA strategy.

Although many DCA strategies based on CIR information have been proposed in recent years, there still exists the problem of how to use the CIR information. The CIR can be the main criterion in channel assignment, but it cannot reach the optimal spectral efficiency if the assignment is based only on the CIR information. Other information is also required in channel assignment strategies. Three strategies that choose channels based only on the CIR information are as follows. The common feature of three cases A, B, and C is that the BS monitors the channel's CIR periodically. When a call is attempted, the BS searches the channels in the channel pool and checks the CIR of each channel. The difference between these strategies is how to select channels from the channel pool according to CIR.

Case A. Always choose the channel with the largest CIR (LDCA). The largest CIR dynamic channel assignment (LDCA) strategy always chooses a free channel with the largest

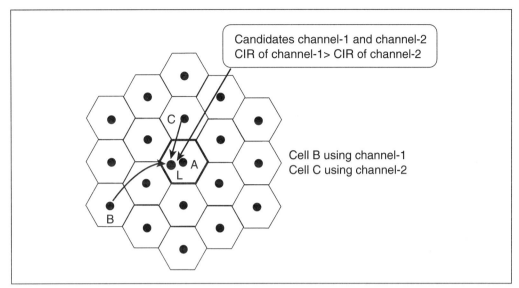

**Figure 5–4** Illustration of Case A

CIR in channel search. Suppose that in cell A, location L, a call is attempted and the candidate, channel 1, has the largest CIR (Figure 5–4). Is it optimal to choose channel 1 in view of its spectral reusability? We can find another candidate, channel 2, which has a smaller CIR but still has a qualified CIR threshold. In the neighboring area, channel 1 has only one cochannel in cell B and channel 2 has only one cochannel in cell C. The distance of location L to BS C can be smaller than the distance of location L to BS B. If we choose channel 2, the reuse distance between channel 2 in cell A and its cochannel in cell C can be smaller than the reuse distance between channel 1 in cell A and its cochannel in cell B. Channel 2 can be more frequently reused than Channel 1. Thus, the LDCA strategy is not the optimal method to increase spectral efficiency.

Case B.  Always choose the channel with the smallest CIR. The smallest CIR dynamic channel assignment (SDCA) strategy always chooses a free channel with the smallest CIR in its channel search. Suppose that in cell A, location L, a call is attempted and candidate channel 1 has the smallest CIR in the channel pool of cell A (Figure 5–5). There can exist another candidate, channel 2, that has a larger CIR. Which channel should be chosen? The following scenario is possible. Channel 1 in cell A has a cochannel used in cell B. We consider the neighboring area of a certain size around cell A. In this entire neighboring area, channel 1 has one cochannel in cell B and channel 2 has two cochannels in cell C and cell D, respectively. The only reason that the channel 1 CIR is smaller than the channel 2 CIR is that the interference source and BS B, are closer for channel 1. If we always choose a candidate channel with the smallest CIR, we can miss an opportunity to reuse more cochannels in the area. In the entire neighboring area, it is better to choose channel 2 than channel 1 to pack more channels. Therefore, the SDCA strategy does not fully explore the spectral efficiency.

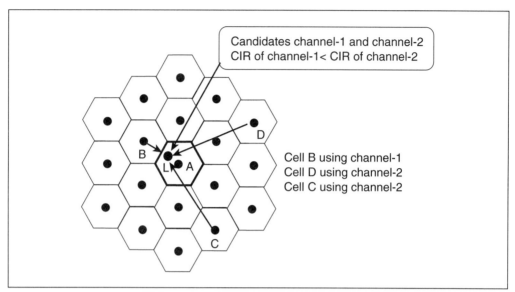

**Figure 5–5** Illustration of Case B

Case C. Choose the first channel that is CIR qualified. The only criterion for this strategy is that the CIR must be equal to or larger than the CIR threshold. This method is not optimal in the view of spectral efficiency.

The CIR criterion is only sufficient to guarantee good communication quality. CIR is a relative value. It does not contain much information about the distribution of channel reuse in a given geographical area. By using CIR as the only criterion for the channel assignment decision we miss information about the distribution of channel reuse. If we choose the distribution method, we certainly need the neighboring information to optimize the channel assignment, thus achieving the optimum packing.

The BS is responsible for channel assignments in wireless networks. Periodically, BSs exchange information with their neighboring BSs. According to the neighboring information, the BS makes a channel assignment function in each cell locally.

The structure of the cellular system is not uniform, thus, in small cell and microcell systems a fixed reuse distance is not suitable for future wireless systems. Reuse distance can be assigned dynamically by using CIR. Each MS and BS detects its CIR periodically and sends this information to its BS. A BS is assigned to an MS according to the received signal strength. A CIR detection function can be designed in the hardware and implemented in the mobile and base stations. The CIR readings can be transferred through the control channel in real time. The channel assignment function implemented in the BS performs a channel search process by using CIR and other information.

Adaptability to traffic has been the initial motivation for using DCA. While a call is in progress, the BS and MS will monitor the channels' CIR. If a channel's CIR is lower than the

CIR threshold, an intracell handover is allowed to keep the quality of communication. A suitable channel with a better quality is searched for among the free channels of the current BSs. If this search fails, the more common intercell handover can be used to search for a channel with a suitable quality in a new BS. By using reconfiguration, the channel assignment pattern can approximate the optimal pattern.

Hać and Mo propose a CDCA strategy that uses neighbor information, especially the number of cochannels, to optimize the channel assignment. Given a cell and a channel in the cell's channel pool, the number of cochannels is defined as the total number of a given channel's cochannels that are used in the given cell's all neighboring cells. While performing a channel search, the channel that has the highest number of cochannels used in the defined neighboring area is chosen. The choice of channel has always the tendency to reach the optimal channel assignment. Highly used channels have the priority to be chosen in the local neighboring area. We always choose the channel with the greater cost. The cost equals the total number of cochannels in the neighboring area. In general, if BSs have different emission powers and the cell sizes are different, there are different sizes of interference areas for different BSs. These factors must be weighted in the cost calculation.

The CDCA keeps track of and checks the neighbor information. For each cell or its BS, the CDCA has its own neighbor list according to the cellular system's topology and its geographical condition. For each cell, the information of channel use in neighboring cells is sent to a given cell. For each channel in the channel pool, the BS calculates the cost, such as the number of cochannels. This information is updated periodically. Generally, the neighboring area of a cell is determined by the structure of the cellular system and implementation. If these factors are changed, the neighboring area is also changed. The neighboring area is represented by a neighbor constant.

We propose the neighbor constant, which is related to each cell. A neighboring area of a given cell is a circle area. The circle is drawn with the neighbor constant as its radius and the BS as its center. All cells that fall in this area are neighboring cells of a given cell. Conceptually, the neighbor constant defines an interference area of a given cell (Figure 5–6).

The concept of distance reuse is inflexible and inefficient in third generation wireless systems. With the CIR test method, we can decide the instant reuse distance dynamically according to the instant environment condition. When the distance between two cells is smaller than the reuse distance in the traditional theory, one cell can still use a cochannel of the other cell. Therefore, we consider cells whose distances are smaller than or equal to the reuse distance. They are defined as the neighbors of a given cell.

The neighbor constant, R, is related to the interference area, the emission power of the BS, and the topology of the cell structure. The neighbor constant indicates an area for channel management. The size of the local view influences the distributed channel assignment strategies. The neighbor constant is directly related to the reuse distance. For evenly distributed cells in a geographical area, we can use the existing theory in cellular radio engineering to estimate the reuse

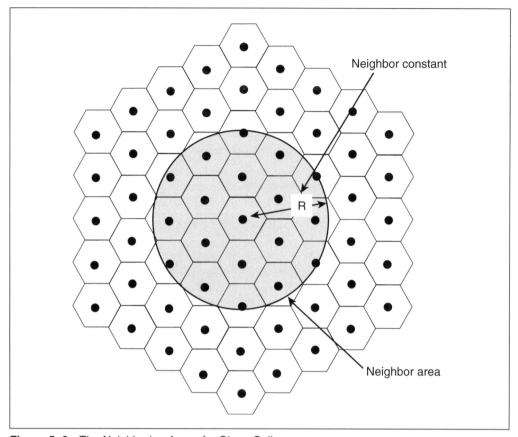

**Figure 5–6**  The Neighboring Area of a Given Cell

distance, and the cochannel reuse ratio to calculate the reuse distance. For an unevenly distributed cell topology, its neighboring area is related to the emission power of the BS.

We assume that the traffic contains only one call type, which is a one–channel call. The CDCA also applies to multichannel traffic. We focus on the DCA traffic adaptability. The basic performance criteria are throughput, a new call blocking rate for new call arrivals, and a forced call termination rate for calls in progress. The CDCA strategy can optimize the channel search and reach the optimum packing in the entire system by optimizing the choice of channels. With the distribution operation, the DCA function can work efficiently.

According to the topology and the environment in the system, each cell has knowledge about its neighboring cells. Neighboring cells exchange the information about channel use periodically. In DCA, all available channels are kept in a pool for the entire system. For each cell, a channel in the system channel pool can be used in this cell if the channel satisfies the reuse criterion. The channel pool for each cell is a copy of the system channel pool, and is called a channel table. The use of channels in the cell is recorded in its channel table. Channel tables are updated when channels are used or released. Suppose a channel table is a one-dimensional array. The record of a channel is an element of the array and can be accessed by an array index. A smaller index channel is a channel with a smaller index in the channel table. A larger index channel is a channel with a larger index in the channel table. The DCA that searches a channel through the channel pool is actually searching the channel table of each cell in the implementation.

The CDCA function is handled in the BS. We assume that an MS and BS have implemented the CIR test function in the hardware and software that tests each channel's CIR locally. The CIR values are read periodically and the CIR values of the MS are sent to neighbor BSs through the control channel. Each cell requests channel use information from neighboring cells periodically. Each BS counts the number of cochannels for each channel.

The CDCA strategy for the new call arrival is as follows.

Step 1: Update each channel's CIR in the channel table and the number of cochannels for each channel.
Step 2: Search the channel table of a given cell. If a channel is free and its CIR is equal to or larger than the CIR threshold, select this channel as the candidate. If no free channel is found, the call is blocked.
Step 3: After finding the candidate in Step 2, update the candidate to the optimum choice based on the neighboring cochannel information. Search the channel table. If the number of cochannels of another free channel is greater than the number of cochannels of the candidate's channel, the other free channel becomes the candidate. Continue the search. Choose the candidate with the largest number of cochannels.
Step 4: Assign the new call to the candidate's channel. Update the channel use status in the channel tables. The channel assignment is completed.

The CDCA strategy for the calls in progress is as follows.

Step 1: Monitor CIR values for calls in progress periodically. If the CIR is lower than the CIR threshold, go to Step 2. Otherwise, discontinue checking the call in progress.
Step 2: Check the signal strength received from the current BS and its neighbor BSs. If the neighbor cell signal strength is higher, reassign the new BS and perform intercell handover using the CDCA strategy. If the channel search fails, the call is terminated.
Step 3: Otherwise, perform an intracell handover using the CDCA strategy to keep the CIR qualified. If the channel search fails, the call is terminated.

If the number of cochannels of a channel is zero, such that no cochannel is used in the entire system or no cochannel is used in the neighboring area, the channel search becomes the search for a channel by sequence and the first channel with a qualified CIR is found.

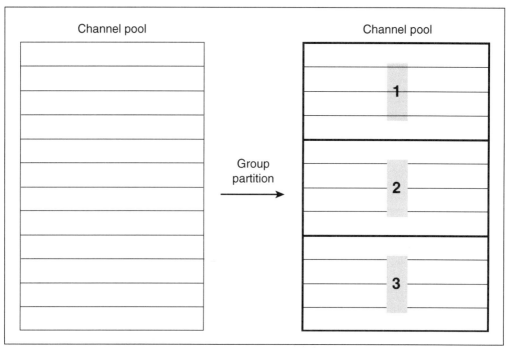

**Figure 5–7**  The Partition of a Channel Pool

### 5.1.2 Group Dynamic Channel Assignment (GDCA) Strategy for Multichannel Traffic

Hać and Mo propose a GDCA strategy that handles multichannel traffic in wireless networks.

A multichannel is required in multimedia to transfer different data types such as voice, data, and video. We consider the multichannel call assignment. We assume that different call types require a different number of channels. A narrowband voice call requires one channel and a wideband image call requires eight channels. Most of the existing channel assignment strategies are proposed for single call type traffic such as one-channel call traffic. Although these strategies can be used for multichannel call traffic, their efficiency greatly degrades in multichannel traffic. The channel assignments in these strategies are less time efficient and less spectrally efficient. The uneven number of channels required in each call is the main reason for DCA degradation. We introduce the group method and the concept of consistency so that the channels can be assigned more efficiently.

We assume that the channel pool of a wireless system is partitioned evenly in groups for multichannel services. The total number of channels in a group is equal to the largest number of channels required by call types in multichannel traffic. The idea of partitioning the channel pool is illustrated in Figure 5–7. For example, if there are 12 channels and the maximum number of channels required by a call is four, we can partition these 12 channels into three groups. We view a group as the basic element in assigning channels.

# Dynamic Channel Assignment in Cellular Mobile Communications Networks

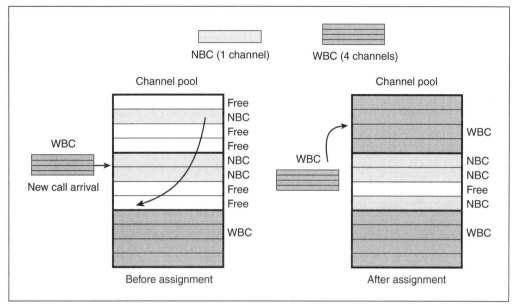

**Figure 5–8** An Illustration of the Consistency Concept

The narrow band call (NBC) is a call type that requires a small number of channels in a group. The wide band call (WBC) is a call type that requires a large number of channels in a group. If one call type requires a larger number of channels as compared to another call type, we can also designate this call type as WBC and another call type as NBC. A used group is a group in which at least one channel is used. A group can be used partially or fully. A partially used group is one in which some channels of the group are used and the other channels of the group are free. A fully used group is a group in which all channels of the group are used. Cogroups are the same groups in the channel pool but used in different cells.

The consistency concept says that the choice of channels for an assignment should always be consistent with the groups. Every channel belongs to a certain group. Channel assignment should always keep channels grouped after a new call access or call handover. The instant distribution of channels used in the system preserves the groups by using optimal packing in the entire system. We can efficiently compact channels in the entire wireless system to reach the optimal spectral efficiency in multichannel traffic.

We always give the WBC a higher priority in taking the channel or group space. The NBC can use a free channel space left by the WBC. If there is not enough channel space for the WBC, the NBC should try a handover to another free channel space to leave this chunk of channel space to the WBC, as shown in Figure 5–8. We assign a WBC according to the consistency concept.

We propose the following consistency rules in a GDCA assignment strategy.

Consistency Rule 1: (Assign a call to a free group or to a partially used group.) The largest call where the number of channels required by the call is the largest in all call types is assigned a free group and uses all channels in the group. For the largest call, there is no free channel space left in the group after the assignment and the group becomes the fully used group. If the call is not the largest, we search partially used groups first. If there is enough free channel space in a partially used group, we assign the new call to this partially used group. If there is no free channel space for the new call in all used groups, we choose a free group and use a part of the channel space of the group for the call. This group becomes a partially used group.

Consistency Rule 2: (Handover to get free channel space.) If a WBC cannot find free channel space and if a group is used by NBCs, we test the feasibility of intracell handover to allow some NBCs a handover to free channel space in other used groups and leave free channel space in this group to the WBC. If this feasibility is allowed, an intracell handover is executed and the WBC is assigned to this free channel space. If the feasibility is not allowed, the NBCs keep the originally assigned channels and the new WBC is blocked. A WBC has a higher priority allowing it to require free channel space and request an NBC to try intracell handover.

By applying the consistency concept, we can treat a group as the basic element in channel packing. If we assign groups with the optimal packing, as DCA strategies reach the optimal packing in one-channel call traffic, we can have the optimal packing for multichannel calls. We give the higher priority to larger WBCs. For example, an eight-channel call has higher priority than a four-channel call, and a four–channel call has higher priority than a one-channel call. If an NBC can make an intracell handover, then this NBC has to handover to leave the space to a WBC requesting a channel space.

We define a qualified CIR as a CIR value which is equal to or larger than a given CIR threshold. A qualified channel has a qualified CIR. The minimum CIR of a group is the minimum CIR of the CIR values of all channels in the group. A qualified group has the qualified minimum CIR of the group and satisfies a certain criterion for a group search, such as the greatest number of cogroups.

The GDCA strategy for the new call arrival is as follows.

Step 1: Assign a cell to the call.
Step 2: With respect to the MS location in a given cell, obtain and update the CIR values for all channels in the channel table of the cell. (The channel table for each cell is a copy of the entire system channel pool. The channel table records the CIR values and use the status of all channels in the cell.)
Step 3: Process group performance data such as the minimum CIR of a group and the other information needed for a group search.
Step 4: Determine the type of arrived call. If the number of channels required by the call is the largest in all call types (the largest WBC), go to Step 5. If the number of channels required by the call is the smallest in all call types (the smallest NBC), go to Step 12. Otherwise (usual WBC), go to Step 8.

Step 5: For the largest WBC, search the channel table for a free qualified group. If a group is found, then assign a free group to the call, and go to Step 15. Otherwise, go to Step 6.

Step 6: If the search of a free qualified group failed, then search the channel table for a qualified group which is used by NBCs. Test the feasibility of an intracell handover that allows the NBCs to switch to the other channel space and leave a free group to the largest WBC. If this test succeeds, then execute an intracell handover, assign the free group to the largest WBC, and go to Step 15. Otherwise, go to Step 7.

Step 7: Search the channel table for a qualified group which is used by the next largest WBC. Test the feasibility of an intracell handover that allows the largest WBC in progress to switch to the other channel space and leave a free group to the largest WBC requested. If the test succeeds, execute an intracell handover, assign the free group to the largest WBC requested, and go to Step 15. Otherwise, the largest WBC is blocked.

Step 8: For the usual WBC, search the channel table and check the partially used groups. If there are enough free qualified channels for the call, assign these channels to the call, and go to Step 15. Otherwise, go to Step 9.

Step 9: Search the channel table and check the used groups for qualified channels that are used by NBCs. Test the feasibility of an intracell handover that allow the NBCs in progress to switch to the other channel space and leave a sufficient number of free qualified channels to the WBC. If the test succeeds, then execute an intracell handover, assign free channels to the WBC, and go to Step 15. Otherwise, go to Step 10.

Step 10: Search the channel table and check used groups for qualified channels that are used by the WBCs. Test the feasibility of an intracell handover that allows the WBCs in progress to switch to another channel space and leave enough number of free qualified channels to the WBC requested. If the test succeeds, then execute an intracell handover, assign these free channels to the new WBC, and go to Step 15. Otherwise, go to Step 11.

Step 11: Search the channel table for a free qualified group. If a qualified group is found, then assign a number of channels of the group to the call, and go to Step 15. Otherwise, the usual WBC is blocked.

Step 12: For the smallest NBC, search the channel table and check the partially used groups. If there are enough free qualified channels for the smallest NBC, assign these channels in the partially used group to the call, and go to Step 15. Otherwise, go to Step 13.

Step 13: Search the channel table and check the used groups for qualified channels that are used by the other smallest NBCs in progress. Test the feasibility of intracell handover that allow the NBCs in progress to switch to other channel spaces and leave enough free qualified channels to the NBC requested. If the test succeeds, then execute an intracell handover, assign these free channels to the new NBC, and go to Step 15. Otherwise, go to Step 14.

Step 14: Search the channel pool and find a free qualified group. If the group is found, then assign a number of channels of the group to the call, and go to Step 15. Otherwise, the call is blocked.

Step 15: Update the group and channel use status. Complete the channel assignment process.

The GDCA strategy for calls in progress is as follows. (This process keeps calls in the required signal quality.)

Step 1: Check the CIR value of the channels being used by calls. If the CIR value of a channel is lower than the CIR threshold, then go to Step 2. Otherwise, complete checking the call in progress.
Step 2: Check the caller's location. If the caller moves to another cell, proceed with an intercell handover. Use the GDCA strategy for the call in another cell.
Step 3: If the caller remains in the same cell, proceed with an intracell handover. Use the GDCA strategy for the call reassignment.

The GDCA solves the problem of multichannel traffic by using a group method. The basic element of channel packing is a group. If the consistency rules are applied, the calls will fill in the groups as compatibly as possible. The goal of keeping channels grouped is to reach the channel packing effect of signal call type, and to compact different call types together with respect to the CIR values. The group method makes channel space in the entire system more manageable. In a given cell, the channels CIR values in the same group tend to be the same, and channels in the group tend to have the same reuse distance. Using search methods to select channels, we can reduce the reuse distance for cogroups to a minimum. With the consistency concept, we can also allow the channels serving each call type to have an even chance of optimal packing.

The group performs as a buffer to prevent CIR fluctuation. Channel assignment is a dynamic process, in which some calls arrive and some leave. If we treat channels as a group, the call leaving a group will leave a chunk of free space for another call to fill in, and the total interference is smaller. When another call occupies this space, the total interference for this group recovers to its original level but does not increase. If cogroups in a group have enough separation distance to avoid interference, introducing and dropping a call from a used group will affect its cogroups to a lesser extent because there is some margin for interference. This margin is the result of the group assignment. The channels for the same call are in the same group. The distribution of the CIR levels of these channels are almost even.

## 5.2 Distributed Dynamic Channel Assignment (DDCA) in Cellular Mobile Communication Networks

Small-cell wireless technologies are needed to support cellular mobile communication in a local environment. The cell size ("microcell" for outdoor and "picocell" for indoor environments) can be several ten times smaller than the cells in existing cellular networks. The number of BSs can, therefore, increase dramatically. In addition, as cellular wireless communication becomes ubiquitous, the propagation environments can become too diverse to be predicted with the desired accuracy. As a result, the conventional channel assignment methodology is no longer a viable approach.

The traditional classification of channel assignment strategies are as follows: FCA, hybrid channel assignment (HCA), DCA, borrowing channel assignment (BCA), and forcible-borrowing channel assignment (FBCA). FCA is used where each cell is assigned a fixed set of nominal channels. If a new call finds that no free nominal channel is available, the call is blocked. The HCA is a part of the FCA and DCA. The channels are divided into two sets. One set of channels is assigned to each cell on a nominal basis and the other set of channels is kept in a central pool for dynamic assignment. Its performance is between FCA and DCA. FCA is not suitable for a cellular mobile network with highly time-varying traffic and a changing propagation condition. This is especially the case in microcellular systems. The DCA methods, which adapt to the instantaneous interference and traffic situations, become more useful. In addition, the DCA removes the need for prior frequency planning, thus increasing system flexibility. As a cellular network evolves, cells become smaller, leading to a microcellular system in MAs. Decreasing the cell size causes a spatial and temporal volatility of traffic and makes optimal frequency planning infeasible. The DCA therefore, plays a more important role in the next generation systems since no centralized coordination is needed in the DCA scheme. Also the DCA adapts to interference, which can achieve high capacity and flexibility in unpredictable propagation environment.

Simulation studies of some DCA schemes show that although they can improve network performance at low to average traffic loads, the heavy load network performance is not better than in the FCA because of the interference of the same channel (i.e., cochannel). The assignment of the channel to a call in a cell can cause degradation of service quality of other calls using the same channel. The interference rate of the channel increases, thus forcing call termination. This always happens in the DCA system, especially under heavy traffic. An optimal DCA scheme should keep the interference probability low as well as the probability of forced call termination.

The function of channel assignment is to divide the total number of channels into subsets and allocate the channels to each cell site. An efficient use of channels determines performance of the system and can be obtained by using different channel assignment techniques. Channel assignment schemes can be classified into two categories: FCA and Nonfixed Channel Assignment (NFCA). The NFCA scheme consists of HCA, BCA, and DCA. There is a remarkable increase in channel occupancy when some of the more efficient DCA schemes are used.

### 5.2.1  Centralized DCA Scheme

The DCA scheme can be classified into two categories: a centralized DCA scheme and a distributed DCA. Centralized DCA schemes are MP, MAXMIN, and the target method.

1. The MP scheme assumes that a new call is blocked only if there is no possible reallocation of the channel to the calls (including reallocating calls in progress, which results in a call being carried). If a new call arrives and finds that there is no available channel, the system reallocates existing calls to the other channels to accommodate the new call. Although the MP scheme can achieve the optimal performance, it has to take into account the "worst

case" prediction of varying traffic and propagation conditions to achieve optimal performance. This can cause a large penalty in system capacity. Furthermore, it requires the systemwide information about channels in use in every cell and the complexity of searching all possible reallocations, which involves considerable system overload.

2. In the MAXMIN scheme (maximum of the minimums CIR of all the calls in the system), a new call is assigned a channel that maximizes the minimum of the CIR of all the calls that are served by the system at one time. This scheme requires a significant computation of existing calls and systemwide coordination.

3. In the target method scheme, a global technique assigns channels so that all mobiles in the system have a CIR close to a defined target. The mean square error (from the target) of the CIR of every mobile on each channel is calculated, and the channel with the smallest error is used. This scheme increases the computation complexity and requires measurement of every cell in the system. Hence, it is not a feasible DCA.

The centralized DCA schemes often require a control center to handle channel allocation or handoff among all cells or MSs. The control center must obtain local status information from all MSs in the cellular system. The current occupancy and interference level of channels, and the propagation condition of MSs within each cell can be measured and transferred to the control center periodically. Based on the actual conditions of the entire system, the decision is made to assign a channel to a mobile. The centralized methods can achieve the optimal performance in the entire system (i.e., maximum capacity and a minimal blocking probability with a certain interference level). However, as the number of cells increases, the centralized computation becomes mathematically intractable. The method is also impractical under highly varying traffic because of the difficulty in measuring the actual conditions. In addition, the measurement and transmission of the current local status information and the control instruction can occupy some spectrum resources and cause a large penalty in the system capacity.

### 5.2.2 DDCA

In the DDCA scheme, centralized control is not required. Each cell can handle a channel assignment or handoff autonomously. Any channel can be allocated to calls in any cells as long as the CIR satisfies the CIR threshold. An allocation or handoff decision is based on the local status information. The exchange of status information with relevant cells is needed to reduce mutual interference. The number of relevant cells is restricted under the computation capability of a cell.

This channel assignment scheme is self-organized and the handoff process is locally controlled. Moreover, in a DCA scheme, each BS measures the interference power of the candidate channels to check to see if the quality of the channel meets the allocation requirement. Based on the result, the BS selects a channel for the call. Measurements of many channels by a BS before a channel allocation increase connection delay. The delay can be reduced if the number of cells needed to exchange information decreases. This is infeasible in the centralized DCA scheme since the system has to know the global use information of the channels. The DDCA scheme reduces the computation complexity because it only considers the local status information. This results in reducing the radio link allocation delay and handoff delay that frequently occurs in the centralized schemes.

The DDCA schemes, such as the compact pattern (CP) based DCA, the geometric DCA, the channel segregation DCA, and the autonomous reuse partitioning DCA, improve the spectrum efficiency (i.e., system capacity). The DDCA scheme also minimizes the interference level if it is possible, as in the first available, the least interference below threshold, and the highest interference below threshold. The DDCA schemes are as follows.

1. The CP based DCA strategy keeps dynamic cochannel cells of any channel in a compact pattern. The compact pattern of a channel is defined as a pattern with the minimal average distance between cochannel cells. This strategy greatly improves traffic-carrying capacity.
2. The geometric DCA strategy exploits the priority information to maintain tight geometric packing of used carriers. The channel, which keeps a tighter pattern, has a higher priority than the other channels.
3. The channel segregation DCA is a self-organized dynamic channel assignment scheme. In this scheme, each cell acquires its favorite channels through learning from past experiences how each channel was used between the current cell and the other cells. As a result, the optimum channel reuse is self-organized, and the spectrum efficiency is improved.
4. The autonomous reuse partitioning DCA is an effective technique to increase capacity by reducing the reuse distance of identical channels. Each BS searches for a candidate channel based on an order common to all BSs. The interference power of channels with a higher priority increases because those channels are used more frequently in many cells. Such channels are likely to be used by the MS close to the BSs since the high signal power received is required to obtain the sufficient CIR level. As a result, each cell is divided into rings. In the inner rings, the channels are reused with distances shorter than those in conventional systems, whereas channels in the outer rings are reused with the same distance as in the conventional system.
5. The first available strategy always selects the first channel with a satisfied CIR. The blocking rate of this strategy is lower, and the interference rate is very high.
6. The least interference below threshold scheme selects the least interfered channel below the interference threshold (i.e., the highest CIR). The scheme also minimizes the overall interference in the system while maintaining the quality of each call above some minimal acceptable level.
7. The highest interference below threshold scheme selects the most interferenced channel below the interference threshold (i.e., CIR still satisfied). The scheme utilizes the frequency spectrum more compactly as long as the CIR is satisfied.

Comparing between the centralized and distributed DCA scheme; the distributed DCA scheme is more flexible and can accommodate an increase in traffic density. The distributed DCA adapts to the growth of the traffic load and varying bursty traffic. This scheme is easier to implement and can avoid allocation delay that frequently occurs in a centralized scheme, especially under heavy and nonuniform traffic load since the distributed DCA scheme is based only on local or near information instead of global status information. In the described distributed DCA schemes, an MS and the BSs estimate the CIR and allocate the channel to a call when the predicted CIR level is above the threshold (i.e., the channel is allocated without checking how the use of that channel use affects the same channel which is used in neighboring cells). However, this allocation can cause the CIR level of a channel used in other cells to drop below the required CIR threshold. In this case, the interference occurs. (Here, interference refers to the condition where the CIR level drops below the CIR threshold.) If an interrupted call cannot select a new channel to continue communication immediately, the call can be terminated, which is called the forced call termination. Even if an interrupted call selects an acceptable new channel, setting a link on the new channel can cause interruption to another established link, which results in a successive interruption. This is called instability. Moreover, it is generally observed in the DDCA schemes that there is a trade-off between the goal of avoiding call blocking and the goal of avoiding interference. Some DDCAs try to achieve the lower blocking rate, others try to minimize the interference as much as possible, but these two purposes cannot be attained at the same time. A lower blocking rate means the system can accept more calls at a time. As a result, the interference increases when more new calls arrive, and the other cochannel call has to handover to the other channel, if needed. The reassignment rate increases dramatically, which introduces the instability to the cellular network. A good channel assignment should introduce a lower reassignment rate. The forced call termination is the most undesirable phenomenon because it causes the calls in progress to be dropped. A good DDCA attains a better forced call termination rate while keeping the blocking rate for the new call at an acceptable level.

### 5.2.3 Priority-Based DDCA (PDCA)

Hać and Xiao propose a DDCA called the PDCA. In this strategy we assign a priority to each channel in the channel pool. The priority considers both the use frequency of the channel in the neighboring cells, and the prediction of the effect of cochannel interference on the other calls in the neighboring cells using the same channel. We define a priority function as one that considers the reuse frequency of a channel as well as the interference on the other cochannel calls.

We measure interference by computing the CIR level of a call. We test the interference of a channel from the candidate channels on the other cochannels in neighboring cells by computing the CIR level of the other cochannel calls in neighboring cells if the channel is assigned to this cell. We compare the interference level of different channels by counting the number of calls to be handed over to the other channel because the CIR level of the channel is lower than the CIR threshold ($CIR_{th}$).

A higher priority is given to the channels used more frequently, and results in less interference to calls handled by the same channels in neighboring cells. The channel with the highest priority in the channel pool is assigned first.

The PDCA strategy has the following features.

1. This is a DDCA strategy similar to the other DDCA strategies. The PDCA strategy considers only the traffic distribution in the neighboring cells when assigning the channel. Each BS in one cell can exchange information with the other BSs in neighboring cells and handle the channel assignment autonomously.
2. This is a dynamic measurement-based DDCA strategy. Because the structure of the cellular system can be nonuniform in the microcell system, the fixed channel reuse distance is not suitable for future mobile communication systems. To decide the reuse distance dynamically, we choose the CIR level as the criterion. When a new call arrives, the local BS checks the CIR level of the candidate channels before making the choice. The CIR depends on real-time information about the calls using the cochannel. From this point of view, the PDCA strategy is a dynamic strategy adaptive to the real-time varying traffic.
3. The PDCA strategy defines the priority order of the available channels based on the channel reuse frequency and interference information. The strategy is optimal in comparison with the other DDCA strategies.
4. Intercell and intracell handover is the solution for an ongoing call if the CIR level of the call is lower than the CIR threshold. The system keeps track of the CIR level of the ongoing call, once the CIR level is lower than the CIR threshold. The system first tries the intercell handover and then the intracell handover.

The following are definitions and assumptions for PDCA strategy.

- The total number of channels in the system is M, and all channels in the system can be allocated to every cell, which is subject to reuse constraint.
- The constraint of channel use is the CIR, which is defined as the Cochannel Interference Ratio, and the threshold of the CIR is denoted as $CIR_{th}$. If the measured CIR level of the channel is below $CIR_{th}$, the channel cannot be used by an MS. Only the cochannel interference is considered; the adjacent channel interference and noise are negligible. The MS and BS can test each channel's CIR locally. The CIR levels are tested periodically and the CIR level of MSs are sent to their BSs through the control channel.
- The cellular network is a regular planar array of hexagonal cells. We define the region of the neighboring cells of cell $i$ as outer $k$ layers of the cells which have the same center as cell $i$. The choice of $k$ depends on the emission power of the base station and the required $CIR_{th}$. The defined region of the neighboring cells is the interference area, which we consider when allocating the channel. Cell $i$ exchanges the information with the outer $k$ layers of cells to get the interference information.
- We assume that the BS in cell $i$ can exchange information with other BSs in the neighboring cells of cell $i$ through the control channel.
- We define $S(i)$ as the set of channels that are used in cell $i$.
- We define $A(i)$ as the set of channels that are used in neighboring cells, the channels' CIR level is greater than the $CIR_{th}$.

- We define $F(i)$ as the set of free channels that are not used in cell $i$ and the neighboring cells.
- We define $N_2(j)$ as the total number of use frequency of channel $j$ in the neighboring cells of cell $i$. When deciding whether to assign channel $j$ to the call in cell $i$, the BS in cell $i$ exchanges information through the control channel to other BSs in the neighboring cells, and inquires about the interference of channel $j$ in cell $i$ with other cochannels in the neighboring cells (i.e., how many calls in $N_2(j)$ meet the threshold $CIR_{th}$ if channel $j$ is used in cell $i$). Other BSs, when receiving the inquiry from the BS in cell $i$, calculate the interference level of channel $j$ which carries the calls in the cell. Those BSs return the information to the BS in cell $i$. We define $N_1(j)$ as the total number of calls using channel $j$ in the neighboring cells of cell $i$, for which CIR is above the $CIR_{th}$ if channel $j$ is assigned to a request call in cell $i$. A larger $N_1(j)$ means a lower interference of a channel $j$ assignment to cochannels' calls within the neighboring cells.
- We define ratio $P(j) = N_1(j)/N_2(j)$ as the priority of channel $j$ in set $A(i)$. If priority $P(j)$ is the highest in set $A(i)$, the selection of channel $j$ is closer to optimal than the other channels in set $A(i)$ based on both the channel reuse frequency and the interference to other cochannels' calls in the neighboring cells.

An algorithm for new call setup in the PDCA strategy is as follows.

Step 1: When a call arrives in cell $i$, cell $i$ searches for all the channels in the channel pool, which are not used in cell $i$ (i.e., the channels not in set $S(i)$). If no such channel is found (i.e., all channels are fully used in cell $i$), the call is blocked. Otherwise, the CIR of those channels is checked at the place where the call arrives in cell $i$. If the CIR of the channel is greater than the threshold $CIR_{th}$, then the channel is available, and set $A(i)$ for cell $i$ is updated. If set $A(i)$ is empty (i.e., no channel with a satisfied CIR is found), then go to Step 4. Otherwise, go to Step 2.

Step 2: For each channel $j$ in set $A(i)$, find the number for $N_2(j)$ and $N_1(j)$ by exchanging the information with the neighboring cells of cell $i$, then calculate priority $P(j)$ for channel $j$, and sort the priority of available channels in set $A(i)$.

Step 3: Find the channel in set $A(i)$ with the highest priority and assign this channel to the new call.

Step 4: If no channel is found in set $A(i)$, choose a new channel in $F(i)$, which is not used in cell $i$ and the neighboring cells, and assign this channel to the call.

Step 5: If no channel is found in $F(i)$, the call is blocked.

A handover is needed to improve the traffic handling load and spectrum use efficiency. If the CIR level of a call is lower than the $CIR_{th}$, the call has to handover to the other channel to keep the communication quality.

The handover in the PDCA strategy is as follows.

Step 1: Periodically check the CIR level of a call in cell *i*. If the CIR is lower than the CIR$_{th}$, attempt a handover, otherwise complete the CIR check.

Step 2: Handover: Check the signal strength of the BS in cell *i* and the neighboring cells to justify whether the intercell handover is needed. If so, assign a new BS to the call and reassign a new channel to the call using the PDCA strategy. If the intercell handover fails, the call is forced to terminate. If the intercell handover is not necessary, try the intracell handover by using the same PDCA channel assignment strategy as used for the new call setup. If the intracell handover failed, the call is forced to terminate.

Multimedia traffic has different types of calls, such as data, audio, and video, which require different bandwidths (i.e., the number of channels to carry the traffic). For example, voice requires one channel to carry a call whereas the video requires eight channels. There are several channel assignment strategies for multimedia traffic. For example, to improve the grade of service of the multichannel calls, we can limit the number of channels available to a single channel call (i.e., voice), while the multichannel calls can dynamically borrow channels from anywhere in the system. This approach is not feasible, since the performance of a single channel call degrades. The other approach is a group channel assignment strategy. The idea of a group DCA is to divide the channel pool into groups, where each has the same number of channels as the maximum number of channels required in multichannel call. When a WBC arrives, the NBC is compacted to make room for the WBC. This strategy has a high reassignment rate and causes high instability in the system. The connection delay is very high because the chance of a handover increases. The optimal DCA scheme for multimedia traffic should assign a group of channels that are consistent (i.e., have a close CIR level). The assignment of channels to a group should result in a lower forced call termination rate and interference rate.

The PDCA strategy is extended to apply to multimedia traffic. The application of PDCA to the multimedia traffic environment sorts the available channels in set $A(i)$ with their CIR level, then divides the available channels in set $A(i)$ into groups by the bandwidth requirement $R(k)$ of multimedia call with type $k$. The CIR level of channels in each group is closer as compared with the CIR level of the channels in other groups. If a group is applied to a type $k$ call, the call is less likely to be handed over to the other channel. The channels for the same call have a close CIR level, so the distribution of the CIR level of these channels is more even and the reuse frequency of the channel becomes more consistent. The question is how to select a group for an arriving call. We use the priority concept in which we assign a priority to each group. Each time a new call arrives, we allocate the channel with the highest priority to the call. We define the priority of a group as the product of the priority of each channel in the group. The priority of each channel is the same as defined in the single channel assignment strategy. Suppose the total number of channels in a group is $n$, then the priority of the group is defined as $\prod_{v=1}^{n} P(v)$, where P(v) is the priority of channel $v$. Since the assignment of the priority is based on the cochannel use information and interference information of the channel in the cellular network, the assignment results in lower interference and reassignment rates, and achieves a good channel use efficiency at the same time.

The PDCA strategy in the multimedia traffic environment for the new call setup is as follows.

Step 1: When a new multimedia call of type $k$ (assume the channel requirement is $R(k)$) arrives in cell $i$, the BS in cell $i$ checks to see if all channels in the system are used. If so, the call is blocked. Otherwise, the BS checks the CIR level of those available channels that are not assigned in cell $i$ but are used in the neighboring cells of cell $i$, and updates $A(i)$. If the total number of channels in $A(i)$ is less than the channel requirement $R(k)$, then go to Step 4, otherwise go to Step 2.

Step 2: Sort the channels in $A(i)$ with the CIR level in the ascendent order, then divide the channels in $A(i)$ by the channel requirement $R(k)$ for a type $k$ call. We get several candidate grouped channels for type $k$ calls.

Step 3: In each group, calculate the priority of all channels in the group by exchanging the information with other BSs in the neighboring cells of cell $i$. Calculate the total priority of the group using formula $\prod_{v=1}^{n} P(v)$. Then select the group with the highest priority, and assign the channels in the group to a type $k$ call.

Step 4: Check to see if the total number of channels in set $A(i)$ plus the channels in set $F(i)$ is greater than $R(k)$. If so, choose all channels in set $A(i)$ and some (the remaining) channels in $F(i)$ to meet the requirement $R(k)$, and assign those channels to type $k$ call. Otherwise, go to Step 5.

Step 5: The call is forced to terminate.

The PDCA strategy in the multimedia traffic environment for handover is as follows.

Step 1: Check the minimum CIR for each type of call. The minimum CIR of the call is defined as the minimum CIR of the channels which are assigned to a multimedia call. If the minimum CIR is less than the $CIR_{th}$, go to Step 2. Otherwise complete checking the call.

Step 2: Check to see whether the intercell handover is necessary. If so, assign a new BS to the call, and reassign the new channels to the multimedia call according to the PDCA strategy. Otherwise, try an intracell handover. For each handover, the call releases all channels it uses, and applies for the new channels to meet the channel requirements.

We use the dynamic group channel assignment in the PDCA strategy to solve the DCA problem in the multimedia traffic environment. The goal of keeping channels grouped is to reach the channel packing effect of the NBC type and to compact different call types together with a less uneven CIR level. The group allocation makes channel space resource manageable in the entire system. The difference between PDCA strategy and the other DCA strategies in the multimedia traffic environment is that the PDCA uses the dynamic group channel allocation. In other DCA schemes, the number of channels in one group is fixed. Instead, in the PDCA, we divide the available channels into groups according to the channel requirement of a call. The number of channels in one group can be adapted to different types of calls. Moreover, the NBC does not handover to the other group if a WBC arrives. The reassignment rate of a call is, thus, kept lower. As a result, the entire system tends to be more stable.

We consider the priority of groups. There is an important difference between our PDCA strategy and other DCA strategies in the multimedia traffic environment. The priority of a group takes care of the channel reuse frequency as well as the interference level to the other cochannels. The interference rate of the entire system is lower than the interference rate of the other group DCA strategies which only consider reuse frequency. The PDCA strategy keeps the channel pattern tighter than other DCA strategies, which only consider the CIR level.

We sort the channels according to their CIR level and divide them into groups. Consequently, the CIR level of channels in a group tends to be closer. The channels in a group are more consistent and have a close interference with the other cochannels. If these channels in a group are used by a multimedia call at the same time and released at the same time, the distribution of the channels' CIR tends to be more even.

## 5.3 The HCA Method for Wireless Communication Networks

Wireless communication networks are assigned to a certain range of the radio spectrum. Usually this spectrum is divided into a set of disjoint radio channels. The channels can be used at the same time, but the adjacent channels used in locations too close to each other can cause interference.

Depending on the radio technologies employed, channels are divided in different ways. There are three major radio technologies that are widely used in wireless networks: frequency division, time division, and code division. In frequency division technology, different users use different frequencies to communicate. The channels are disjointed frequency bands. In time division technology, all the communication is carried on the same frequency band, but different users are assigned different time periods during which they can communicate. The communication time is divided into time slots. Each recurring time slot is a channel. The code division technology separates channels by using different modulation codes. A combination of these technologies results in more methods with which to divide channels. Regardless of the method used, the division of a given radio spectrum cannot be done indefinitely. A certain level of quality must be maintained for the radio signal in each channel.

Due to the propagation path loss characteristic of the radio signal, the average power received by a receiver is proportional to $P \times d^{-a}$, where $P$ is the transmitter power, $d$ is the distance between receiver and transmitter, and $a$ is the number between three and five as determined by the physical environment. By controlling the transmitter power level and taking advantage of the geographical characteristics of the wireless communication network's service area, the same channel can be used simultaneously in different cells, providing that the CIR is over a certain threshold. Here, the carrier is the received signal power in a channel from the transmitter (i.e., BS) in the cell where the receiver is located, and the interference is the sum of received signal powers in the same channel from transmitters in all the other cells, plus some environmental noise. There are two basic ideas behind the channel allocation methods. One is physical separation, that is, controlling the distance between cochannel cells (cells that use the same channel) to get a good balance between channel utilization efficiency and the high carrier-to-interference ratio. The other is power control, which means achieving the required CIR by reducing power at the interfering transmitters and increasing the desired signal's power level.

The HCA methods combine strategies from both FCA and DCA. In HCA, all the available channels are divided into two groups, fixed and dynamic. The channels in the fixed group are assigned to cells as nominal channels, similarly as in the FCA methods. Channels in the dynamic group are shared among all the cells. Nominal channels are preferred for use in their respective cells. When a call request comes into a cell, it is assigned a nominal channel if it is available. When no nominal channel is available, a channel from a dynamic group is assigned.

The HCA has characteristics of both the FCA and DCA. Under a light traffic load, the HCA performs like the FCA method. When the traffic load increases, almost all the nominal channels are in use. New calls are served by channels in the dynamic group. Similar to the DCA, the HCA does not perform well under a heavy traffic load.

Hać and Chen introduce an HCA method, which emphasizes not only the DCA part, but also the nominal channel allocation part. Minimizing the cochannel distance is used as a criterion in both parts of the methods for consistency. There is no strict separation of the nominal and dynamic channels. The channels already assigned as nominal channels can be assigned to calls in other cells as dynamic channels, as long as the channel separation constraints are satisfied. The first part of the method is the allocation of nominal channels for each cell. This is carried out in the planning stage of the wireless communication network. The second part is the allocation of channels to ongoing call requests while the wireless network is in use. This is carried out dynamically (i.e., when a call originates in a cell without free nominal channels). We will discuss nominal channel allocation strategies: the frequency exhaustive and requirement exhaustive methods, and show the importance of the nominal channel slot order in the nominal channel allocation.

### 5.3.1 Definition of the Problem

We define the wireless network by using the following parameters.

> A set of $n$ cells.
> A set of $m$ channels.
> An $n \times n$ symmetric constraint matrix

$$C = \begin{bmatrix} c_{11} & c_{12} & \cdots & c_{1n} \\ c_{21} & c_{22} & \cdots & c_{2n} \\ \cdots & \cdots & \cdots & \cdots \\ c_{n1} & c_{n2} & \cdots & c_{nn} \end{bmatrix}$$

where $c_{ij}$ is the minimum separation between an arbitrary pair of channels, one used in cell $i$, the other in cell $j$.

An $n \times n$ symmetric cell distance matrix

$$D = \begin{bmatrix} d_{11} & d_{12} & \cdots & d_{1n} \\ d_{21} & d_{22} & \cdots & d_{2n} \\ \cdots & \cdots & \cdots & \cdots \\ d_{n1} & d_{n2} & \cdots & d_{nn} \end{bmatrix}$$

where $d_{ij}$ is the distance between cells $i$ and $j$.

By introducing matrix $C$, both the cochannel and adjacent channel interference are considered in the channel allocation. If $c_{ij} = 0$, cells $i$ and $j$ are cochannel cells. If $c_{ij} = 1$, there is a cochannel constraint between cells $i$ and $j$. If $c_{ij} > 1$, there is an adjacent channel constraint between cells $i$ and $j$. The diagonal element $c_{ii}$ represents cosite constraint.

Let $S_i$ be the set of cells that are assigned channel $i$. The objective of our algorithm is to minimize the cochannel distance, which is defined by

$$\sum_i \sum_{j,k \in S_i} d_{jk} \quad 1 \le i \le m, \; 1 \le j, k \le n.$$

If channel $i$ is assigned to only one cell, we define its corresponding cochannel distance as the largest element in the cell distance matrix $D$. That is

$$\sum_{j,k \in S_i} d_{jk} = \underset{p,q}{MAX}(d_{pq}) \quad 1 \le p, q \le n.$$

This means if channel $i$ is not assigned to any cell originally, assigning $i$ to a new call request makes its corresponding cochannel distance change from zero to the longest distance between any pair of cells. If we assign a channel already in use to the new call request, it gives a smaller cochannel distance increase. This way, channel reuse is encouraged.

In the wireless network planning stage, we decide how many nominal channels are needed for each cell. This is determined by user density, user calling patterns, and user movement patterns. The result is the nominal channel requirement vector

$$R = \begin{bmatrix} r_1 & r_2 & \cdots & r_n \end{bmatrix}$$

where $r_i$ denotes the number of nominal channels required by cell $i$.

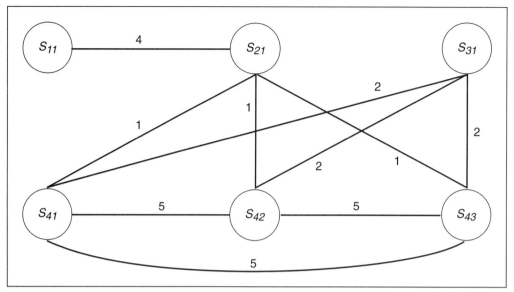

**Figure 5–9** Graph Representing an Example of Adjacent Cells

The channel planning problem can be modeled as a generalized graph-coloring problem. Before actually being assigned nominal channels, a cell has a channel slot for each nominal channel it needs. We can create a graph such that each vertex of the graph represents a channel slot. There is an edge between the two vertices if two channel slots cannot have the same channel. The label of an edge is the minimum separation between two channels assigned to two channel slots, which is $c_{ij}$ assuming one channel slot belongs to cell $i$, and the other is in cell $j$.

For example, assume that we have a wireless network with four cells, and its constraint matrix is

$$C = \begin{bmatrix} 5 & 4 & 0 & 0 \\ 4 & 5 & 0 & 1 \\ 0 & 0 & 5 & 2 \\ 0 & 1 & 2 & 5 \end{bmatrix}$$

and nominal channel requirement vector is

$$R = \begin{bmatrix} 1 & 1 & 1 & 3 \end{bmatrix}.$$

We have a total of six nominal channel slots. They are denoted by $s_{11}$, $s_{21}$, $s_{31}$, $s_{41}$, $s_{42}$, and $s_{43}$. In $s_{ij}$, $i$ is the cell number and $j$ is the nominal channel slot number in cell $i$. Each nominal channel slot represents a place where we need to assign a nominal channel. The nominal channel allocation problem for the network can be represented by the graph shown in Figure 5–9.

Assume a graph is denoted by $(V, E)$, where $V$ is the set of vertices and $E \subset V^2$ is the set of edges. Now the nominal channel allocation is equivalent to assigning positive integers between 1 and $m$ to the vertices of the graph (i.e., find a mapping $V \to I$ where $I$ is the set of positive integers between 1 and $m$), under the following conditions.

If two vertices $v_1$ and $v_2$ are connected by an edge (i.e., $(v_1, v_2) \in E$), the absolute value of the difference between the integer assigned to $v_1$ and $v_2$ is greater than or equal to the label of the edge. This condition corresponds to constraints represented by matrix $C$. Let $V_i \subset V$ be the set of vertices which are assigned the integer $i$. The following value is the minimum.

$$\sum_i \sum_{j, k \in V_i} d_{jk} \quad 1 \le i \le m, \ 1 \le j, k \le n.$$

This is a generalized graph-coloring problem, which is known to be NP-complete. Thus, channel planning problem is also NP-complete. We will use approximation techniques in nominal channel allocation.

### 5.3.2 Nominal Channel Assignment Strategy and Channel Slot Order

#### 5.3.2.1 Frequency Exhaustive and Requirement Exhaustive Strategies

Given a nominal channel requirement vector $R$, the set of cells, and the set of available channels, there are several ways to assign nominal channels to each cell. We discuss the frequency exhaustive strategy and requirement exhaustive strategy.

The frequency exhaustive strategy takes a nominal channel slot, scans all the available channels, finds the first channel that satisfies the constraints specified in matrix $C$ while considering all the nominal channels which are already filled, and assigns this channel to the nominal channel slot. This procedure is repeated until all the nominal channel slots are filled.

The requirement exhaustive strategy takes the first channel in the list, and checks each nominal channel slot. The channel is assigned to a nominal channel slot if it does not violate the constraints specified in matrix $C$. When the first channel has been assigned to all possible slots, the second channel is processed in the same way; followed by the third channel, the fourth channel, and so on. This continues until every nominal channel slot has a channel assigned to it.

In the example in Figure 5–9 described by matrices $C$ and $R$, we assume that the cell distance matrix is

$$D = \begin{bmatrix} 0 & 2 & 4 & 5 \\ 2 & 0 & 2 & 3 \\ 4 & 2 & 0 & 1 \\ 5 & 3 & 1 & 0 \end{bmatrix}$$

We use $s_{11}, s_{21}, s_{31}, s_{41}, s_{42}$, and $s_{43}$ to denote the six channel slots.

Using the frequency exhaustive strategy, and processing the nominal channel slots in the following order: $s_{11}, s_{21}, s_{31}, s_{41}, s_{42}$, and $s_{43}$ (we call this slot order 1), we first assign channel 1 to $s_{11}$. Next we consider $s_{21}$. According to constraint matrix $C$, the separation between channels assigned to cells 1 and 2 must be at least four. The first channel that satisfies this constraint, channel 5, is assigned to $s_{21}$. When selecting a channel for $s_{31}$, we must consider the two channel assignments that are already made. The status of the current nominal channel assignment is as follows.

| Cell | Channels Assigned |
|------|-------------------|
| 1    | 1                 |
| 2    | 5                 |

The minimum separation between channels in cells 3 and 1 is 0. So is the minimum separation between channels in cells 3 and 2. This means that the nominal channel assignment in cell 3 is not affected by the nominal channel assignments in cell 1 or 2. Thus, we can use channel 1 again and assign it to $s_{31}$. Now we have a channel assignment scheme as follows.

| Cell | Channels Assigned |
|------|-------------------|
| 1    | 1                 |
| 2    | 5                 |
| 3    | 1                 |

According to the fourth row of constraint matrix $C$, the minimum separations between the channels in cells 4 and 2, and cell 3 are 1 and 2, respectively, and the minimum separation between cosite channels within cell 4 is 5. For $s_{41}$, channel 3 is selected. Similarly, channels 8 and 13 are assigned to $s_{42}$ and $s_{43}$, respectively. The final nominal channel assignment in slot order 1 is as follows.

| Cell | Channels Assigned |
|------|-------------------|
| 1    | 1                 |
| 2    | 5                 |
| 3    | 1                 |
| 4    | 3, 8, 13          |

Using a notation where $S_i$ is the set of cells that are assigned channel $i$ as one of the nominal channels, for this nominal channel assignment scheme, the cochannel distance is

$$\sum_i \sum_{j,k \in S_i} d_{jk} \quad d_{jk} = 24$$

Using the frequency exhaustive strategy with the order of nominal channel slots to be processed changed to $s_{41}, s_{42}, s_{43}, s_{21}, s_{31}$, and $s_{11}$ (slot order 2), we have another nominal channel assignment scheme whose cochannel distance is 25.

| Cell | Channels Assigned |
|------|-------------------|
| 1    | 6                 |
| 2    | 2                 |
| 3    | 3                 |
| 4    | 1, 6, 11          |

Similarly, for the frequency exhaustive strategy and nominal channel slot order $s_{21}$, $s_{41}$, $s_{42}$, $s_{43}$, $s_{31}$, and $s_{11}$ (slot order 3), the cochannel distance is 30, and the nominal channel assignment is as follows.

| Cell | Channels Assigned |
|------|-------------------|
| 1    | 6                 |
| 2    | 1                 |
| 3    | 4                 |
| 4    | 2, 7, 12          |

Using the requirement exhaustive strategy on the nominal channel slot order 1, 2, and 3, respectively, we have three more nominal channel assignment schemes. They are summarized as follows, along with the results from using the frequency exhaustive strategy. FE is the frequency exhaustive strategy, and RE is the requirement exhaustive strategy. Assignments (*a*), (*b*), (*c*), and (*d*, *e*, *f*) mean that cell 1 is assigned channel *a*, cell 2 is assigned channel *b*, cell 3 is assigned channel *c*, and cell 4 is assigned channels *d*, *e*, and *f*.

| Slot Order | Strategy | Assignment | Cochannel Distance |
|------------|----------|------------|--------------------|
| 1 | FE | (1), (5), (1) (3, 8, 13) | 24 |
| 2 | FE | (6), (2), (3), (1, 6, 11) | 25 |
| 3 | FE | (6), (1), (4), (2, 7, 12) | 30 |
| 1 | RE | (1), (5), (3), 1, 6, 11) | 25 |
| 2 | RE | (1), (5), (3), (1, 6, 11) | 25 |
| 3 | RE | (5), (1), (1), (3, 8, 13) | 22 |

This example shows that by using different channel assignment strategies for processing nominal channel slots in different orders, we have nominal channel assignment schemes with different cochannel distances. The problem is to determine which channel assignment strategy or nominal channel slot order gives us better channel assignment schemes with smaller cochannel distance.

### 5.3.2.2 Importance of the Nominal Channel Slot Order

The nominal channel slot order and channel assignment strategy are two key factors in good nominal channel assignment schemes. They are also two aspects of the nominal channel assignment problem.

The nominal channel slot order and channel assignment strategy are not completely independent. There is an interaction between them. Using different channel assignment strategies on the same nominal channel slot order results in different nominal channel assignment schemes. On the other hand, given a channel assignment strategy, the quality of the nominal channel allocation schemes it generates is affected by the different nominal channel slot orders used.

The effectiveness of channel assignment strategies is much more difficult to measure than the nominal channel slot order. Most of the research focuses on the nominal channel slot order. The channel assignment strategy and nominal channel slot order are closely related in the process of looking for a good nominal channel assignment scheme. Thus, if we consider all the possible orders of nominal channel slots, it is not important which channel assignment strategy we use. The quality of the resulting nominal channel assignment scheme is largely determined by the nominal channel slot order in which the channels are assigned.

A concept that can be used in sorting nominal channel slots is the degree of cells and nominal channel slots. Because of the geographical characteristics of the wireless communication network's service area, cells are different from each other. When assigning nominal channels, there are different restrictions in different cells. For cells with more restrictions, finding appropriate channels is more difficult than in cells with fewer restrictions. If we sort cells according to the difficulty of channel assignment, from the most difficult to the least difficult, and assign nominal channels to cells in that order, heuristically, there is a better chance that we will have a good nominal channel assignment scheme.

Based on this idea, a degree of cell was introduced by Sivarajan, et al. The degree of cell $i$ is defined as

$$d_i = \left( \sum_{j=1}^{n} r_i c_{ij} \right) - c_{ii}$$

where $n$ is the number of channels, $r_i$ comes from the nominal channel requirement vector $R$, and $c_{ij}$ comes from the constraint matrix $C$. This is a heuristic measure of the difficulty of assigning a nominal channel to the cell. The degree of a nominal channel slot is the degree of the cell to which it belongs.

There are two cell ordering methods based on the definition of the degree of cell: node-degree order and node-color order. If we arrange cells in a decreasing manner, this is called the node-degree order. The node-color order is obtained by computing degrees of $n$ cells in the system. The cell with the smallest degree is put in the last place in the list. We take this cell out and recompute degrees of the other $(n-1)$ cells. We put the cell with the smallest degree at the $(n-1)$th place in the list. We continue this procedure until there is only one cell left, and put it in the first place in the list.

### 5.3.3 Nominal Channel Allocation in the Network Planning Stage

The nominal channel assignment methods we described are noniterative, and are based on the sequential assignment of channels, sometimes according to the heuristic ranking of the channel assignment difficulty in cells and nominal channel slots. In these methods, once a channel assignment is made, or a cell or nominal channel slot's position in the order is decided, it is final.

There will be no change thereafter, even if it is proven to be a bad choice in the later stages of the algorithm.

Contrary to the noniterative nominal channel assignment methods, the iterative methods return to the previous channel assignment or cell/channel slot placement decisions to reevaluate them. While these decisions are good at the time they are made, later additional assignments or placements affect the original decision. Based on this idea, during each iteration an iterative method assigns new channels or places a new nominal channel slot. At the same time, the iterative method checks previous decisions under the new situation and changes those decisions if it gives a better solution. Iterative method techniques include simple iteration, neural network algorithm, simulated annealing, pattern-based optimization, and local search.

Box proposed a simple iterative channel assignment method, which arranges cells in a descending order of channel assignment difficulty based on the channel requirement. During each iteration, channels are assigned to one cell according to channel assignment difficulty. If there is a denial of assignment, the assignment order is changed.

Simulated annealing uses a cost function to measure the efficiency of channel assignments. The cost function considers channel-separation constraints and channel demand. When all the constraints are satisfied, the cost function reaches its minimum of zero. Both Duque-Aton and Mathar tried to solve the channel allocation problem using simulated annealing, although different models were used.

The neural network algorithm proposed by Kunz creates a neuron for each channel at each cell. There is a cost function considering both the channel separation and channel demand, which is minimized by the algorithm. However, adjacent channel constraints are not considered by Kunz. Funabiki's algorithm employs a parallel two-dimensional neural network model, which considers general adjacent-channel and channel-separation constraints. The results given by this algorithm are better than those proposed by Kunz.

Algorithms using simulated annealing and neural network technologies are usually more complicated, and take a long time to execute. Furthermore, their results, according to simulations, fail to give good solutions to some hard cases of channel allocation problems. These methods usually do not have systematic ways to find appropriate parameters.

Hać's and Chen's nominal channel assignment method uses local search techniques. The channel assignment problem is an NP-complete problem, and it is computationally intractable. Therefore, it is difficult to find an efficient algorithm to solve it. Because channel allocation is an optimization problem, one of the widely used methods for a problem in this category is the local search approach.

Given an optimization problem, an objective function $f$ can be defined so that better solutions to the problem result in smaller values of $f$. Now the problem becomes how to minimize function $f$. Let $T$ be the state space of the problem. The objective function is a mapping

$$T \to R$$

where $R$ is the set of real numbers. It means for every state $t$ in $T$, there is a value $f(t)$ associated with it. Our goal is to find the state with the smallest value, or an approximation of the state. We define the neighborhood of state $t$ as the set of states that are reachable from $t$ with one transition. It is denoted by $N(t)$. $N(t)$ is a subset of $T$.

A local search algorithm starts from the current state $t_n$. It searches $t_n$'s neighborhood $N(t_n)$ for a state, $t_{n+1}$, representing a better solution. For example, it looks for a state, $t_{n+1}$, satisfying the following condition

$$t_{n+1} \in N(t_n) \text{ and } f(t_{n+1}) < f(t_n).$$

If such a state, $t_{n+1}$, exists, it is made the new current state, and the process is repeated. Otherwise, we say the local optimum has been reached. As the first step of the algorithm, a feasible solution represented by the initial state $t_0$ is generated before the iteration begins.

To apply the local search technique to the nominal channel allocation problem, let us redefine the problem, specify what the state space $T$ is, and formulate the objective function $f$.

The performance of a nominal channel assignment method is primarily determined by the nominal channel slot order in which channels are assigned. Because of this, we define the state space for the channel assignment problem using a nominal channel slot order. Theoretically, by processing all of the possible orders of the nominal channel slots, we can find good nominal channel assignment schemes. In practice, this is not efficient, and almost impossible for large problems.

For the nominal channel assignment problem, we define the state space $T$ as a set that includes all the possible orders of nominal channel slots. For the example in Figure 5–9, as described in matrices $C$, $R$, and $D$, its state space includes all possible permutations of $s_{11}$, $s_{21}$, $s_{31}$, $s_{41}$, $s_{42}$, and $s_{43}$. That is

$$T = \{(s_{11}, s_{21}, s_{31}, s_{41}, s_{42}, s_{43}), (s_{11}, s_{21}, s_{31}, s_{41}, s_{43}, s_{42}), (s_{11}, s_{21}, s_{31}, s_{42}, s_{41}, s_{43}),$$
$$(s_{11}, s_{21}, s_{31}, s_{42}, s_{43}, s_{41}), (s_{11}, s_{21}, s_{31}, s_{43}, s_{41}, s_{42}), (s_{11}, s_{21}, s_{31}, s_{43}, s_{42}, s_{41}), \ldots\}$$

The size of $T$ is 720 (i.e., $|T| = 720$. For a wireless network with a nominal channel requirement vector

$$R = \begin{bmatrix} r_1 & r_2 & \ldots & r_n \end{bmatrix}$$

if we define

$$r = \sum_{1 \leq i \leq n} r_i$$

the size of its state space is $|T| = r!$.

When the size of the wireless network increases, so does $r$. According to the above formula, the size of the state space increases much faster than $r$. This is why it is almost impossible to search the entire state space for the solution.

We define the objective function for the nominal channel assignment problem. Given a nominal channel assignment scheme, the sum of the distance between each pair of cochannels is called a cochannel distance. It is the criterion we use to measure the quality of a nominal channel assignment scheme. When the cochannels are used at a shorter distance, they can be used in more cells simultaneously. This leads to a more efficient channel use. The smaller the cochannel distance is, the better the nominal channel assignment scheme.

Using our notation, let $S_i$ be the set of cells which are assigned channel $i$, and $D = [d_{ij}]$ the cell distance matrix, the objective function for the nominal channel assignment problem is

$$MIN_{t \in T} \left( \sum_i \sum_{j, k \in S_i} d_{jk} \right) \quad 1 \leq i \leq m, \ 1 \leq j, \ k \leq n.$$

where $n$ is the number of cells, and $m$ is the number of channels.

The neighborhood of state $t$ (a list of nominal channel slots) is defined as

$$N(t) = \{t' | \forall t'(t' \in T \wedge H(t, t'))\}$$

where $H(t, t')$ is a logical expression. The neighborhood of $t$ includes all the members of $T$ which satisfy $H(t, t')$. By using different $H(t, t')$, there are many different ways of defining a neighborhood.

We use the definition that if $t'$ and $t$ both as sorted lists differ in two places in the list, $t'$ is in $t$'s neighborhood. Now the definition of $t$'s neighborhood becomes

$$N(t) = \{t' | \forall t'(t' \in T \wedge D(t, t') = 2)\}$$

where $D(t, t')$ is the number of places lists $t$ and $t'$ differ from each other. That is, if we arbitrarily select two elements from list $t$ and switch them, the resulting new state $t'$ belongs to $t$'s neighborhood. For example, if

$$t = (s_{11}, s_{21}, s_{31}, s_{41}, s_{42}, s_{43})$$

then the neighborhood of $t$ is

$$N(t) = \{(s_{21}, s_{11}, s_{31}, s_{41}, s_{42}, s_{43}), (s_{31}, s_{21}, s_{11}, s_{41}, s_{42}, s_{43}), (s_{41}, s_{21}, s_{31}, s_{11}, s_{42}, s_{43}),$$
$$(s_{42}, s_{21}, s_{31}, s_{41}, s_{11}, s_{43}), (s_{43}, s_{21}, s_{31}, s_{41}, s_{42}, s_{11}), (s_{11}, s_{31}, s_{21}, s_{41}, s_{42}, s_{43}), \ldots\}$$

The number of elements in $N(t)$ is 15. Generally, if $r$ is the total number of nominal channel slots, the size of any state's $t$ neighborhood is

$$|N(t)| = \frac{r(r-1)}{2}.$$

The local search algorithm used to assign nominal channels is as follows.

Initialize the algorithm.
Find an initial state and assign it to *current–state*.
Assign channels to a nominal channel slot list represented by *current–state*.
Compute the cochannel distance for the resulting nominal channel assignment scheme, and assign it to *distance*.

Set *limit* to $\frac{r(r-1)}{2}$, where $r$ is the total number of nominal channel slots.

Assign list {*current–state*} to *visited–states*.

Assign an empty list to *visited–neighbors*.

While *limit* > 0, do the following,

    Randomly select a number $i$ from 1, 2, 3, ..., $r$.

    Randomly select a number $j$ from 1, 2, 3, ..., $i$–1, $i$+1, ..., $r$.

    Swap the $i$th and $j$th elements in *current–state*, and assign the result to *state'*.

    If *state'* is in *visited–neighbors*, go back to the beginning of the loop. Otherwise, add *state'* to list *visited–neighbors*.

    If *state'* is in *visited–states*, reduce *limit* by 1, and go back to the beginning of the loop. Otherwise, add *state'* to list *visited–states*.

    Assign channels to nominal channel slot list represented by *state'*.

    Compute the cochannel distance for the resulting nominal channel assignment scheme, and assign it to *distance'*.

    If *distance'* < *distance*, assign *state'* to *current–state*, *distance'* to *distance*, reset *visited–neighbors* to { }, and reset *limit* to $\frac{r(r-1)}{2}$. Otherwise, reduce *limit* by 1.

Assign channels to nominal channel slot list represented by *current–state*, and return the channel assignment scheme as the result.

    In the algorithm initialization, parameters such as the number of cells $n$, the number of available channels $m$, the constraint matrix $C$, the distance matrix $D$, and the nominal channel requirement vector $R$ are inputed.

    The initial state can be randomly selected. To speed up the algorithm, it is better to choose a state with a good nominal channel assignment scheme. This way, the local search has a better chance of finding a good solution. Using the concept of cell degree, we can create an initial state by putting the nominal channel slots in node-degree order and node-color order.

    When assigning channels to a list of nominal channel slots represented by a selected state, either the frequency exhaustive or requirement exhaustive strategy can be used. In the channel assignment algorithms, the frequency exhaustive strategy is used.

    List *visited–states* is used to keep track of those states which are already processed. At first, it only contains the initial state. Every time a neighbor of the current state is created, the algorithm checks to see if it is in *visited–states*. The neighbor is skipped if it is in *visited–states*, and the next neighbor is created. If the neighbor is not in *visited–list*, the algorithm adds it to *visited–states*, and checks if it is a state providing a better solution than the current state.

    Variable *limit* puts an upper limit on the time the loop can run. Initially, it is set to $\frac{r(r-1)}{2}$ because it is the number of the state's neighbors. This represents the worst case of how long the loop runs.

    During the early stages in the execution of the algorithm, it is usually not very difficult to find a neighbor giving a better solution. This is common in many local-search optimization algorithms. Usually it takes very few iterations to find the next state. As the optimization continues, finding a better state among neighbors becomes more difficult. More neighbors are already visited (appearing in the list *visited*). More iterations are needed to find the next state. Many of them create neighbors which are already in *visited*. The algorithm stops when all the neighbors of the current state are checked and none of them is better than the current state.

While *visited–states* remember those states which are already visited for the entire algorithm, *visited–neighbors* contains current state's neighbors which have been checked. Because $i$ and $j$ are generated randomly, it is possible that the same neighbor can be created more than once. When this happens, using *visited–neighbors* prevents it from being processed again. The *visited–neighbors* is reset to empty when new current state is determined.

The algorithm terminates when all the neighbors of the current state either produce worse solutions than the current state or are already visited.

### 5.3.4 Dynamic Channel Allocation in the Network Running Stage

In the HCA method, the assignment of nominal channels is the first part of the solution. The second is the DCA to call requests. To be consistent with our nominal channel assignment method, we use the cochannel distance minimization as the criterion when dynamically assigning channels to call requests.

For those cells to which channel $i$ is assigned, we define the cochannel distance of channel $i$ as

$$cd_i = \sum_{j,k \in S_i} d_{jk} \quad 1 \leq i \leq m, \ 1 \leq j, k \leq n$$

where $m$ is the number of channels, and $n$ is the number of cells in the system. $S_i$ is the set of cells to which channel $i$ is assigned. Pair $(j, k)$ is unordered (i.e., $(j, k)$ is considered the same as $(k, j)$), which prevents the distance between the two cells from being counted twice.

When a call comes in, a channel needs to be assigned to it, and the overall co-channel distance increases inevitably. Because our goal is to minimize cochannel distance, we need to find which channel results in the smallest cochannel distance increase, provided that this channel can be assigned to the incoming call without violating the channel-separation constraints specified in constraint matrix $C$.

It seems that we can achieve our goal by finding channel $i$ which, when assigned to the call request, causes the smallest increase of its cochannel distance $cd_i$. However, the change of channel's $i$ cochannel distance $cd_i$ is greatly influenced by the number of cells already assigned channel $i$ (i.e., the size of $S_i$).

For example, channel $i$ is already assigned to calls in 20 cells, and if we assign it to the new call in another cell, the increase of its cochannel distance is the sum of the distance from the new cell to each of the 20 cells in $S_i$. In the same system, there is another channel $j$, which is already assigned to two cells only. The increase of its co-channel distance caused by assigning channel $j$ to the new call in another cell is the sum of the distance between the new cell and each of the other two cells. Thus, if the new cell is far away from the other two cells, the increase in channel's $j$ cochannel distance can still be smaller than the increase in channel's $i$ cochannel distance. Based on this, channel $j$ can be selected to serve the new call. This is not the result we are interested in because our objective is to assign a channel to cells which are close to each other. This way, the channel can be used in more cells, and more efficient channel use can be achieved. Thus, we introduce average cochannel distance.

Let us define

$$Q_i = \{(j, k) | j, k \in S_i \text{ and } j < k\}$$

as the set of all different unordered pairs of cells from $S_i$. Using condition $j < k$ assures that the same two cells are counted only once. The definition of channel's $i$ cochannel distance now becomes

$$cd_i = \sum_{j, k \in Q_i} d_{jk} \quad 1 \leq i \leq m, \ 1 \leq j, \ k \leq n.$$

The number of elements in $Q_i$ is defined as

$$p_i = |Q_i| = \frac{|S_i|(|S_i| - 1)}{2}.$$

Channel's $i$ average cochannel distance is

$$acd_i = \frac{cd_i}{p_i}.$$

The cochannel distance for the entire network is

$$cd = \sum_i acd_i$$

If channel $i$ is not assigned to any cell, its average cochannel distance is zero.

When channel $i$ is assigned to the new call, an increase in cochannel distance $cd$ is the same as the increase in channel's $i$ average cochannel $acd_i$ since the other channel's average cochannel distances are not affected. Computing the increase of $i$'s average cochannel distance requires calculating $acd_i$ both before and after its new assignment. To find the best channel, we need to compute this for all the candidate channels, which is time consuming.

If channel $i$ is assigned to the new channel request in cell $k$, the new average cochannel distance of channel $i$ is

$$acd_i' = \frac{cd_i + \sum_{j \in S_i} d_{jk}}{p_i + |S_i|}.$$

Here $S_i$ does not include cell $k$. We compare this with the definition of $acd_i$, and the change of channel's $i$ average cochannel distance is primarily determined by $\sum_{j \in S_i} d_{jk}$. We define

$$\Delta acd_i = \frac{\sum_{j \in S_i} d_{jk}}{|S_i|} \quad S_i \neq \{\ \}$$

where $S_i$ is the set of cells that have channel $i$ before $i$ is assigned to the new request. When channel $i$ is assigned to only one cell (i.e., $S_i$ is empty), $\Delta acd_i$ is defined as the longest distance between any two cells. This is a penalty for assigning an unused channel to incoming channel requests. It serves as an incentive to use channels that are already assigned to the other cells. We use $\Delta acd_i$ as the measurement of the increase of $acd_i$ when channel $i$ is assigned to cell $k$.

The following is the procedure of DCA which is invoked whenever there is a channel request. Among all the candidate channels, find a channel $i$ that meets the following conditions. If channel $i$ is assigned to the request, all the channel-separation constraints indicated in constraint matrix $C$ are satisfied.

The increase in cochannel distance caused by assigning channel $i$ is the smallest, that is

$$\forall j (j \neq i \wedge \Delta acd_i)$$

The resulting channel, $i$, is assigned to the new channel request. Unlike regular hybrid channel allocation methods, this method does not divide channels into fixed and dynamic groups. When there is no nominal channel available, all the channels are considered for dynamic allocation, including those used in other cells as nominal channels. This improves the efficiency of channel use.

## 5.4 QoS in Wireless Microcellular Networks

The third generation mobile cellular networks (e.g., high speed mobile networks) support multimedia traffic (voice, video, data, and image). Multimedia traffic puts a heavy bandwidth demand on the cellular network. Bandwidth is the most critical resource in cellular networks and thus, requires mechanisms to efficiently use the available bandwidth. To increase the spectrum utilization, microcellular and picocellular networks are deployed. The use of smaller cells gives rise to the problem of frequent handoffs. The handoff rate increases as calls are handed off more frequently from one cell to another. Cellular networks carrying multimedia traffic have to provide a predefined QoS to the calls being serviced. We discuss the problem of providing a guaranteed QoS to the existing calls in a microcellular network.

Research in the area of wireless communications focused on multiple access techniques like TDMA, CDMA or higher layer control issues like handoff support, hierarchical network design, and channel allocation. Since wireless mobile networks are designed to support voice and data traffic, research efforts are directed towards these traffic types. Multimedia traffic support in mobile cellular networks is a new area for research. Multimedia traffic requires different bit rate services at the same time. This requires adaptive multicarrier systems transmitting multimedia signals by changing the number of carriers according to the capacity of signals. New CDMA techniques for mapping different bit rate services onto the same allocated bandwidth can handle multimedia communication. A brief introduction to advanced wireless technologies for achieving high-speed different bit rate services is given by Morinaga.

In cellular networks carrying multimedia traffic, it becomes necessary to provide a QoS guarantee for multimedia traffic calls supported by the network. Resource management becomes difficult as smaller cells are used to increase the bandwidth efficiency. The smaller the cell, the higher the rate is at which traffic statistics in the network change. The microcellular and picocellular architectures increase the handoff rate due to the small coverage area of each BS. Frequent changes in the network traffic make the provision of guaranteed QoS more difficult.

The QoS provision for multimedia traffic in nonwireless networks such as B-ISDN networks has been extensively studied. Research in the area of QoS provisioning in high-speed wireless networks focuses on the integration of resource allocation and admission control policies. An adaptive call admission control mechanism, in combination with a resource sharing scheme, guarantees a predefined quality of service to each class of traffic.

Broadly, there are two different ways to handle the problem of rapid handoffs in microcellular networks. The first method is based on the concept of hierarchical network architecture. A geographical area is covered with a large number of microcells with overlaying macrocells. The calls attempting handoff from one cell site to another are handed off to the macrocell covering both cell sites. The macrocell can then handoff the call back to the microcell when the required bandwidth is available or it can continue to service the call. In the second approach, the cell cluster concept is used. When a call is initially accepted by a cell site, an appropriate amount of bandwidth is reserved in the adjoining cells. The cell site serving the call together with its adjoining cells form a group that is referred to as a cell cluster. We use the cell cluster concept to handle the problem of guaranteeing a predefined QoS to existing calls in the network. The QoS is defined as a set of parameters that are to be maintained at a certain specified value. We define the QoS on the basis of the new call blocking rate, handoff dropping rate and forced call termination rate.

Armstrong and Hać propose an adaptive resource allocation scheme that uses call admission control for new calls and a bandwidth readjustment of existing calls for handoff calls. The call admission control becomes necessary to guarantee a certain grade of service to calls accepted in the network. The resource allocation scheme provides a predefined QoS to the existing calls.

The operational limitations of conventional mobile telephone systems led to the development of mobile cellular networks. The traditional mobile systems suffered from three main drawbacks—limited service capability, poor service performance, and inefficient frequency utilization. In discussing the developments of cellular networks, we generally refer to a different generation of systems.

The first generation systems comprise today's analog cellular telephone systems. These systems provide a certain amount of mobility to users. However, the mobility is confined to the service area of a cellular company. The multiple access technique used in these systems is FDMA. Because the propagation environment for cellular networks introduces noise and interference in the transmitted signal, these systems are not very rugged. These systems use the concept of frequency reuse for efficient utilization of available frequency spectrum.

Digital transmission is used in second generation networks. The digital transmission allows for correction of impairments introduced in the medium. This regeneration process provides the primary advantages for digital transmission. In addition, the digital systems provide a cost advantage. The power consumption in mobile equipment is low and the size of the equipment is small. However, the bandwidth requirement per call is higher in these systems. The multiple access technique used in these systems is either TDMA or CDMA. Some other multiple access techniques, like PRMA and CDPD (cellular digital packet data), are also being developed and deployed on an experimental basis.

The integration of wired and wireless networks underlie the concept of third generation cellular networks. The idea is to create a single network infrastructure that allows users to exchange any kind of information between any desired locations. This network merges the existing first and second generation networks and encompasses other means of wireless access such as paging, cordless phones, and wireless LANs. The third generation cellular networks offer enough bandwidth for multimedia services.

The multiple access techniques used in this network are either TDMA, CDMA, or a combination of both. Active research is going on for the search of a suitable multiple access scheme. The third generation cellular networks that support multimedia applications rely on the provision of a highly reliable radio link to provide good quality service.

### 5.4.1 Multimedia Traffic in Microcellular Networks

The frequency spectrum available for mobile communication is limited. The frequency reuse concept has been used to increase the system capacity. The mobile cellular networks experience a severe bandwidth limitation as they support multimedia applications. These bandwidth intensive applications require the use of micro/picocellular architectures to obtain a greater spectrum efficiency. The microcellular networks are currently deployed in high user density areas to increase capacity. These networks have the inherent problem of rapid handoffs due to a smaller coverage area of cells. This problem leads to network congestion and higher call dropping rates. A solution to this problem in microcellular networks is to apply a distributed call admission control at the connection setup time. The spectrum efficiency in these networks can only be realized if there are strict rules for admission control and flexible resource allocation.

In multimedia, the transmitted information comes from a variety of sources with different source characteristics (e.g., data rates, burstiness, and activity ratio) and different qualities (e.g., BER and delay sensitivity). Multimedia applications impose stringent QoS demands on the network. Multimedia traffic is a combination of both real-time and non-real-time traffic. Real-time traffic includes voice and video while data and graphics make up non-real-time traffic. Real-time traffic is delay sensitive and non-real-time traffic is error sensitive. The different service applications (e.g., videophone, fax, and email) of these traffic classes differ in their bandwidth requirement and call duration times.

Multimedia traffic requires a predefined QoS and hence, the resource management schemes need to be designed to maintain a predefined QoS to each class of traffic. QoS is defined as a set of performance measuring parameters, such as call blocking rate and handoff dropping rate, required to maintain an acceptable service to the user. Each class of traffic can have its own QoS. Multimedia traffic imposes a demand for variable transmission rate. Real-time traffic requires a constant bit transmission rate and is highly sensitive to delay and delay jitter. On the other hand, non-real-time traffic requires a variable bit transmission rate and has stringent BER requirements. Consequently, resource management schemes become more complex. It is argued that either DCA or FCA can be suitable for multimedia traffic support in cellular networks. However, for providing a predefined QoS to calls accepted in the system, some kind of fixed channel allocation is needed. We focus on a scheme in which each cell site is allowed to use the complete frequency spectrum. The frequency spectrum is not divided into band limited channels but each call request is assigned a bandwidth as per its requirement. This allows for a variable transmission rate for different kinds of traffic.

Resource management in microcellular networks supporting multimedia applications must resolve the problem of frequent handoffs. Resource management, thus, involves frequency reservation for handoff calls or distributed call admission control. Resource management in such networks largely depends on the amount of information available at the BS for each call in service. Resource management schemes can apply handoff queuing and delaying policies to provide a better performance. The unpredictable mobility patterns of mobile users make resource management difficult. A brief discussion on the prediction of mobility patterns for users with different mobility rates is given by Sohraby. Resource management schemes can be more effective if information about users' movement is made available. Queuing and delaying of handoff results in better forced termination rates. A priority-based handoff queuing scheme is discussed by Lin, et al. Handoff queuing or delaying cannot be effectively employed to support delay sensitive traffic.

### 5.4.2 Resource Allocation in Microcellular Networks

The support for bandwidth intensive (multimedia) services in mobile cellular networks increases the network congestion and requires the use of micro/picocellular architectures in order to provide higher capacity in regard to radio spectrum. Micro/picocellular architectures introduce the problem of frequent handoffs and make resource allocation difficult. As a result, the availability of wireless network resources at the connection setup time does not necessarily guarantee that wireless resources are available throughout the lifetime of a connection. Multimedia traffic imposes the need to guarantee a predefined QoS to all calls serviced by the network.

In microcellular networks supporting multimedia traffic, the resource allocation schemes have to be designed such that a call can be assured of a certain QoS once it is accepted into the network. The resource allocation for multimedia traffic becomes quite complex when we consider the different classes of traffic comprising multimedia traffic. These classes of traffic have different delay and error rate requirements. Resource allocation schemes must be sensitive to traffic characteristics and adapt to rapidly changing load conditions. From a service point of view, multimedia traffic can be categorized into two main categories: real-time traffic with stringent time delays and relaxed error rates, and non-real-time traffic with relaxed time delays and stringent error rates.

It is important to note providing QoS to different classes of traffic necessitates a highly reliable radio link between the mobile terminal and its access point. This requires efficient communication techniques to mitigate the problems of delay sensitivity, multipath fading, shadow fading, and cochannel interference. Some methods like array antennas and optimal combining can be used to combat these problems. We focus on the mobility aspect of mobile networks and consider the resource allocation with reliable radio links.

Schemes have been proposed in the literature that address the problem of resource allocation for multimedia traffic support in microcellular networks. In these schemes, real-time traffic, because it is more delay sensitive, is given priority over non-real-time traffic.

In these schemes, the central approach used is call admission control (CAC). CAC imposes a limit on the number of calls accepted into the network. Each cell site only supports a predetermined number of call connections. This call threshold is periodically calculated depending on the number of existing calls in the cell in which the call arrives, its adjoining cells, and the resources utilized by all calls in the cell. Once the threshold is reached, all subsequent requests for new call connections are refused.

#### 5.4.2.1  Admission Threshold Based Scheme

In this scheme, resource management is done by periodically calculating the admission threshold and by blocking all new call connection requests once the threshold is reached. The call admission decision is made in a distributed manner whereby each cell site makes a decision by exchanging status information with adjoining cells periodically. A cell with a BS and a control unit is referred to as cell site.

When a new call connection is requested, the system estimates the state of the cell in which the call admission request is made, in addition to the state of its adjacent cells $T$ units of time after the call admission request. If a call connection requests arrives at time $t_0$, then in order to accept the new call connection, the following three admission conditions must be satisfied:

1. At time $t_0 + T$, the handoff dropping probability in the current cell as well as the adjacent cells should not exceed the highest tolerable handoff dropping probability ($P_{HD}$). $P_{HD}$ forms a part of the QoS to be guaranteed to the calls accepted in the network.

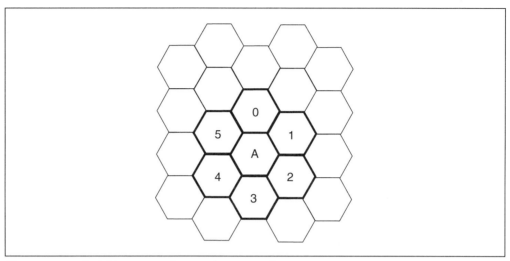

**Figure 5–10** Cell Cluster as a Group of Adjacent Cells

2. At time $t_0 + T$, the overload probability (the likelihood that the number of calls in the concerned cell goes beyond the admission threshold) should not exceed the highest tolerable overload probability ($P_{OL}$). $P_{OL}$ is another measure of a QoS guarantee.
3. At time $t_0 + T$, the overload probability in the adjacent cells due to traffic handoffs must be smaller than $P_{OL}$.

This scheme relies on the prediction of future events in the network. As a result, the performance of this scheme could be unpredictable in rapidly changing traffic environments which are likely to exist in microcellular networks.

### 5.4.2.2 Resource Sharing Based Scheme

To support traffic classes with different delay and error requirements, resource sharing provides a mechanism to ensure a different grade of service to each class of traffic. This scheme employs a resource sharing mechanism that reacts to rapidly changing traffic conditions in a cell. An adaptive call admission control policy that reacts to changing new call arrival rates can be used to keep the handoff dropping rate and forced call termination rate acceptably low.

The CAC scheme differentiates the new call on the basis of its traffic class and a decision is made based on the traffic class of the new call connection request and number of call connections of each class already being serviced in the cell cluster. A cell cluster is defined as the group of cells surrounding a cell as shown in Figure 5–10.

For real-time call connections, a new call is blocked if no bandwidth is available to service the request. A similar algorithm is used to service a handoff request. The QoS metrics for real-time calls are handoff dropping probability and forced call termination probability. For non-real-time calls, the available bandwidth is shared equally among all non-real-time call connections in the cell. Handoff queuing or delaying is not used in this scheme. CAC keeps the probability of a call being terminated before its lifetime acceptably low. Resource sharing algorithms provide a better performance for a particular class of traffic.

Resource sharing schemes can be implemented using different mechanisms based on the bandwidth available to a particular class of traffic. The three different ways in which resources can be shared among the various classes of traffic are described below. We describe these mechanisms here for two classes of multimedia traffic. Real-time traffic calls are referred to as Class I calls and non-real-time calls as Class II calls.

1. Complete Partition — The total bandwidth $B$ is divided among the various classes of traffic. For the case with two classes of traffic, $B_I$ ($<B$) is assigned to Class I calls and the remaining bandwidth is available to Class II calls. This scheme is the most inefficient of the three resource sharing schemes described here. A fixed partition of resources is inefficient in changing traffic conditions since resources not utilized by one class of traffic are wasted and cannot be used by the other class.
2. Complete Access — Class I calls have access to the total bandwidth $B$ of the cell with preemptive priority over Class II calls. The Class II calls can only use the bandwidth not utilized by Class I calls. This scheme provides a better performance for Class I calls at the expense of poor service to Class II calls.
3. Restricted Access — Class I calls have access to a portion $B_I$ ($<B$) of the bandwidth with a preemptive priority over Class II calls. The remaining bandwidth ($B - B_I$) is reserved for Class II calls only. Class II calls can also use the bandwidth not used by Class I calls. This scheme gives the best performance among all schemes.

In Naghshineh and Acampora, it is shown through simulations that the restricted access scheme gives a better performance than the complete access or partition scheme in regards to the handoff dropping probability and forced calls termination probability. The bandwidth utilization has been found to be the best with the complete access scheme. To provide better QoS to Class I calls due to their delay sensitivity, a trade-off is made between QoS guaranteed to Class I calls and Class II calls. These schemes can be extended for resource allocation for multiple classes of traffic.

### 5.4.2.3 Resource Reservation (RR) Based Scheme

In the above schemes, CAC is used such that the call admission threshold is periodically determined based on the traffic conditions in the network, call duration time statistics, and the QoS requirements of different classes of traffic. In this scheme, the decision to accept a call depends on the bandwidth available in the cell in which the call arrives and in the adjacent cells, at the time of a call connection request. The CAC mechanism uses the information regarding the network state at the time a call arrives. The call admission procedure for Class I calls is different from the one for Class II calls. The call admission procedure for new calls and handoff calls is also different. Handoff calls are given priority over new call connections.

This scheme uses the idea of a bandwidth reservation in the adjacent cells to provide a better performance in regards to the handoff dropping rate. For Class I calls, a bandwidth reservation is performed in all cells adjacent to the cell in which the call arrives. The amount of bandwidth to be reserved in adjacent cells can be calculated according to various policies. Some of the policies are suggested in Olivera, et al. They either depend on the total number of calls of the particular class or on the maximum bandwidth used by a call. When a Class I call connection is requested, the cell site in which the call arrives makes decision based on the bandwidth available in the cell and bandwidth available in the adjacent cells.

A new call connection is accepted by a cell site if the requested bandwidth is available in the cell and sufficient bandwidth is available in adjacent cells to guarantee a required QoS to the call. A connection is refused if either the bandwidth is not available in the cell or sufficient bandwidth cannot be reserved in the adjacent cells. For handoff calls, a similar procedure is followed. However if the call can be serviced with a lower transmission rate, an attempt is made to service the handoff by assigning it the minimum bandwidth needed.

For Class II calls, there is no bandwidth reservation in the adjacent cells. Class II calls are accepted by a cell site if some bandwidth is available in the cell. Class II calls are generally data calls and can be serviced with variable transmission rates. Each Class II call has a requested bandwidth and a minimum bandwidth. If enough bandwidth is available in cell, then the requested bandwidth is assigned. If the requested bandwidth cannot be assigned, then an attempt is made to assign the minimum bandwidth. A new Class II call request is blocked if the minimum bandwidth cannot be assigned. For Class II handoff calls, the procedure is similar except that if the minimum bandwidth is not available, the call is still serviced and whatever bandwidth is available is assigned to the call. A handoff call is only dropped if the cell is being used at its full capacity and no bandwidth is available.

When a Class I call hands off, the bandwidth reserved in the adjacent cells is released and a reservation is performed in adjacent cells of the new cell to which the call hands off. The reserved bandwidth is also released when a Class I call terminates or is dropped by the network.

The RR scheme can provide better service but there is a trade-off between service quality and processing overhead. There are several issues that need to be resolved to efficiently apply the RR scheme. One of the issues is to determine in which cells the reservation needs to be performed. If information about a mobile users' mobility pattern and direction is available in advance, a reservation can be made in the cells in which the mobile user is expected to handoff in the future. If this information is not available, a reservation needs to be performed in all surrounding cells. The next issue is to decide how much bandwidth needs to be reserved in surrounding cells. If reservation is to be performed in all the cells, then only a fraction of the requested bandwidth of the call needs to be reserved in the surrounding cells as the call will be handing off to one of the surrounding cells. This is an important parameter that affects the efficiency of the RR schemes.

Another issue is to determine when to release and reserve the bandwidth in the surrounding cells. When a new call connection is requested, the bandwidth reservation can be made in all

QoS in Wireless Microcellular Networks 273

the surrounding cells and when a call finishes or hands off, the reserved bandwidth is released. The maximum amount of bandwidth that can be reserved in the surrounding cells to support handoff is also an important parameter. A careful choice of this parameter results in lower call blocking rates and lower handoff dropping rates. All these parameters also depend on the traffic load in a cell. To solve this problem, the cell size has to be carefully chosen depending on the traffic density in a region. A simple method to estimate cell size depending on traffic density is given by Naghshineh and Acampora.

Multimedia traffic is comprised of different service applications having varied bandwidth requirements. We refer to these service application as call type. Each call type differs from the other bandwidth requirement and average call duration times. Bandwidth reservation becomes more complex when we consider different service applications.

### 5.4.3 RR with Renegotiation

Armstrong and Hać propose a resource allocation scheme that provides a QoS guarantee to real-time traffic and at the same time guarantees a better performance to non-real-time traffic. The resource allocation scheme uses resource reservation in surrounding cells for real-time calls and a renegotiation of bandwidth assigned to non-real-time calls. The resource allocation scheme is simple enough and can be implemented in a distributed manner to ensure fast decision making. We refer to this scheme as resource reservation and renegotiation (RRN) scheme.

In microcellular networks supporting a number of service applications, determining the amount of bandwidth to be reserved is difficult. The reservation scheme described by Olivera, et al. does not consider the bandwidth differences between the various service applications for one class of traffic. This leads to an unfair treatment of one type of service application. For instance, if the maximum amount of bandwidth that can be reserved in the surrounding cells is $r_{max}$, then if a service application has a bandwidth requirement greater than $r_{max}$, this service application will have a poorer performance. This also affects the policy that can be employed for determining the amount of bandwidth to be reserved. If we use a common pool of bandwidth for a reservation with the largest requested bandwidth to be reserved, then calls arriving in the system when the largest of the bandwidth can not be reserved in the surrounding cells, are dropped. This can result in an unacceptably high call blocking rate.

In the RRN scheme, for service applications requiring smaller bandwidths, a shared pool of bandwidth is used for reservation. For applications requiring greater bandwidth, the largest requested bandwidth is reserved. This helps in keeping the call blocking rate low and does not affect the handoff dropping rate.

The RRN scheme uses the following assumptions.

- The cellular network consists of a two-dimensional array of hexagonal cells of equal size. The cells are uniformly distributed in a geographical area and consist of a BS with omni-directional antenna. We refer to a cell with a BS as a cell site.

- Efficient techniques to combat multipath and shadow fading, and delay variations are employed to provide a reliable radio link between the mobile users and BS. We assume the presence of a wireless MAC layer responsible for providing reliable data and control channels is used to transmit and receive all messages between the BS and mobile users.
- We assume that every cell can use the full available bandwidth spectrum. We do not consider the adverse propagation environment, and assume that efficient techniques are employed to provide a reliable multi-bit rate communication between mobile users and the cell site.
- The cellular network supports both real-time traffic, referred to as Class I calls, and non-real-time traffic referred to as Class II calls. The multimedia traffic carried by the network is comprised of these two classes of traffic. Each class of traffic is further categorized into three service types, referred to as call type depending on their average call duration times and bandwidth requirement.
- The total bandwidth available to a cell is $B$ Mbps. A fraction $R_l$ of the total bandwidth $B$ is reserved for handoff calls. The bandwidth $(1 - R_l)B$ is shared by the Class I calls and Class II calls. Another fraction $R_h$ of this bandwidth can be used for reservation purposes.
- The mobile user requesting a connection to the network provides information regarding the class and type of traffic, required bandwidth, minimum bandwidth and speed of movement.
- The bandwidth reservation is performed using the cell cluster concept. Figure 5–10 shows a cell cluster with cluster size of seven. Cells numbered zero to five belong to the cell cluster of cell A. Every cell exchanges information with all other cells in a cell cluster.
- There can be one of three different mobile characteristics for each call type depending on the mobile users' mobility pattern. A mobile user can be stationary, slow moving or fast moving. The mobility pattern affects the handoff rate.
- Every call has an associated dwell time, which is the time a call spends in a cell before handing off to another cell. The dwell time of a call depends on its handoff rate. Every call has an initial handoff rate, $h$, which is based on users' mobility pattern and call class.
- The handoff rate is assumed to be changing with the number of handoffs. The effective handoff rate every time a call requests a handoff is given by $h/2^n$, where $n$ is the number of handoffs already experienced by a call.
- Handoffs are assumed to be mobile assisted. A mobile user spends a certain amount of time equal to the dwell time in a cell and then hands off to another cell in the cell cluster. After a call hands off, the dwell time is recalculated based on the new handoff rate.
- All bandwidth occupied by a call, reserved or in use, is released when a call departs from the system or is forced to terminate.

In microcellular networks, calls require handoffs at much faster rates in comparison to networks with larger cells. On the other hand, microcellular networks provide a higher system capacity. The RRN scheme supports real-time calls and non-real-time calls along with a variety

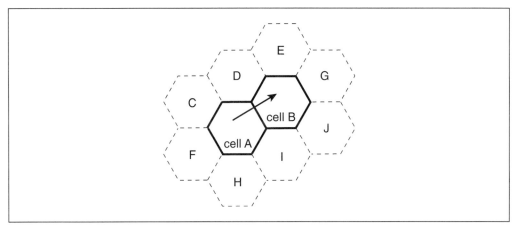

**Figure 5–11** Bandwidth Reservation in a Cell Cluster

of service type for each class. Real-time calls are delay sensitive and hence cannot be queued or delayed. Resources must be available when a handoff is requested. To guarantee that real-time calls are not forced to terminate at the time of handoff, we use the RR mechanism. By using RR, we can guarantee an acceptably low handoff dropping rate and forced call termination rate for real-time traffic.

For a real-time call, bandwidth is reserved in all cells adjacent to the cell in which the call arrives. When a call hands off to another cell, if enough bandwidth is not available to service the handoff, it uses the bandwidth reserved in the target cell and thus, the likelihood that a call will be dropped is reduced. When a call is successfully handed off to another cell, we release the bandwidth reserved for the call earlier (in the cell cluster of the old cell) and reserve bandwidth in the cell cluster of a new cell.

Figure 5–11 shows the process of bandwidth reservation. When a call arrives in cell A, bandwidth is reserved in cell cluster of A (i.e., cells B, C, D, F, H, and I). When the call hands off to cell B, we release the bandwidth reserved in cell B, C, D, F, H and I, and then reserve bandwidth in cell cluster of B (i.e., cells A, D, E, G, I, and J).

Non-real-time calls are more tolerant to delay as compared to real-time calls. Delay tolerance is equivalent to accepting a variable service rate. This property of data traffic makes resource renegotiation possible in microcellular networks. Non-real-time calls receive higher service rates under low traffic conditions, while under heavy traffic conditions, the service rate available is kept at a minimum. Thus, the RRN scheme adapts to changing traffic conditions in the network.

RNR can be implemented in a distributed manner and each cell site can make decisions as to how to assign bandwidth to Class II calls within its service area. RNR is further facilitated by keeping an ordered list of calls that will undergo bandwidth readjustment. This reduces the overhead imposed by the mechanism. The calls are selected on a priority basis for bandwidth reassignment. The bandwidth assignment mechanism is reversible. If the service rate of a call is reduced under heavy traffic load conditions, then under low traffic conditions, the same call can receive higher service rates. There is a trade-off between the call processing overhead and the service quality that can be provided.

RNR overcomes some of the drawbacks associated with RR. RNR is only possible with Class II calls. In the RRN scheme, to provide a better quality of service to Class II calls we use a parameter called the renegotiation factor, which helps to determine if a handed-off call can be serviced through readjustment. The renegotiation factor is defined as the ratio of nonnegotiated calls to the total calls being serviced by a cell at any time.

$$Renegotiation\ Factor = \frac{Nonnegotiated\ Class\ II\ calls}{Total\ Class\ II\ calls\ in\ service}$$

For handoff calls to be accepted in a cell, the renegotiation factor must be greater than a certain value that is a design parameter, depending on the traffic density and cell size.

New calls arriving in the system can be Class I or Class II calls. The resource allocation scheme for establishing a new call connection is different for Class I and Class II calls.

Class I calls: For new Class I calls, RR is performed. When a call connection is requested, the BS determines whether the bandwidth $B$ is available to service the calls. If $B$ is more than the requested bandwidth $B_r$, the BS then calculates the bandwidth $B_{res}$ it needs to reserve in the adjacent cells. The information from other cells in the cell cluster is gathered and if $B_{res}$ can be reserved, the call is accepted. If $B_{res}$ is not available in all cells of the cell cluster, the call is blocked.

Class II calls: There is no bandwidth reservation for Class II calls. When a new call connection is requested, the BS determines the available bandwidth $B$. If $B$ is greater than $B_r$, the call is accepted in the network. The call is assigned to the requested bandwidth. When $B$ is less than $B_r$, the call is blocked.

In microcellular networks, due to the smaller coverage area of cells, the handoffs are very frequent. Keeping the handoff dropping rate to a minimum is an important issue in designing resource allocation schemes. In the RRN scheme, different policies are used to service the handoffs depending on the class of traffic.

Class I calls — For handoff of Class I calls, initially we attempt to assign bandwidth to the call using the available bandwidth. If enough bandwidth is not available to service the handoff request, we try to assign the bandwidth already reserved in the target cell. If the reserved bandwidth in the target cell has been used by calls from other cells, we try to make available enough bandwidth to service the handoff by re-assigning bandwidth to existing Class II calls in the cell. A Class II call is chosen for re-assignment according to the following policy:

1. Schedule the calls on the basis of bandwidth $B_{asn}$ being used by the call, and the mobility pattern of the call.
2. Select the call with the largest used bandwidth and slow mobility.
3. Reduce the bandwidth assigned to the selected call by the amount of bandwidth needed.
4. If needed, readjust the bandwidth for more than one Class II calls.

This way, the bandwidth is made available to service a handoff request. One important parameter that is used is the bandwidth that is dedicated to Class II calls. If the bandwidth utilized by Class II calls goes below the threshold value, then no re-adjustment is made to service a Class I handoff. The handoff request is refused and the call is forced to terminate.

The use of RRN can provide a better handoff dropping rate. When the RR is performed for new calls, it is likely that for a call handoff request, the RR in the cell cluster exceeds the reservation limit. A handoff is refused, even if enough bandwidth is available in the target cell to service the request. Using renegotiation can improve the handoff dropping rate. Handoff calls are accepted only if the renegotiation factor in all cells in the cell cluster is more than a minimum value. This value is selected depending on the traffic load in each cell.

Class II calls — Handoff of Class II calls is simple if enough bandwidth is available in the target cell to service the handoff request. If enough bandwidth is not available, then the following steps are taken to service the handoff:

1. If the available bandwidth $B$ is greater than the minimum bandwidth $B_{min}$ of the handoff call, service the call by assigning $B_{min}$ to it.
2. If $B$ is less than $B_{min}$, select calls for readjustment. The calls for bandwidth readjustment are selected based on bandwidth utilized by the call, and the mobility pattern of the call.
3. Select a call with a maximum bandwidth and slow mobility. If slow mobility calls are not present, then a fast-moving call is selected.
4. In this case, a bandwidth readjustment is performed for only one call to keep the procedure simple. The handoff calls and the bandwidth readjusted calls share the bandwidth.

The performance evaluation and simulation results for the RRN scheme are as follows.

1. The RRN scheme improves performance of the handoff dropping rate and the forced call termination rate while keeping the blocking rate within acceptable limits. Under a uniform traffic environment, the RRN scheme provides the lowest handoff dropping rate and forced call termination rate. For traffic loads up to 700 calls per hour per cell, these parameters are almost close to zero.
2. The RRN scheme can provide a predefined QoS for both real-time and non-real-time traffic. This can be achieved by carefully selecting the design parameters of the scheme.
3. The performance of the RRN scheme depends on the fraction $f$, and improves as the value of $f$ is reduced. However, it is possible that an exceedingly low value of $f$ can result in poor performance.
4. The RR policy used in the RRN scheme does not depend on the cell size, but on the offered traffic load and the available bandwidth in the cell.

# References

1. N. Abramson, "Multiple Access in Wireless Digital Networks," *Proceedings of the IEEE*, 82:99(Sep.), 1360–1369, 1994.
2. A. S. Acampora and J. H. Winters, "A Wireless Network for Wide-Band Indoor Communications," *IEEE Journal of Selected Areas in Communications,* 5:(Jun.), 1987.
3. A. S. Acampora and M. Naghshineh, "Control and Quality of Service Provisioning in High Speed Microcellular Networks," *IEEE Personal Communications*, 1:2(2nd Quarter), 36–43, 1994.
4. H. Ahmadi and W. Denzel, "A Survey of Modern High-Performance Switching Techniques," *IEEE Journal of Selected Areas in Commununications*, 7:7(Sep.), 1091–1030, 1989.
5. A. Aho, J. Hopcroft, and J. Ullman, *Data Structures and Algorithms*, Reading, MA: Addison-Wesley, 1983.
6. S. R. Ahuja and J. R. Ensor, "Coordination and Control of Multimedia Conferencing," *IEEE Communication Magazine*, May, 38–43, 1992.
7. Y. Akaiwa and H. Andoh, "Channel Segregation—A Self-Organized Dynamic Channel Allocation Method: Application to TDMA/FDMA Microcellular System," *IEEE Journal on Selected Areas in Communications*, 11:6(Aug.), 949–954, 1993.
8. W. D. Ambrosch, A. Maher, and B. Sasscer, *The Intelligent Network: A Joint Study by Bell Atlantic, IBM, and Siemens*, Berlin Heidelberg, Germany: Springer-Verlag, 1989.
9. N. Amitay, "Distributed Switching and Control with Fast Resource Assignment/Handoff for Personal Communications Systems," *IEEE Journal of Selected Areas of Communications*, 11:6(Aug.), 842–849, 1993.
10. H. Arai and M. Kawarasaki, "Analysis of Call Admission Control in the ATM Network," *Electronics and Communications in Japan*, 73:10, 42–51, 1990.

11. Y. Argyropoulos, S. Jordan, and S. P. R. Kumar, "Dynamic Channel Allocation Performance under Uneven Traffic Distribution Conditions," *IEEE International Conference on Communication*, 1855–1859, 1995.
12. A. Armstrong and A. Hać, "A New Resource Allocation Scheme for QoS Provisioning in Microcellular Networks," *Proceedings of the IEEE ICC,* Atlanta, GA, 1685–1689, 1998.
13. The ATM Forum, *ATM User—Network Interface Specification Version 3.0*, Englewood Cliffs, NJ: Prentice Hall, 1993.
14. ATM Forum, *LAN Emulation over ATM Version 2 LNNI Specification*, Feb. 1999. URL: http://www.atmforum.com/atmforum/specs/approved.html.
15. B. Awerbuch and D. Peleg, "Concurrent on Line Tracking of Mobile Users," *Proceedings ACM SIGCOMM Symposium on Communication, Architectures and Protocols*, Oct., 1991.
16. B. Awerbuch and D. Peleg, "Concurrent Online Tracking of Mobile Users, " *Proceedings ACM SIGCOMM Symposium on Communications, Architectures and Protocols, September 3–6, 1991*, 221–233, 1991.
17. E. Ayanoglu, K. Y. Eng, and M. J. Karol, "Wireless ATM: Limits, Challenges, and Proposals," *IEEE Personal Communications*, 3:4, 18–34, 1996.
18. A. Aziz, "A Scalable and Efficient Intra-Domain Tunneling Mobile-IP Scheme," *Computer Communication Review*, 24:1(Jan.), 12–20,1994.
19. B. R. Badrinath, T. Imielinski, and A. Virmani, "Locating Strategies for Personal Communication Networks," *IEEE GLOBECOM 1992 Workshop on Networking of Personal Communications Applications*, Orlando, Florida, Dec., 1992.
20. B. R. Badrinath, A. Acharya, and T. Imielinski, "Structuring Distributed Algorithms for Mobile Hosts," *Proceedings of the 12th International Conference on Distributed Computing Systems, June 1994*, Washington, DC: IEEE Computer Society Press, 21–28, 1994.
21. A. Baiocchi, F. Delli Priscoli, F. Grilli, and F. Sestini, "The Geometric Dynamic Channel Allocation as a Practical Strategy in Mobile Networks with Bursty User Mobility," *IEEE Transaction on Vehicular Technology*, 44:1(Feb.), 14–23, 1995.
22. F. Baker, "OSPF Version 2 Management Information Base," *Internet Request for Comments*, Jul., 1248, 1991.
23. R. Beck and H. Panzer, "Strategies for Handover and Dynamic Channel Allocation in Micro-Cellular Mobile Radio Systems," *Proceedings of the IEEE Vehicular Technical. Conference,* San Francisco, CA, VTC-89, 178–185, 1989.
24. J. Bennett and G. T. Des Jardins, "Comments on the July PCRA Rate Control Baseline," *ATM Forum Contribution*, 94–0682, Jul.,1994.
25. D. Bertsekas and R. Gallager, *Data Networks*, Englewood Cliffs, NJ: Prentice-Hall, 1992.
26. P. Bianchini and H. S. Kim, "The Tera Project: A Hybrid Queuing ATM Switch Architecture for LAN," *IEEE Journal on Selected Areas in Communications*, 13.4(May), 673–685, 1995.
27. U. Black, *TCP/IP and Related Protocols*, New York: McGraw-Hill, 1992.
28. F. Box, "A Heuristic Technique for Assigning Frequencies to Mobile Radio Nets," *IEEE Transactions on Vehicular Technology*, VT-27:2, May, 57–64, 1978.
29. E. Buitenwerf, G. Colombo, H. Mitts, and P. Wright, "UMTS: Fixed Network Issues and Design Options," *IEEE Personal Communications*, Feb., 30–37, 1995.

30. D. C. A. Bulterman and R. Van Liere, "Multimedia Synchronization and UNIX," *Proceedings of the Second Inernational. Workshop on Network and Operating System Support for Digital Audio and Video, 1991*. Lecture Notes in Computer Science, New York: Springer-Verlag, Vol. 614, 108–119, 1992.
31. R. Caceres, P. B. Danzig, S. Jamin, and D. J. Mitzel, "Characteristics of Wide-Area TCP/IP Conversations," *Proceedings of the SIGCOMM 91 Conference: Communications Architectures and Protocols, ACM*, 101–112, Sep. 1991.
32. R. Caceres and L. Iftode, "The Effects of Mobility on Reliable Transport Protocols," *Proceedings of the 12th International Conference on Distributed Computing Systems, June 1994*, Washington, DC: IEEE Computer Society Press, 12–20, Jun. 1994.
33. A. Cayley "A Theorem on Trees," *Journal of Math.*, 23, 376–378, 1989.
34. Y. Chang, N. Golmie, and D. Su, "Study of Interoperability Between EFCI and ER Switch Mechanism for ABR Traffic in an ATM Network," National Institute of Standards and Technology Report, 1994.
35. M. M.-L. Cheng and J. C.-I. Chuang, "Performance Evaluation of Distributed Measurement-Based Dynamic Channel Assignment in Local Wireless Communication," *IEEE Journal on Selected Areas in Communication*,14:4(May), 698–710, 1996.
36. J. C. S. Cheung, M. A. Beach, and J. P. McGeehan, "Network Planning for Third-Generation Mobile Radio Systems," *IEEE Communications Magazine*, 32:11(Nov.), 54–59, 1994.
37. L. Chew and A. Hać, "A Multiple Access Protocol for Wireless ATM Networks," *Proceedings of the IEEE Vehicular Technology Conference*, Amsterdam, The Netherlands, 1715–1719, Sept. 19–22, 1999.
38. G. Cho and L. F. Marshall, "An Efficient Location and Routing Scheme for Mobile Computing Environments," *IEEE Journal on Selected Areas in Communications*, 13:5(Jun.), 868–879, 1995.
39. W. J. Clark, "Multipoint Multimedia Conferencing ," *IEEE Communication Magazine*, May, 44–50, 1992.
40. D. Comer, *Internetworking with TCP/IP, Principles, Protocols, and Architecture*, Englewood Clifs, NJ: Prentice Hall, 1988.
41. D. C. Cox and D. O. Reudink, "A Comparison of Some Channel Assignment Strategies in Large Scale Mobile Communications Systems," *IEEE Trans. Comm.*, COM-20:2, Apr., 190–195, 1972.
42. L. J. Cimini and G. J. Foschini, "Distributed Dynamic Channel Allocation Algorithms for Microcellular Systems," in *Wireless Communications: Future Directions*, Jack M. Holtzman and David J. Goodman, eds., Boston: Kluwer Academic Publishers, 1993.
43. L. G. Cuthbert and J.-C. Sapanel, *ATM: The Broadband Telecommunications Solution*, The Institution of Electrical Engineers, Telecommunications Series, No. 29, London, U.K., 1993.
44. Y. K. Dalal and R. M. Metcalfe, "Reverse Path Forwarding of Broadcast Packets," *Communications ACM*, 21:12(Dec.), 1040–1048, 1978.
45. S. Deering and D. Cheriton, "Multicast Routing in Datagram Internetwork and Extended LANs," *ACM Trans. Comput. Syst.*, 8:2(May), 85–110, 1990.

46. E. Del Re, R.Fantaccii, and G. Giambene, "Handover and Dynamic Channel Allocation Techniques in Mobile Cellular Networks," *IEEE Transaction on Vehicular Techn.*, 44:2(May), 229–237, 1995.
47. M. Diaz and P. Senac, "Timed Stream Petri Nets: A Model for Timed Multimedia Information," *Proceedings of the 15th International Conference on Application and Theory of Petri Nets, 1994*. Lecture Notes in Computer Science, Vol. 815, New York: Springer-Verlag, 219–238, 1994.
48. E. Dijkstra, "A Note on Two Problems in Connection with Graphs," *Numerische Mathematik*, 1, 269–271, 1959.
49. S. Dixit and P. Skelly, "MPEG-2 over ATM for Video Dial Tone Networks: Issues and Strategies," *IEEE Network*, Sep., 30–40, 1995.
50. B. Doshi, S. Dravida, P. Johri, and G. Ramamurthy, "Memory, Bandwidth, Processing and Fairness Considerations in Real Time Congestion Controls for Broadband Networks, Teletraffic and Datatraffic in a Period of Change," *IAC*, ITC-13, 153–159, 1991.
51. M. Duque-Aton, D. Kunz, and B. Ruber, "Channel Assignment for Cellular Radio Using Simulated Annealing," *IEEE Transactions on Vehicular Technology*, 40:1(Feb), 14–21, 1991.
52. A. E. Eckberg, "B-ISDN/ATM Traffic and Congestion Control,"*IEEE Networks*, Sep., 28–37, 1992.
53. D. E. Everitt, and N. W. Macfadyen, "Analysis of Multicellular Mobile Radiotelephone Systems with Loss," *British Telecom Technol. J.*, 1:2, 37–45, 1983.
54. D. Everitt and D. Manfield, "Performance Analysis of Cellular Mobile Communication Systems with Dynamic Channel Assignment," *IEEE Journal on Selected Areas in Communication*, 7:8(Oct.), 1172–1180, 1989.
55. D. Everitt, "Traffic Engineering of the Radio Interface for the Cellular Mobile Networks," *Proceedings of the IEEE*, 82:9(Sep.), 1371–1382, 1994.
56. L. Fedaoui, A. Seneviratne, and E. Horlait, "Implementation of a End-to-End Quality of Service Management Scheme," *Proceedings. of International COST 237 Workshop on Multimedia Transport and Teleservices, 1994*. Lecture Notes in Computer Science, Vol. 882, New York: Springer-Verlag, 124–144, 1994.
57. D. Ferrari, "Delay Jitter Control Scheme for Packet-Switching Internetworks," *Computer Communications*, 15:6(Jul.), 367–373, 1992.
58. D. Ferrari, "Design and Applications of a Delay Jitter Control Scheme For Packet-Switching Internetworks," *Proceedings of the Second International Workshop on Network and Operating System Support for Digital Audio and Video, 1991*. Lecture Notes in Computer Science, Vol. 614, New York: Springer-Verlag, 72–83, 1992
59. S. Feit, *TCP/IP Architecture, Protocols, and Implementation*, New York: McGraw-Hill, 1993.
60. M. Fujioka, S. Sakai, and H. Yagi, "Hierarchical and Information Handling for UPT," *IEEE Network*, Nov., 50–60, 1990.
61. N. Funabiki and Y. Takefuji, "A Neural Network Parallel Algorithm for Channel Assignment in Cellular Radio Networks," *IEEE Transactions on Vehicular Technology*, 41:4(Nov.), 430–437, 1992.

62. H. Furukawa and Y. Akaiwa, "Self-Organized Reuse Partitioning, a Dynamic Channel Assignment Method in Cellular System," *43th IEEE Vehicular Technology Conferenc. '93*, 524–527, 1993.
63. J. Gait, "A Kernel for High-Performance Multicast Communication," *IEEE Transactions on Computers*, 38:2, 218–226, Feb. 1989.
64. R. G. Gallager, "A Perspective on Multiaccess Channels," *IEEE Trans. Info. Theory*, IT-31, Mar., 124–142, 1985
65. M. Galletti and F. Grossini, "Performance Simulation of Congestion Control Mechanisms for Intelligent Networks," *International Zurich Seminar on Digital Communications*, 391–406, 1992.
66. A. Gamst and W. Rave, "On Frequency Assignment in Mobile Automatic Telephone Systems," *Proceedings of the IEEE, GLOBECOM '82*, 309–315, 1982.
67. M. R. Garey and D. S. Johnson, *Computers and Intractability: A Guide to the Theory of NP–Completeness*, San Francisco, CA: Freeman, 1979
68. J. J. Garrahan et. al., "Intelligent Network Overview," *IEEE Communications Magazine*, 31:3(Mar.), 30–36, 1993.
69. M. Gerla and L. Kleinrock, "Flow control: A Comparative Survey," *IEEE Communications*, COM-28, Apr., 553–573, 1980.
70. D. J. Goodman, R. A. Balenzuela, K. T. Gayliard, and B. Ramamurthi, "Packet Reservation Multiple Access for Local Wireless Communications," *IEEE Transaction of Communications*, 37:8(Aug.), 885–890, 1989.
71. D. J. Goodman, "Trends in Cellular and Cordless Communications," *IEEE Communication Magazine*, Jun., 1991.
72. D. J. Goodman, S. A. Grandhi, and R. Vijayan, "Distributed Dynamic Channel Assignment Scheme," *43th IEEE Vehicular Technology Conference '93*, 532–535, 1993.
73. W. J. Goralski, "Introduction to ATM Networks," McGraw-Hill Series on Computer Communications, New York: McGraw-Hill, 1995.
74. L. J. Greenstein, "Microcells in Personal Communications Systems," *IEEE Communications Magazine*, 30:12(Dec.), 76–88, 1992.
75. F. Guillenmin, P. Boyer and A. Dupus, "Peak Rate Enforcement in ATM Networks," *Proceedings of the IEEE INFOCOM*, Florence, Italy, 753–758, 1992.
76. Z. Haas, "Adaptive Admission Congestion Control," *ACM SIGCOMM, Computer Communications Review*, 1992, 58–76.
77. A. Hać, "Congestion Control and Switch Buffer Allocation in High-Speed Networks," *Proceedings of the IEEE INFOCOM*, Miami, FL, 314–322, 1991.
78. A. Hać, "Distributed Multicasting Algorithm with Congestion Control and Message Routing," *Journal of Systems and Software*, 24:1, 49–65, 1994.
79. A. Hać, "Resource Management in a Hierarchical Network," *Proceedings of the SBT/IEEE International Telecommunications Symposium*, 563–567, Sep. 3–6, 1990.
80. A. Hać and Z. Chen, "Hybrid Channel Allocation in Wireless Networks," *Proceedings of the IEEE Vehicular Technology Conference*, Amsterdam, The Netherlands, 2329–2333, Sept. 19–22, 1999.
81. A. Hać and L. Gao, "Analysis of Congestion Control Mechanisms in Intelligent Network," *International Journal of Network Management*, 8:1, 18–41, 1998.

82. A. Hać and L. Guo, "Mobile Host Protocols for the Internet," *Proceedings of the IEEE Vehicular Technology Conference*, Amsterdam, The Netherlands, 2790–2794, Sept. 19–22, 1999.
83. A. Hać and A. Hossain, "Virtual LAN Supporting Quality of Service in Wireless ATM Networks," *Proceedings of the IEEE Vehicular Technology Conference*, Amsterdam, The Netherlands, 2686–2690, Sept. 19–22, 1999.
84. A. Hać and Y. Huang "Hierarchical Location Update and Routing Scheme for Mobile Computing Environment," *Proceedings of the IEEE International Conference on Universal Personal Communications*, San Diego, CA, 728–732, Oct. 12–16, 1997.
85. A. Hać and H. Lin, "Congestion Control for ABR Traffic in ATM Network," *International Journal of Network Management*, 9:4, 249–264, 1999.
86. A. Hać and B. Liu, "Database and Location Management Schemes for Mobile Communications," *IEEE/ACM Transactions on Networking*, 6:6, 851–865, 1998.
87. A. Hać and D. Lu, "Architecture, Design, and Implementation of a Multimedia Conference System," *International Journal of Network Management*, 7: 2, 64–83, 1997.
88. A. Hać and Y. Ma, "A Rate-Based Congestion Control Scheme for ABR Service in ATM Network," *International Journal of Network Management*, 8:5, 292–317, 1998.
89. A. Hać and C. Mo, "Dynamic Channel Assignment Strategy for Microcell Systems," *Proceedings of the IEEE Pacific Rim Conference on Communications, Computers and Signal Processing*, Victoria, BC, Canada, 346–349, Aug. 20–22, 1997.
90. A. Hać and C. Mo, "Dynamic Channel Assignment in Wireless Communication Networks," *International Journal of Network Management*, 9:1, 44–66, 1999.
91. A. Hać and J. On, "Multimedia Virtual Circuit Model for MPEG on ATM Networks," *International Journal of Network Management*, 8:6, 325–342, 1998.
92. A. Hać and C. Sheng,"User Mobility Management in the PCS Network Through the Placement of Hierarchical Databases," *International Journal on Wireless Information Networks*, 5:3, 235–256, 1998.
93. A. Hać and D. Wang, "A Multicast Routing Algorithm Reducing Congestion," *International Journal of Network Management*, 7:3, 147–163, 1997.
94. A. Hać and C. Xiao, "Improving Performance of Channel Assignment in Cellular Mobile Communication Network," *Proceedings of the IEEE Pacific Rim Conference on Communications, Computers and Signal Processing*, Victoria, BC, Canada, 358–361, Aug. 20–22, 1997.
95. A. Hać and C. Xiao, "A New Distributed Dynamic Channel Assignment Strategy in a Cellular Mobile Communication Network," *International Journal on Wireless Information Networks*, 6:1, 49–58, 1999.
96. A. Hać and C. X. Xue, "Synchronization in Multimedia Data Retrieval," *International Journal of Network Management*, 7:1, 33–62, 1997.
97. A. Hać and X. Zhou,"Locating Strategies for Personal Communication Networks: A Novel Tracking Strategy,"*IEEE Journal on Selected Areas in Communications*, 15:8, 1425–1436, 1997.
98. A. Hać and K. Zhou, "Multicasting Algorithm in Multimedia Communication Network," *Proceedings IEEE GLOBECOM Global Communications Conference*, Sydney, Australia, 747–752, Nov. 8–12, 1998.

99. A. Hać and K. Zhou, "A New Heuristic Algorithm for Finding Minimum Cost Multicast Trees with Bounded Path Delay," *International Journal of Network Management*, 9:4, 265–279, 1999.
100. A. Hać and Z. Zhu, "Efficient Routing in Wireless Personal Networks," *Proceedings of. IEEE Pacific Rim Conference on Communications, Computers and Signal Processing*, Victoria, BC, Canada, 342–345, Aug. 20–22, 1997.
101. A. Hać and Z. Zhu, "Performance of Routing Schemes in Wireless Personal Networks," *International Journal of Network Management*, 9:2, 80–105, 1999.
102. J. Hagouel, "Source Routing and a Distributed Algorithm to Implement It," *Proceedings of the IEEE INFOCOM*, Apr., San Diego, CA, 115–123, 1983.
103. W. K. Hale, "Frequency Assignment: Theory and Applications," *Proceedings of the IEEE*, 68:12(Dec.), 1497–1514, 1980.
104. H. Heeke, "A Traffic-Control Algorithm for ATM Networks," *IEEE Transactions on Circuits and Systems for Video Technology*, 3:3(Jun.), 182–189, 1993.
105. J. S. M. Ho, H. Uzunalioglu, and I. F. Akyildiz, "Cooperating Leaky Bucket for Average Rate Enforcement of VBR Video Traffic in ATM Networks," *Proceedings of the IEEE INFOCOM*, Boston, MA, 1995.
106. W. Hoffman and R. Pavley, " A Method for the Solution of the $N$th Best Path Problem,"*J. Assoc. Computing Machinery*, 6:4, 1959.
107. J. Homa and S. Harris, "Intelligent Network Requirements for Personal Communications Services," *IEEE Communications Magazine*, 30:2(Feb.), 70–76, 1992.
108. N. F. Huang, Y. T. Wang, B. Li, and T. L. Liu, "Virtual LAN Internetworking over ATM Networks for Mobile Stations," *IEEE INFOCOM*, 1397–1403, 1997.
109. D. Hughes, "Fairness Share in the Context of MCR," *ATM Forum Contribution*, 94–0977, Oct., 1994.
110. C. Huitema, *Routing in the Internet*, Englewood Cliffs, NJ:Prentice Hall-PTR, 1995.
111. T. Imielinski and B. R. Badrinath, "Querying Locations in Wireless Environments," in *Wireless Communications: Future Directions*, J. M. Holtzman and D. J. Goodman, eds., Boston: Kluwer, 85–108, 1993.
112. J. Ioannidis, D. Duchamp, and G. Q. Maguire, Jr., "IP-Based Protocols for Mobile Internetworking," *Proceedings of the ACM SIGCOMM Symposium on Communication, Architectures and Protocols*, 235–245, Sep. 1991.
113. ITU, Joint Technical Committee ISO/IEC JTC 1, *Information Technology—Coding of moving pictures and associated audio for digital storage media at up to about 1.5 Mbit/s*, ISO/IEC 11172.
114. ITU, "Narrow-Band Visual Telephone Systems and Terminal Equipment," *ITU-T Recommendation H.320*, Mar., 1993.
115. B. Jabbari et al., "Intelligent Network Concepts in Mobile Communications," *IEEE Communications Magazine*, 30:2(Feb.), 64–69, 1992.
116. B. Jabbari et al., "Network Issues for Wireless Communications," *IEEE Communications Magazine*, 33:1(Jan.), 88–98, 1995.
117. V. Jacobson, "VAT-X11 Based Audio Conferencing Tool," *UNIX Manual Pages*, Feb., 1993.

118. R. Jain, "Congestion Control in Computer Networks: Issues and Trends," *IEEE Network Magazine*, May, 24–30, 1990.
119. R. Jain, "Reducing Traffic Impacts of PCs Using Hierarchical User Location Databases," *Proceedings of the IEEE INFOCOM*, San Francisco, CA, 2, 1153–1157, 1996.
120. R. Jain, Y.-B. Lin, C. Lo, and S. Mohan, "A Forwarding Strategy to Reduce Network Impacts of PCS," *Proceedings of the IEEE INFOCOM*, Boston, MA, 2, 481–489, 1995.
121. X. Jiang, "Distributed Path Finding Algorithm for Stream Multicast," *Computer Communications*, 16:12, 767–775, Dec. 1993.
122. D. B. Johnson, "Scalable and Robust Internetwork Routing for Mobile Hosts," *Proceedings of the 14th International Conference on Distributed Computing Systems, June 1994*, Washington DC: IEEE Computer Society Press, 2–11, 1994.
123. D. B. Johnson and C. Perkins, "Route Optimization In Mobile IP," *Internet Draft*, Nov. 1996. URL: http://www.monarch.cs.cmu.edu/internet-drafts/draft-ietf-mobileip-optim-0.5txt.
124. T. Kanai, "Autonomous Reuse Partitioning in Cellular System". *IEEE Vehicular Technology Conference '92*, 782–785, 1992.
125. M. Karol, Z. Liu, and K. Y. Eng, "Distributed-Queuing Request Update Multiple Access (DQRUMA) for Wireless Packet (ATM) Networks," *Proceedings of the International Conference on Communication 1995, Seattle, WA*, 1224–1231, Jun., 1995.
126. R. Kawahara, T. Asaka, and S. Sumita, "Overload Control for the Intelligent Network and Its Analysis by Simulation," *IEICE Transactions on Communications*, E78–B:4(Apr.), 494–503, 1995.
127. L. Kleinrock and F. A. Tobagi, "Packet Switching in Radio Channels, Part 1: Carrier Sense Multiple Access Modes and Their Throughput Delay Characteristics," *IEEE Transaction of Communications*, COM-23, Dec., 1400–1416, 1975.
128. D. Kohler and H. Muller, "Multimedia playout synchronization using buffer level Control, *Proceedings of the Second International Workshop on Multimedia: Advanced Teleservices and High-Speed Communication Architectures, 1994*. Lecture Notes in Computer Science, Vol. 868, New York: Springer-Verlag, 167–180, 1994.
129. V. P. Kompella, J. C. Pasquale, and G. C. Polyzos, "Multicast Routing for Multimedia Communication," *IEEE/ACM Transactions on Networking*, 1:3, 286–292, 1993.
130. L. Kou, G. Markowsky, and L Berman, "A Fast Algorithm for Steiner Trees," *Acta Informatica*, 15, 141–151, 1981.
131. O. Kubbar and H. T. Mouftah, "Multiple Access Control Protocols for Wireless ATM: Problems Definition and Design Objectives," *IEEE Communications Magazine*, 35:11(Nov.), 93–99, 1997.
132. K. Kubota, M. Murata, and H. Miyahara, "Congestion Control for Bursty Video Traffic in ATM Networks," *Electronics and Communications in Japan, Part 1*, 75:4(Feb.), 13–22, 1991.
133. D. Kunz, "Channel Assignment for Cellular Radio Using Neural Networks," *IEEE Transactions on Vehicular Technology*, 40:1(Feb.), 188–193, 1991.
134. J. Kurose, "Open Issues and Challenges in Providing Quality of Service Guarantees in High Speed Networks," *Computer Communication Review*, 23:1(Jan.), 234–39, 1993.

135. K. Kvols and S. Blaabjerg, "Bounds and Approximation for the Periodic On/Off Queue," *Proceedings of the IEEE INFOCOM*, Florence, Italy, 487–494, 1992.
136. B. Lague, C. Rosenberg, and F. Guillemin, "A Generalization of Some Policing Mechanisms," *INFOCOM*, 767–775, 1992.
137. S. S. Lam and G. G. Xie, "Burst Scheduling: Architecture and Algorithm for Switching Packet Video," *Proceedings of the IEEE INFOCOM*, Boston, MA, 8a.1.1–8a.1.11, 1995.
138. L. Lamont and N. D. Georganas, "Synchronization Architecture and Protocols for a Multimedia News Service Application," *Proceedings of the IEEE International Conference on Multimedia Computing and Systems*, 3–8, 1994.
139. F. Langlois and J. Regnier, "Dynamic Congestion Control in Circuit-Switched Telecommunications Networks," *Teletraffic and Datatraffic in a Period of Change*, ITC-13, 127–132, 1991.
140. T. F. LaPorta et al., "Distributed Call Processing for Personal Communications Services," *IEEE Communication Magazine*, Jun., 1995.
141. G. S. Lauer, "IN Architectures for Implementing Universal Personal Telecommunications," *IEEE Network*, Mar./Apr., 6–16, 1994.
142. W. C. Y. Lee, "New Cellular Schemes for Spectral Efficiency," *IEEE Trans. Veh. Tech.*, VT-36:4(Nov.), 188–192, 1987.
143. W. C .Y. Lee, *Mobile Cellular Telecommunications: Analog and Digital System*, New York: McGraw-Hill, Inc., 1995.
144. W. -H. F. Leung, T. J. Baumgartner, and Y. H. Hwang, "A Software Architecture for Workstations Supporting Multimedia Conferencing in Packet Switching Networks," *IEEE Journal on Selected Areas in Communications*, 8:3(Apr.), 380–389, 1990.
145. W.-H. F. Leung, T. J. Baumgarner, and S.-C. Tu, "A Software Architecture for Workstations Supporting Multimedia Conferencing in Packet Switching Networks," *IEEE Journal on Selected Area of Communications*, 68:91(Apr.), 1991.
146. A. K. Li and N. D. Georganas, "Multimedia Segment Delivery Scheme and Its Performance for Real-Time Synchronization Control," *Proceedings of the IEEE INFOCOM*, Toronto, Canada, 3, 1734–1738, 1994.
147. B. Li and P. Vankwikelberge, "Virtual LAN Configuration and Address Resolution in an ATM Network," *Second International Symposium on Internetworking, INTERNETWORKING '94,* Sophia Antipolis, 179–190, May 1994.
148. Y. Li and S. Andressen, "Traffic Management Principles for Guaranteed Wireless Network Performance," *Third Annual International Conference on Universal Personal Communications*, San Diego, CA, 325–29, September 1994.
149. Y. B. Lin, S. Mohan, and A. Noepel, "Queueing Channel Assignment Strategies for PCS Hand-off and Initial Access," *Third Annual International Conference on Universal Personal Communications, San Diego, CA, September 1994*, 365–70, 1994.
150. T. D. C. Little and A. Ghafoor, "Network Considerations for Distributed Multimedia Objects Composition and Communication," *IEEE Network Magazine*, 4:6, 1990.
151. T. D. C. Little and A.. Ghafoor, "Synchronization and Storage Models for Multimedia Objects," *IEEE Journal on Selected Areas in Communications*, 8:3(Apr.), 413–427, 1990
152. T. D. C. Little and A. Ghafoor, "*Network Considerations for Distributed Multimedia Object Composition and Communication, IEEE Network*, Nov., 32–49, 1990.

153. T. D. C. Little and A. Ghafoor, "Multimedia Synchronization Protocols for Broadband Integrated Services," *IEEE Journal on Selected Areas in Communications*, 9:9(Dec.), 1368–1381, 1991.
154. T. D. C. Little and A. Ghafoor, "Scheduling of Bandwidth-Constrained Multimedia Traffic," *Computer Communications*, 15, 381–387, 1992.
155. T. D. C. Little and F. Kao, "An Intermedia Skew Control System for Multimedia Data Presentation," *Proceedings of the Third International Workshop on Network and Operating System Support for Digital Audio and Video, 1992*. Lecture Notes in Computer Science, Vol. 712, New York: Springer-Verlag,, 130–141, 1993.
156. T. D. C. Little and D. Venkatesh, "Prospects for Interactive Video-on-Demand," *IEEE Multimedia, Fall*, 14–2, 1994.
157. N. C. Lobley, "Intelligent Mobile Networks," *British Telecom Journal*, Apr., 21–29, 1995.
158. H. Lu, K. Pung, T. S. Chua, and S. F. Chan, "Temporal Synchronization Support for Distributed Multimedia Information Systems," *Computer Communications*, 17:12( Dec.), 852–862, 1994.
159. A. G. MacInnis, "The MPEG Systems Coding Specification," *Signal Processing: Image Communication*, 4:2(Apr.), 153–159, 1992.
160. J. MacLellan and C. Rose, "Resource Allocation for Wireless Network," *Proceedings of the 44th IEEE Vehicular Technology Conference*, 2, 804–808, 1994.
161. U. Madhow, M. L. Honig, and K. Steiglitz, "Optimization of Wireless Resources for Personal Communications Mobility Tracking," *IEEE/ACM Transactions on Networking*, 3:6(Dec.), 698–707, 1995.
162. N. M. Marafih, Y-Q. Zhang, and R. L. Pickholtz, "Modeling and Queueing Analysis of Variable-Bit-Rate Coded Video Sources in ATM Networks," *IEEE Transactions on Circuits and Systems for Video Technology*, 4:2(Apr.), 121–127, 1994.
163. R. Mathar and J. Mattfeldt, "Channel Assignment for Cellular Radio Networks," *IEEE Transactions on Vehicular Technology*, 42:4(Nov.), 647–656, 1993.
164. D. E. McDysan and D. L. Spohn, *ATM: Theory and Applications*, McGraw-Hill Series on Computer Communications, New York: McGraw-Hill, 1994.
165. D. Minoli, *Dialtone Technology* (Video), New York: McGraw-Hill, 1995.
166. D. Minoli and M. Vitella, *ATM and Cell Relay Service for Corporate Environments*, McGraw-Hill Series on Computer Communications, New York: McGraw-Hill, 1994.
167. S. Mohan and R. Jain, "Two User Location Strategies for Personal Communications Services," *IEEE Personal Communications*, 1st Quarter, 42–50, 1994.
168. N. Morinaga, "Advanced Wireless Communication Technologies for Achieving High-Speed Mobile Radios," *IEICE Transactions on Communications*, E78-B:8(Aug.), 1089–93, 1993.
169. J. Moy, "OSPF Version 2," *Internet Request for Comments*, 1247, Jul. 1991.
170. J. Moy, "Multicast Routing Extensions for OSPF," *Communications of the ACM*, 37: 8, 61–66, Aug. 1994.
171. J. Moy, "OSPF Version 2," *Internet Request forComments*, 1583, Mar., 1994.
172. M. Murata and H. Miyahara, "Survey on Traffic Control in ATM Networks Based on Traffic Theory," *Electronics and Communications in Japan*, 73:10, 28–40, 1990.

173. A. Myles and D. Skellern, "Comparing four IP Based Mobile Host Protocols," *Computer Networks and ISDN Systems*, 26:3(Nov.), 349–355, 1993.
174. A. Myles and D. Skellern, "Comparison of Mobile Host Protocols for IP," *Journal of Internetworking Research and Experience*, Dec. 1993.
175. A. Myles, D. B. Johnson, and C. Perkins, "A Mobile Host Protocol Supporting Route Optimization and Authentication," *IEEE Journal on Selected Areas in Communications*, 13:5(Jun.), 839–849, 1995.
176. M. Naghshineh and A. S. Acampora, "Design and Control of Microcellular Networks with QOS Provisioning for Real Time Traffic," *Third Annual International Conference on Universal Personal Communications, San Diego, CA, September 1994*, 376–81, 1994.
177. M. Naghshineh and A. S. Acampora, "QOS Provisioning in Micro-Cellular Networks Supporting Multimedia Traffic," *IEEE INFOCOM,* Boston, MA, 1075–84, Apr. 1995.
178. M. Naghshineh and M. Schwartz, Distributed Call Admission Control in Mobile/Wireless Networks," *IEEE Journal on Selected Areas in Communications*, 14:4(May), 711–716, 1996.
179. K. Nahrstedt and R. Steinmetz, "Resource Management in Networked Multimedia Systems," *Computer*, Vol. 28:5(May), 52–63, 1995.
180. A. Nakajima, Y. Yamamoto, and H. Sawada, "Automatic Pursuit Routing for Mobile Communications Network," *Electronics and Communications in Japan*, Part 1, 73:3, 81–91, 1990.
181. P. Newman, "Backward Explicit Congestion Notification for ATM Local Area Networks," *Proc. IEEE*, 1993.
182. L. J. Ng et al., "Distributed Architectures and Databases for Intelligent Personal Communication Networks," *Preceeding of the IEEE International Conference on Selected Topics in Wireless Communications*, June 1992, 300–304, 1992.
183. C. Nicolaou, "An Architecture for Real-Time Multimedia Communication Systems," *IEEE Journal on Selected Areas in Communications*, 8:3(Apr.), 391–400, 1990.
184. B. D. Noble and M. Satyanarayanan, "A Research Status Report on Adaptation for Mobile Data Access," *Proceedings of the IEEE International Conference on Network Protocols*, Dec. 10–15, 1995.
185. T. Norp and A. J. M. Roovers, "UMTS Integrated with B-ISDN," *IEEE Communications Magazine*, Nov., 60–65, 1994.
186. A. B. Noy and I. Kessler, "Tracking Mobile Users in Wireless Communications Networks," *Proceedings of the IEEE INFOCOM*, San Francisco, CA, 1232–1239, 1993.
187. H. Ohsaki et al., "Rate-Based Congestion Control for ATM Networks," *Computer Communication Review*, 25:2(Apr.), 60–71, 1995.
188. T. Ohsawa, K. Okanoue, M. Madihian, K. Lshii, and Y. Kakura, "Seamless Networked Mobile Computing Realized by Wireless Internet Access," *Proceedings of the IEEE INFOCOM*, San Francisco, CA, 2, 792–796, 1996.
189. C. Oliviera, J. Bae Kim, and T. Suda, "Quality of Service Guarantee in High Speed Multimedia Wireless Networks," *Proceedings of the IEEE INFOCOM,* San Francisco, CA, 728–34, 1996, Mar. 1996.
190. K. Pahlavan and A. H. Lavesque, *Wireless Information Networks*, New York: John Wiley and Sons, 1995.

191. P. Pancha and M. El Zarki, "Bandwidth-Allocation Schemes for Variable-Bit-Rate MPEG Sources in ATM Networks," *IEEE Transactions on Circuits and Systems for Video Technology*, 3:3(Jun.), 190–198, 1993.
192. P. Pancha and M. El Zarki, "Leaky Bucket Access Control for VBR MPEG Video," *Proceedings of the IEEE INFOCOM*, Boston, MA, 1995.
193. R. Pandya, "Emerging Mobile and Personal Communication Systems," *IEEE Communications Magazine*, 33:6(Jun.), 44–51, 1995.
194. C. E. Perkins and P. Bhagwat, "A Mobile Networking System Based on Internet Protocol," *IEEE Personal Communications*, 1:1, 34–41, 1994.
195. C. Perkins, A. Myles, and D. B. Johnson, "IMHP: A Mobile Host Protocol for the Internet," *Computer Networks and ISDN Systems*, 27:3(Dec.), 479–491, 1994.
196. C. Perkins, "IP Mobility Support," *Internet Draft,* Nov. 1995. URL: http://www.it.kth.se/~d91-fta/exjobb/draft/mobileip-protocol-13.txt.
197. X. H. Pham and R. Betts, "Congestion Control for Intelligent Networks," *Computer Networks and ISDN Systems,* 26, 511–524,1994.
198. G. P. Pollini, "Signaling System No. 7. Messaging in North American Cellular," *WINLAB, Rutgers University Technical Report*, TR-73, May, 1994.
199. G. P. Pollini, K. S. Meier-Hellstern, and D. J. Goodman, "Signaling Traffic Volume Generated by Mobile and Personal Communications," *IEEE Communication Magazine*, Jun. 1995.
200. G. P. Pollini and K. S. Meier-Hellstern, "Efficient Routing of Information Between Interconnected Cellular Mobile Switching Centers," *IEEE/ACM Transactions on Networking*, 8:6(Dec.), 765–774, 1995.
201. J. B. Postel, "User Datagram Protocol," *Internet Request for Comments,* 768, Aug. 1980.
202. J. B. Postel, "Internet Protocol," *Internet Request for Comments*, 791, Sep. 1981.
203. J. B. Postel, "Internet Control Message Protocol," *Internet Request for Comments*, 792, Sep. 1981.
204. L. Pouzin, "Methods, Tools, and Observations on Flow Control in Packet-Switched Data Networks," *IEEE Transactions on Communications*, COM-29, Apr., 413–426, 1981.
205. N. B. Pronios and T. Bozios, "A Scheme for Multimedia and Hypermedia Synchronization," *Proceedings of the International COST 237 Workshop on Multimedia Transport and Teleservices, 1994.* Lecture Notes in Computer Science, Vol. 882, New York: Springer-Verlag, 340–345, 1994.
206. S. Rajagopalan and B. R. Badrinath, "An Adaptive Location Management Strategy for Mobile IP," *First ACM Mobilcom 95*, Nov., 1995.
207. V. Rangan, S. Ramanathan, H. M. Vin, and T. Kaeppner, "Techniques for Multimedia Synchronization in Network File Systems," *Computer Communications*, 16:3(Mar.), 168–176, 1993.
208. V. Rangan, S. Ramanathan, H. M. Vin, and T. Kaeppner, "Performance of Inter-media Synchronization in Distributed and Heterogeneous Multimedia Systems," *Computer Networks and ISDN Systems*, 27, 549–565, 1995.
209. D. Raychaudhuri, "Wireless ATM: An Enabling Technology for Multi-Media Personal Communications," *Second International Workshop on Mobile Multi-Media Communications*, Apr., 1995.

210. S. Robert, *Algorithms in C*, Reading, MA: Addison-Wesley, 1990.
211. R. Rom and M. Sidi, "Multiple Access Protocols: Performance and Analysis," New York: Springer-Verlag, 1990.
212. H. Saito, "Call Admission Control in an ATM Network Using Upper Bound of Cell Loss Probability," *IEEE Transactions on Communications*, 40:9(Sept), 1512–1521, 1992.
213. S. Sakata, "Development and Evaluation of an In-House Multimedia Desktop Conferencing System," *IEEE Journal on Selected Area of Communications*, 68: 91(Apr.), 1991.
214. J. Sanchez, R. Martinez, and M. W. Marcellin, "A Survey of MAC Protocols Proposed for Wireless ATM," *IEEE Network*, 11:6(Nov.), 52–62, 1997.
215. S. Sarin and I. Grief. "Computer-Based Real-Time Conferencing Systems," *IEEE Computer*, 18:10, 1985.
216. S. Sathaye, "ATM Forum Traffic Management Specification," *ATM Forum Contribution*, Dec., 1994.
217. M. Satyanarayanaan, B. Noble, P. Kumar, and M. Price, "Application-Aware Adaptation for Mobile Computing," *Proceedings of the Operating System Review*, Jan. 52–55, 1995.
218. R. Saunders and L. Lopes, "Performance Comparison of Global and Distributed Dynamic Channel Allocation Algorithms," *44th IEEE Vehicular Technology Conference '94*, 2, 799–803, 1994.
219. M. Schwartz, "Network Management and Control Issues in Multimedia Wireless Networks," *IEEE Personal Communications*, 2:3(Jun.), 8–16, 1995.
220. M. Sengoku, K. Itoh, and T. Matsumoto., "A Dynamic Frequency Assignment Algorithm in Mobile Radio Communication Systems," *Trans. TEICE, Japan*, E61:7(Jul.), 527–533, 1978.
221. M. Sengoku, "Efficient Utilization of Frequency Spectrum for Mobile Radio Communication Systems Algorithms for Channel Assignments," *J. IEICE*, 69:4(Apr.), 350–356, 1986.
222. M. Serizawa and D. J. Goodman, "Instability and Deadlock od Distributed Dynamic Channel Allocation". *43th IEEE Vehicular Technology Conference '93*, 528–531, 1993.
223. Z.-Y. Shae, P.-C. Chang, and M. Chen, "Capture and Playback Synchronization in Video Conferencing," *Proceedings on Multimedia Computing and Networking, 1995, SPIE*, Vol. 2417, 90–101, 1995.
224. C. Shim, I. Ryoo, J. Lee, and S. Lee, "Modeling and Call Admission Control Algorithm of Variable Bit Rate Video in ATM Networks," *IEEE Journal on Selected Areas in Communications*, 12:2(Feb.), 332–344, 1994.
225. K.-Y. Siu and H.-Y. Tzeng, "Intelligent Congestion Control for ABR Service in ATM Networks," *Computer Communication Review*, 24:5(Oct.), 81–106, 1994.
226. K. N. Sivarajan, R. J. McEliece, and J. W. Ketchum, "Channel Assignment in Cellular Radio," *Proceedings of the 39th IEEE Vehicular Technology Conference*, 846–850, May 1–3, 1989.
227. K. N. Sivarajan, R. J. McEliece, and J. W. Ketchum, "Dynamic Channel Assignment in Cellular Radio," *40th IEEE Vehicular Technology Conference*, 631–637, 1990.
228. D. E. Smith, "Ensuring Robust Call Throughput and Fairness for Scp Overload Controls," *IEEE/ACM Transactions on Networking*, 3:5(Oct.), 538–548, 1995.
229. K. Sohraby, "Mobility Calculus in Cellular Wireless Networks," *Proceedings of the IEEE INFOCOM*, San Francisco, CA, 1258–62, March 1996.

230. M. Spreitzer and M. Theimer, "Architecture Considerations for Scalable, Secure, Mobile Computing with Location Information," *Proceedings of the 12th IEEE International Conference on Distributed Computing Systems, June 1994*, Washington, DC, 29–38, 1994.
231. R. Steinmetz, "Synchronization Properties in Multimedia Systems," *IEEE Journal on Selected Areas in Communications*, 8:3(Apr.), 401–412, 1990.
232. R. Steinmatz, "Synchronization Properties in Multimedia Systems," *IEEE Selected Areas in Communication*, 6:3, 1990.
233. R. Steinmetz and C. Engler, *Human Perception of Media Synchronization,* IBM European Networking Center, Technical Report, 43.9310, 1993.
234. R. Steinmetz, "Analyzing the Multimedia Operating System," *IEEE Multimedia*, 2:1(Spring), 68–84, 1995.
235. T. Steven, "Multimedia Over ATM? It's a Matter of Timing," *Data Communication*, Jun., 21–22, 1994.
236. D. L. Stone and K. Jeffay, "Queue Monitoring: A Delay Jitter Management Policy," *Fourth International Workshop on Network and Operating System Support for Digital Audio and Video, 1993.* Lecture Notes in Computer Science, Vol. 846, New York: Springer-Verlag, 149–160, 1994.
237. M. N. S. Swamy. and K. Thulasiraman, *Graphs, Networks, and Algorithms*, New York: John Wiley and Sons, 494, 1988.
238. E. D. Sykas, I. Ch. Paschalidis and G. K. Moutzinous, "Congestion Avoidance in ATM Networks," *Proceedings of the IEEE INFOCOM*, Florence, Italy, 905–913, 1992.
239. S. Tabbane, "An Alternative Strategy for Location Tracking," *IEEE Journal on Selected Areas in Communications*, 13:5(Jun.), 880–892, 1995.
240. H. Tamura, M. Sengoku, S. Shinoda, and T. Abe, "Channel Assignment Taking into Account Interference in a Cellular Mobile System and Coloring Problem of a Graph," *IEICE Technical Report*, CAS89-43, Aug., 1989.
241. A. S. Tanenbaum, *Computer Networks*, Upper Saddle River, NJ: Prentice Hall-PTR, 1996.
242. F. Teraoka, K. Claffy, and M. Tokoro, "Design, Implementation, and Evaluation of Virtual Internet Protocol," *Proceedings of the 12th International Conference on Distributed Computing Systems, June 1992*, Washington, DC: IEEE Computer Society Press, 170–177, Jun. 1992.
243. F. Teraoka and M. Tokoro, "Host migration Transparency in IP Networks: The VIP Approach," *Computer Comm. Review, ACM*, 45–65, Jan. 1993.
244. F. Teraoka, Y. Yokote, and M. Tokoro, "A Network Architecture Providing Host Migration Transparency," *Proceedings of the SIGCOMM 91 Conference: Communications Architectures and Protocols, ACM*, 209–220, Sep. 1991.
245. C. -K Toh, "Protocol Aspects of Mobile Radio Networks," PhD Thesis, Cambridge University Computer Laboratory, Aug. 1996.
246. C. -K Toh, *Wireless ATM and AD-HOC Networks—Protocols and Architectures*, Boston: Kluwer Academic Publisher, 1996.
247. C. -Y. Tse and S. C. Liew, "Video Aggregation: An Integrated Video Compression and Multiplexing Scheme for Broadband Networks," *Proceedings of the IEEE INFOCOM*, Boston, MA, 1995.

248. N. Tsolas, G. Ado, and R. Botthem, "Performance and Overload Considerations When Introducing IN into an Existing Network," *International Zurich seminar on digital communications*, 407–414, 1992.
249. M. Veeraraghavan, T. F. La Porta, and R. Ramjee, "A Distributed Control Strategy for Wireless ATM Networks," *ACM Wireless ATM Journal (WINET)*, Sep., 323–339, 1995.
250. M. Veeraraghavan, M. J. Karol, and K. Y. Eng, "Mobility and Connection Management in a Wireless ATM LAN," *IEEE Journal on Selected Areas in Communications*," Jan., 50–68, 1997.
251. R. J. Vetter, "ATM Concepts, Architectures, and Protocols," *Communications of the ACM*, 38:2(Feb), 31–38, 1995.
252. Y. Viniotis and R. O. Onvural, *Asynchronous Transfer Mode Networks*, New York: Plenum Press, 1993.
253. J. Walrand, *Communication Networks: A First Course, 2nd edition*, Boston: McGraw-Hill, 1998.
254. C. Wang, J. Wang, and Y. Fan, "Performance Evaluation of a Hierarchical Location Registration and Call Routing for Personal Communications," *Proceedings, International Conference on Communication*, 1722–1726, 1992.
255. J. Z. Wang, "A Fully Distributed Location Registration Strategy for Universal Personal Communication Systems," *IEEE Journal on Selected Areas in Communications*, 11:6(Aug.), 850–860, 1993.
256. W. Wang and C. K. Rushforth, "An Adaptive Local-Search Algorithm for the Channel-Assignment Problem (CAP)," *IEEE Transactions on Vehicular Technology*, 45:3(Aug.), 459–466, 1996.
257. B. Waxman, "Routing of Multipoint Connections," *IEEE Journal on Selected Areas in Communication*, 6(Dec.), 1617–1622, 1988.
258. W. Wei and B. H. Soong, "Distributed Algorithms for Dynamic Channel Allocation in Cellular Mobile Systems," *Proceedings of the IEEE Vehicular Technical. Conference*, 548–551, 1994.
259. N. D. Wilson et al., "Packet CDMA vs. Dynamic TDMA for Multiple Access in an Integrated Voice/Data PCN," *IEEE JSAC*, 11:6(Aug), 870–883, 1993.
260. P. E. Wirth, "Teletraffic Implications of Database Architectures in Mobile and Personal Communications," *IEEE Communications Magazine*, 33:6(Jun.), 54–59, 1994.
261. P. E. Wirth, "Teletraffic Implications of Database Architectures in Mobile and Personal Communications," *Computer Networks and ISDN Systems*, 28, 1996.
262. M. Woo, N. U. Qazi, and Arif Ghafoor, "A Synchronization Framework for Communication of Pre-Orchestrated Multimedia Information, *IEEE Network*, Jan./Feb., 52–61, 1994.
263. G. M. Woodruff and R. Kositpaiboon, "Multimedia Traffic Management Principles for Guaranteed ATM Network Performance," *IEEE Journal on Selected Areas in Communications*, 8:3(Apr.), 437–446, 1990.
264. M. D. Yacoub, *Foundations of Mobile Radio Engineering*, Boca Raton, FL: CRC Press, Inc., 1993.
265. S. Yazid and H. T. Mouftah, "Congestion Control Methods for BISDN," *IEEE Communications Magazine*, Jul., 42–47, 1992.

266. J. Ye and S-q. Li, "Analysis of Multimedia Traffic Queues with Finite Buffer and Overload Control," *Proceedings of the IEEE INFOCOM*, Florence, Italy, 848–859, 1992.
267. W. Yen and I. F. Akyildiz, "On the Synchronization Mechanisms for Multimedia Integrated Services Networks," *Proceedings of the International COST 237 Workshop on Multimedia Transport and Teleservices, 1994,* Lecture Notes in Computer Science, Vol. 882, New York: Springer- Verlag, 168–184 1994.
268. K. L. Yeung and T. S. P. Yum, "Compact Pattern Based Dynamic Channel Assignment for Cellular Mobile System," *IEEE Transaction on Vehicular Techn.*, 43:4(Nov.), 892–896, 1996.
269. N. Yin, "Fairness Definition in ABR Service Model," *ATM Forum Contribution*, 94–0928R2, Nov., 1994.
270. R. Yuan, "An Adaptive Routing Scheme for Wireless Mobile Computing," in *Wireless and Mobile Communications*, J. M. Holtzman and D. J. Goodman, eds., Boston: Kluwer, 39–50, 1994.
271. R. Yuan, "Traffic Pattern Based Mobile Routing Scheme," *Computer Communications*, 18:1(Jan.), 32–36, 1995.
272. W. Yue, "Analytical Methods to Calculate the Performance of a Cellular Mobile Radio Communication System with Hybrid Channel Assignment," *IEEE Transaction on Vehicular Techn.*, 40:2, 453–460, 1991.
273. T. Yum and M. Chen, "Multicast Source Routing in Packet-Switched Networks," *Proceedings of the IEEE INFOCOM*, San Francisco, CA, 1284–1288, 1990.
274. J. Zander, "Generalized Reuse Partitioning In Cellular Mobile Radio," *43th IEEE Vehicular Technology Conference '93*, 181–184, 1993.
275. J. Zeft, G. Rufa, "Congestion and Flow Control in Signaling System No.7—Impacts of Intelligent Networks and New Services," *IEEE Journal on Selected Areas in Communications*, 12:3(Apr.), 501–509, 1994.
276. T. Znati and B. Field, "A Network Level Channel Abstraction for Multimedia Communication in Real-Time Networks, *IEEE Transactions on Knowledge and Data Engineering*, 5:4(Aug.), 590–599, 1993.
277. J. A. Zoellner and C. A. Beall, "A Breakthrough in Spectrum Conserving Frequency Assignment Technology," *IEEE Transactions on Electromagnetic Compatibility*, EMC-19, Aug., 313–319, 1977.

# Glossary of Acronyms

## A

| | |
|---|---|
| AAL | adaptation layer |
| ABR | available bit rate |
| AC | access component |
| ACK | acknowledgment |
| ACR | allowed cell rate |
| AMPS | advanced mobile phone system (s) |
| API | applications programming interface |
| APU | audio presentation units |
| ARCMA | adaptive request channel multiple access |
| ARP | address resolution protocol |
| ARP | autonomous reuse partitioning |
| AS | alternative strategy |
| ATM | asynchronous transfer mode |

# B

| | |
|---|---|
| B-ISDN | broadband integrated services digital network |
| BCA | borrowing channel assignment |
| BEB | binary exponential broadcast |
| BECN | backward explicit notification system |
| BER | bit error rate |
| BR | border router |
| BS | base station (s) |
| BSC | base station controller |
| BUS | broadcast and unknown service |

# C

| | |
|---|---|
| CAC | call admission control |
| CBR | constant bit rate |
| CCAF | call control agent function |
| CCF | call control function |
| CCR | current cell rate |
| CD-I | compact disk interface |
| CDCA | cochannel information band dynamic channel assignment |
| CDMA | code division multiple access |
| CDPD | cellular digital packet data |
| CDV | cell delay variation |
| CER | cell error rate |
| CG | call gapping |
| CHS | channel server |
| CIR | carrier-to-interference ratio |
| CLB | cooperating leaky bucket |
| CLP | cell loss priority |
| CLR | cell loss ratio |
| CMR | cell misinsertion rate |
| CP | compact pattern |

| | |
|---|---|
| CPU | central processor unit |
| CS | convergence sublayer |
| CSMA/CD | carrier sense multiple access with collision detection |
| CTD | cell transfer delay |
| CX | crossover switch |

# D

| | |
|---|---|
| DAF | direct access function |
| DAMA | demand assignment multiple access |
| DB | delay bound |
| DBF | delay bound function |
| DC | data compression |
| DCA | dynamic channel assignment |
| DCT | discrete cosine transform |
| DD | data downstream |
| DDBMA | dynamic delay bounded multicast algorithm |
| DDCA | distributed dynamic channel assignment |
| DF | rate decrease factor |
| DMIS | distributed multimedia information system |
| DMST | delay-bounded minimum Steiner tree |
| DPST | dynamic priority spanning tree |
| DQRUMA | distributed queuing request update multiple access |
| DSDSP | dynamic state dependent shortest path |
| DVI | digital video interface |

# E

| | |
|---|---|
| EACR | estimated allowable cell rate |
| EFCI | explicit forward congestion indicator |
| EIR | equipment identity register |
| ER | explicit rate |

| ESN | electric series number |
| ESP | extended shortest path |
| EUA | end user agent |

# F

| FA | foreign agent |
| FBCA | forcible-borrowing channel assignment |
| FCA | fixed channel assignment |
| FDMA | frequency division multiple access |
| FECN | forward explicit congestion notification |
| FQS | frame quality service |

# G

| GDCA | group dynamic channel assignment |
| GFC | generic flow control |
| GSM | global system for mobile communication |

# H

| HA | home agent |
| HCA | hybrid channel assignment |
| HDB | home database |
| HEC | header error control |
| HLR | home location register |
| HLS | home location server |
| HOL | head of the line |
| HOS | handoff switch |
| HT | high threshold |

# I

| | |
|---|---|
| IAM | initial address message |
| ID | identifier |
| IETF-MIP | internet engineering task force-mobile IP |
| IF | rate increase factor |
| ILMI | interim local management interface |
| IN | intelligent network |
| IP | internet protocol (not intelligent peripheral) |
| IS | interim standard |
| ISDN | integrated services digital network |
| ITU-T | International Telecommunications Union – Telecommunication |

# J

| | |
|---|---|
| JPEG | Joint Picture Experts Group |

# L

| | |
|---|---|
| LA | location area |
| LAN | local area network |
| LB | leaky bucket |
| LDCA | largest CIR dynamic channel assignment |
| LE | LAN emulation |
| LEC | LE client(s) |
| LECS | LE configuration server |
| LES | LE server(s) |
| LSRIP | loose source routing IP |
| LSRR | loose source and record route |
| LT | low threshold |
| LU | location updating |

# M

| | |
|---|---|
| M-QoS | mobile QoS |
| MA | metropolitan area |
| MAC | media access control |
| MACR | modified ACR |
| MAN | metropolitan area network |
| MCB | multimedia conference bridge |
| MCP | multimedia conference protocol |
| MCR | minimum cell rate |
| MH | mobile host(s) |
| MHB | mobile handling backbone |
| MHP | mobile host protocol |
| MICP | mobile internet control protocol |
| MIN | mobile identification number |
| MP | maximum packing |
| MPEG | motion picture experts group |
| MS | mobile station(s) |
| MSC | mobile switching center |
| MSR | mobile subnet router(s) |
| MSRN | mobile station roaming number |
| MSS | mobile support stations |
| MST | minimum spanning tree |
| MSU | mobile switching unit |
| MT | mobile terminal(s) |
| MTD | membership tracking database |
| MV | motion vector |
| MVC | multimedia virtual circuit |

# N

| | |
|---|---|
| NBC | narrow band call |
| NFCA | nonfixed channel assignment |
| NIU | network interface unit |
| nrt-VBR | non-real-time VBR |

# O

| | |
|---|---|
| OCPN | object composition Petri net |
| OSI | open system interconnection |
| OSPF | open shortest path first |
| OUT | outstanding messages |

# P

| | |
|---|---|
| PCN | personal communication network |
| PCR | peak cell rate |
| PCS | personal communication service |
| PDCA | priority based dynamic channel assignment |
| PDU | protocol data unit |
| PFSSP | priority first search for the shortest path |
| PGBK | piggyback |
| PHY | physical layer |
| PMD | physical medium-dependent |
| PRCA | proportional rate control algorithm |
| PRMA | packet reservation multiple access |
| PSTN | public switched telephone network |
| PTI | payload type identifier |
| PVC | permanent virtual connection |

# Q

QoS       quality of service

# R

| | |
|---|---|
| RA | request access |
| RAMA | resource allocation multiple access |
| RF | radio frequency |
| RM | resource management |
| RR | resource reservation |
| RRN | resource reservation and renegotiation |
| rt-VBR | real-time variable bit rate |

# S

| | |
|---|---|
| SAR | segmentation and reassembly |
| SB | shared ABR bandwidth |
| SCF | service control function |
| SCP | service control point |
| SDF | service data function |
| SDH | synchronous digital hierarchy |
| SDP | service data point |
| SDCA | smallest CIR dynamic channel assignment |
| SECBR | severely-errored cell block ratio |
| SH | source host |
| SIU | synchronization information unit |
| SLB | shared leaky bucket |
| SMDS | switched multimegabit data service |
| SMS | service management system |
| SONET | synchronous optical network |
| SPST | shortest path spanning tree |
| SRF | specialized resource function |
| SS | signaling system |

| | |
|---|---|
| SS7 | signaling system number 7 |
| SSF | service switching function |
| SSP | service switching point |
| STP | signaling transfer point |
| STS-1 | synchronous transport signal level 1 |
| SVC | switched virtual connection |

# T

| | |
|---|---|
| TA | transmission access |
| TC | transmission convergence |
| TCP | transmission control protocol |
| TDD | time division duplex |
| TDM | time division multiplexing |
| TDMA | time division multiple access |
| TIA | temporary internet address |
| TMSI | temporary mobile station identity |
| TMTI | temporary mobile terminal identity |
| TPN | timed Petri net |
| TTL | time to live |
| TX | transit exchange |

# U

| | |
|---|---|
| UBR | unspecified bit rate |
| UCB | University of California, Berkeley |
| UDP | user datagram protocol |
| UI | update interval |
| UMTS | universal mobile telecommunication system |
| UNI | user network interface |

# V

| | |
|---|---|
| VAU | video access unit(s) |
| VBR | variable bit rate |
| VC | virtual channel (not virtual circuit) |
| VCC | virtual channel connection |
| VCI | virtual channel identifier |
| VCN | virtual circuit negotiator |
| VDB | visitor database |
| VIP | virtual IP |
| VLAN | virtual LAN |
| VLR | visitor location register |
| VLS | visitor location server |
| VMS | video multiplexing server |
| VP | virtual path |
| VPI | virtual path identifier |
| VPU | video presentation unit(s) |
| VSOAIC | video mosaic |

# W

| | |
|---|---|
| WAN | wide area network |
| WATM | wireless ATM |
| WBC | wide band call |
| WIN | window size variable |
| WLAN | wireless LAN |

# Index

## A

acknowledgment (ACK), 29-30, 57-58, 215
adaptive request channel multiple access (ARC-MA) protocol, 32
admission threshold, 269
advanced mobile phone system (AMPS), 78
ALOHA, 26-27, 30
allowed cell rate (ACR), 20, 42-44, 50-51
application programming interface (API), 62
asynchronous transfer mode (ATM), 1-8, 81
ATM adaptation layer (AAL), 2-8, 18, 73, 76, 104
ATM cell format, 2
ATM cell header, 8
audio access unit (AAU), 13-14
audio presentation unit (APU), 13
audio SIU (AIU), 86
autonomous reuse partitioning (ARP) DCA, 245
available bit rate (ABR), 14-21, 24, 31, 41-43

## B

backward explicit congestion notification (BECN), 20, 45-47
bandwidth window, 62
base station (BS), 22, 30, 102-103, 113, 117-118, 125, 129, 131-133, 177-189, 225, 228-237, 244-250, 252, 268-269, 273
controller (BSC), 114
beat down problem, 47-48
binary exponential backoff (BEB), 32,
bit error rate (BER), 25, 267-268
border router (BR), 196-209, 211-215
borrowing channel assignment (BCA), 243
broadband ISDN (B-ISDN), 1-3, 81, 84-85, 103-104, 176-177
broadband networks, 1-17, 175
broadcast and unknown server (BUS), 60, 71-79
buffer adjustment, 101
buffer configuration, 91

## C

call admission control (CAC), 269-270
call control agent function (CCAF), 104
call control function (CCF), 105
call gapping (CG), 54-59
carrier sense multiple access (CSMA), 26
carrier sense multiple access with collision detection (CSMA/CD), 26
carrier-to-interference ratio (CIR), 230-237, 240-242, 244-252
cell delay variation (CDV), 42

cell error ratio (CER), 42
cell loss priority (CLP) bit, 8
cell loss ratio (CLR), 42
cell misinsertion rate (CMR), 42
cell transfer delay (CTD, 42
cellular digital packet data (CDPD), 267
centralized DCA, 243-244
central processor unit (CPU), 55, 85
channel reuse, 23, 227
channel segregation DCA, 245
channel server (ChS), 64
cochannel information based dynamic channel assignment (CDCA), 232-237
cochannel interference, 23
code-division multiple access (CDMA), 27, 29, 78, 265-267
Columbia MHP, 211
compact disk interactive (CD-I), 11
compact pattern (CP) based DCA, 245
compression technology, 11-14
constant bit rate (CBR), 1, 7, 12, 17-18, 24, 31, 41-43, 77
convergence sub-layer (CS), 2
cooperating leaky bucket (CLB) model, 10-11, 15
credit-based flow control, 18
crossover switch (CX), 69-71
current cell rate (CCR), 21

distributed multimedia information system (DMIS), 81-85, 96
distributed queuing request update multiple access (DQRUMA), 30-32
dogleg routing, 128
dynamic channel assignment (DCA), 227-232, 243-246, 249-252, 263, 265, 268
dynamic delay bounded multicast algorithm (DDB-MA), 147-159
dynamic neighbor zone, 131
dynamic priority spanning tree (DPST), 161-174
dynamic state-dependent shortest path (DSDSP), 177, 182, 188-192

## E

electric series number (ESN), 126
end user agent (EUA), 36-37, 40
equipment identity register (EIR), 126
estimated allowed cell rate (EACR), 50-51
Ethernet, 37
explicit forward congestion indication (EFCI), 20, 44-47
explicit rate (ER), 21, 50-51
extended shortest path (ESP), 177, 182, 184-188, 191-192

## D

data downstream (DD), 29
data transmission access (TA), 28
delay-bounded minimum Steiner tree (DMST), 147-159
delay bound function (DBF), 148-149
delay jitter, 82, 84
demand assignment multiple access (DAMA), 28-32
demand assignment demultiplexing, 4
destination node, 147
digital video interactive (DVI/Indeo), 11
direct access function (DAF), 106
discrete cosine transform (DCT), 13, 16-17
distributed database, 109
distributed dynamic channel assignment (DDCA), 242-247

## F

file transfer protocol (FTP), 39
fixed channel assignment (FCA), 227, 243, 252, 268
forcible-borrowing channel assignment (FBCA), 243
foreign agent (FA), 195-209, 211-224
forward explicit congestion notification (FECN), 20, 43-45
forwarding pointers, 134
frame, 3
frame quality scaling (FQS), 16
frequency reuse, 24, 230
frequency-division multiple access (FDMA), 26, 267

## G

generic flow control (GFC), 4
geometric DCA, 245
global scope, 128. 136
global system for mobile communications (GSM), 66, 108, 124, 126, 131, 133
group dynamic channel assignment (GDCA), 238-242

## H

handoff (handover), 61-63, 68, 79, 181, 184-185, 188-192, 240-242, 269, 271-272
handoff switch (HOS), 66
header error control (HEC), 3
head of line (HOL), 164
hierarchical database, 110-111
highway cell, 103
home agent (HA), 195-211, 219
home database (HDB), 115
home location register (HLR), 105, 109, 114, 126-128, 136-137, 177, 217
home location server (HLS), 64, 128, 136-137
hybrid channel assignment (HCA), 243, 251-252, 263

## I

initial address message (IAM), 126
integrated services digital network (ISDN), 42, 125, 127
intelligent network (IN), 52-58, 104-107, 114-115
intelligent peripheral, 105
intermedia synchronization, 81-83, 98
interim local management interface (ILMI), 75
interim standard 41 (IS-41), 108
international standards organization (ISO), 12
Internet engineering task force mobile IP (IETF-MIP), 193, 195-196, 200, 203
internet protocol (IP), 38-39, 136-137, 193-210
intramedia synchronization, 81-83, 98

## J

jitter, 82, 84
joint picture experts group (JPEG), 11

## K

KMB algorithm, 147-159, 160, 172
KPP algorithm, 147-159

## L

LAN emulation (LE), 59, 71-74
largest CIR dynamic channel assignment (LDCA), 232-233
LE client (LEC), 60, 71-78
LE configuration server (LECS), 75
LE server (LES), 60, 72-79
leaky bucket (LB), 8, 15, 55
    model, 8-10
link cost function, 148
link delay function, 148
local area network (LAN), 37-38, 161, 178, 211, 267
local scope, 129-130
location area (LA), 114, 117
location database, 108
location server (LS), 140
location tracking, 116-119
location updating (LU), 114-115, 117
loose source and record route (LSRR), 194
loose source routing IP (LSRIP), 193-196

## M

macrocell, 103
maximum packing (MP), 231, 243
media access control (MAC), 21-32, 60, 73-77, 218, 222, 274
membership tracking database (MTD), 79, 243
metropolitan area (MA), 115-118
    network (MAN), 115-116, 178
microcell, 103

microcellular networks, 265, 267-269
minimum cell rate (MCR), 42, 49-50
minimum cost multicast tree, 145-147
minimum spanning tree (MST), 160-163, 171-174, 182-183
mobile handling backbone (MHB), 61, 77-79
mobile host (MH), 130, 176-182, 193-224
    protocol (MHP), 211
mobile identification number (MIN), 126
mobile internetworking control protocol (MICP), 130
mobile QoS (M-QoS), 60-63
mobile station (MS), 27, 113, 115, 118, 230, 234, 244-247
    roaming number (MSRN), 114
mobile subnet router (MSR), 211
mobile support station (MSS), 130
mobile switching center (MSC), 66, 103, 109, 113, 115, 125-127, 131, 229
mobile terminal (MT), 103, 107, 125-133, 136-143
mobility support border router (MSBR), 211
modified ACR (MACR), 20-21
motion joint picture experts group (MJPEG), 11
motion picture experts group (MPEG), 8, 12-17
    standard 1 (MPEG-1), 11, 13
    standard 2 (MPEG-2), 11
multicast tree, 145, 148
multimedia conference bridge (MCB), 36-41
multimedia conference protocol (MCP), 39-40
multimedia conference system, 36-41
multimedia synchronization, 81-101
multimedia virtual circuit (MVC) model, 14-17
multimedia virtual circuit queue, 14
multiplex stream, 13
multiplexer, 14
multiplexing, 4

## N

narrow band call (NBC), 239-241
network interface unit (NIU), 66
nominal channel, 253-163
nominal channel slot order, 257-258
nominal value (NV), 100
non fixed channel assignment (NFCA), 243

non real-time variable bit rate (nrt-VBR), 24
NP-complete, 122, 147, 259

## O

object composition Petri net (OCPN), 82-83
    model, 82, 86-89
open shortest path first (OSPF), 196, 211-224
open system interconnection (OSI), 2
optical carrier signal level 1, 3, and 12 (OC-1, OC-3, OC-12), 12
outstanding messages (OUT), 57

## P

packet reservation multiple access (PRMA), 30, 267
paging, 108
path searching, 186-188, 192
patron area, 212-216
patron host, 212, 216
payload type identifier (PTI), 20, 44
    field, 20
peak cell rate (PCR), 42, 50
permanent virtual connection (PVC), 1, 7, 75
personal communication network (PCN), 66, 113-115, 124
personal communication service (PCS), 101-111, 113-114
physical layer (PHY), 2
physical medium-dependent (PMD), 2
    sub-layer, 2
picocell, 103
piggyback (PGBK) bit, 30
playback, 97
playout synchronization, 88
priority based DDCA (PDCA), 246-251
priority first search for the shortest path (PFSSP), 164-169
proportional rate control algorithm (PRCA), 20, 47
protocol data unit (PDU), 2
public switched telephone network (PSTN), 108, 125-128, 132

## Q

quality of service, 21, 60-61, 77, 84, 146, 176, 225, 265-277
    guarantee, 21, 146, 270
query, 124

## R

radio frequency (RF), 225
random assignment, 26
rate-based control, 19-21
real-time variable bit rate (rt-VBR), 24
redirection agent, 212, 217, 220
relay node, 148
reporting cell, 118
reporting center, 118, 121-124
request access (RA), 29
rerouting path, 184, 187-192
resource auction multiple access (RAMA), 29
resource management (RM) cell, 19-21, 44-52
resource reservation (RR), 271-273, 275-277
resource reservation and renegotiation (RRN), 273-277
resource sharing, 270

## S

scalable mobile host IP (SMIP), 196-209
segmentation and reassembly (SAR) layer, 2
service control function (SCF), 105-106
service control point (SCP), 53-57, 105, 107
service data function (SDF), 105-107
service data point (SDP), 105
service management system (SMS), 53-57
service switching function (SSF), 105
service switching point (SSP), 53-57, 105-106
severely-errored cell block ratio (SECBR), 42
shared ABR bandwidth (SB), 21
shared leaky bucket (SLB), 14-16
shortest path, 148-149, 152, 182-185, 189-190
shortest-path-spanning-tree (SPST), 161-163, 170-171
signaling network, 56
signaling system number 7 (SS7), 55, 107, 113, 126, 177
signaling transfer point (STP), 114
smallest CIR dynamic channel assignment (SD-CA), 253
source host (SH), 139-142, 211
source node, 147
specialized resources function (SRF), 105
state-dependent shortest path, 183
static partitioning, 131
Steiner tree, 147-149
superpath, 152-154
switched multimegabit data service (SMDS), 7
switched virtual connection (SVC), 1, 7
synchronization information unit (SIU), 86, 89-92
synchronous digital hierarchy (SDH), 2
synchronous optical network (SONET), 3-4
synchronous transport signal level 1 (STS-1), 3

## T

TCP/IP protocol, 38, 193
temporary IP (TIP) address, 193-194
temporary mobile station identity (TMSI), 114
temporary mobile terminal identity (TMTI), 105, 126-127
time division duplex (TDD) system, 28
time division multiple access (TDMA), 26-27, 229, 265, 267
time division multiplexing (TDM), 29, 229
time to live (TTL) field, 138
timed Petri nets (TPN) model, 86
tracking strategy, 116-123
transmission access (TA), 29
transmission control protocol (TCP), 38-41
transmission convergence (TC), 2
    sub-layer, 2
transmitter, 26
triangle routing, 128, 216-217
tunnel packet, 220-222

## U

umbrella cell, 103
universal mobile telecommunication system

(UMTS), 68, 114
unspecified bit rate (UBR), 24, 31, 41, 43
update interval (UI), 20
user datagram protocol (UDP), 38-41
user-network interface (UNI), 4, 74

## V

variable bit rate (VBR), 1, 7, 12, 17-18, 31, 41-43, 77
video access unit (VAU), 14
video mosaic (VSOAIC), 63
video multiplexing server (VMS), 14
video presentation units (VPUs)
video SIU (VIU), 86
virtual channel (VC), 18, 44, 47-49, 66
    connection (VCC), 59, 72-78
    identifier (VCI), 4-5, 64-65, 69

virtual circuit negotiator (VCN), 14-15
virtual IP (VIP), 193-194, 196
virtual LAN, 59-60
virtual path (VP), 18, 176
    identifier (VPI), 4-5, 64
visitor database (VDB), 115
visitor location register (VLR), 105, 109, 114, 126-129, 177, 217
visitor location server (VLS), 64, 128

## W

wide area network (WAN), 18, 38, 50, 161, 195, 213
wide band call (WBC), 239-241
window size variable (WIN), 57
wireless ATM (WATM), 21
wireless integration, 25
wireless local area network (WLAN), 22, 64

# About the Author

**Dr. Anna Hac´** is Professor in the Department of Electrical Engineering, University of Hawaii at Manoa, Honolulu, and a member of the Editorial Board of the *IEEE Transactions on Multimedia*. She has been a Visiting Scientist at the Imperial College, University of London, England; a Post-Doctoral Fellow at UC Berkeley; an Assistant Professor of Electrical Engineering and Computer Science at The Johns Hopkins University, and a Member of Technical Staff at AT&T Bell Laboratories. Her research contributions include system and workload modeling, performance analysis, reliability, modeling process synchronization mechanisms for distributed systems, congestion control, and wireless networking.

# PRENTICE HALL
## Professional Technical Reference
*Tomorrow's Solutions for Today's Professionals.*

## *Keep Up-to-Date with*
# PH PTR Online!

We strive to stay on the cutting-edge of what's happening in professional computer science and engineering. Here's a bit of what you'll find when you stop by **www.phptr.com**:

- **Special interest areas** offering our latest books, book series, software, features of the month, related links and other useful information to help you get the job done.

- **Deals, deals, deals!** Come to our promotions section for the latest bargains offered to you exclusively from our retailers.

- **Need to find a bookstore?** Chances are, there's a bookseller near you that carries a broad selection of PTR titles. Locate a Magnet bookstore near you at www.phptr.com.

- **What's New at PH PTR?** We don't just publish books for the professional community, we're a part of it. Check out our convention schedule, join an author chat, get the latest reviews and press releases on topics of interest to you.

- **Subscribe Today!** **Join PH PTR's monthly email newsletter!**

    Want to be kept up-to-date on your area of interest? Choose a targeted category on our website, and we'll keep you informed of the latest PH PTR products, author events, reviews and conferences in your interest area.

    Visit our mailroom to subscribe today! **http://www.phptr.com/mail_lists**

| DATE DE RETOUR | L.-Brault | |
|---|---|---|
| 1 4 AVR. 2004 | | |
| 2 0 AVR. 2007 | | |
| 1 5 NOV. 2007 | | |